Differential Geometry in Euclidean and Minkowskian Spaces

Paul E. Ehrlich and Seon-Bu Kim

2014년도
대한민국학술원 선정
우수학술도서

이 도서는 교육부의 지원으로
대한민국학술원에서 선정한
"2014년도 우수학술도서"임.

KYUNG MOON SA

Preface

This book is about differential geometry for curves and surfaces in three dimensional Euclidean and Minkowskian spaces.

In chapter 1, we sum up the history of Euclidean and non-Euclidean Geometry and introduce the vector fields which are the basic tools to study the theory of curves and surfaces in differential geometry. In addition, we study the basic geometric properties of Minkowskian space and the hyperbolic trigonometry which plays a similar role in Minkowskian space to the usual trigonometry in Euclidean space.

In chapter 2, we study the Frenet-Serret apparatus for the reparametrization of a regular curve by arc length and introduce the curvature and torsion along the curve. Moreover, we classify curves by their curvature and torsion in three cases, Euclidean curves and both timelike and spacelike curves in Minkowskian space.Further, we need to study the rotation index which will be used as a theorem of turning tangents in the proof of the Gauss-Bonnet Theorem in chapter 5.

In chapter 3, we formulate the concept of regular surface covered by parametrized patches and give examples of surfaces with Maple figures. To measure the curvature of a patch for a surface in Euclidean and Minkowskian spaces, we need to compute the first and second fundamental forms of Euclidean surfaces and both timelike and spacelike surfaces.

In chapter 4, we study the shape operator and compute the normal curvatures, principal curvatures, Gaussian and mean curvatures of a patch for Euclidean surfaces and both timelike and spacelike surfaces. Moreover, we observe the invariant properties of curvatures by isometries between two surfaces. Further, we study the parallel translation and geodesics of the surfaces in Euclidean and Minkowskian spaces.

In chapter 5, we study the Gauss-Bonnet theorem, one of the most important theorems in mathematics, which shows that the geometrical property "Gaussian curvature" on a compact surface in Euclidean space is related to the topological property "Euler number". But in Minkowskian space we just apply the theory for an interior region of a unit speed timelike or spacelike n-polygonal curve.

Acknowledgements

In 2004, the second author visited the mathematics department of the University of Florida for research with the first author to prove volume comparison theorems for Lorentzian manifolds and they decided to write this book for the undergraduate and graduate students. After the second author came back to Chonnam National University in Korea, they continued their writing by using email, with the second author sending the original drafts of the materials for revisions to the first author and then typed his returned corrections by the LaTex program. The second author received some financial aid for publishing from

Chonnam National University in 2012.

The authors wish especially first to thank the patience of both their families and of the University of Florida and Chonnam National University for many years of encouragement as this book was developed, and secondly to Kyungmoon Publishers in Seoul in Korea for bringing the book to fruition.

Contents

Chapter 1
Preliminaries

1.1 A Brief History of Euclidean and Non-Euclidean Geometry

Euclidean geometry is a plane geometry derived from Euclid's work "The Elements" published around 300 B.C. Writing at his academy in Alexandria, Egypt, Euclid drew on results from previous centuries of classical Greek mathematics. Especially, Euclid himself followed a convention of Aristotle (384-322 B.C.) and distinguished between axioms as "truths applicable to all sciences" and postulates as applying only to geometry. An example of an axiom or "common notion" familiar to many generations of students is "Things which are equal to the same thing are also equal to one another."

The celebrated Parallel Postulate is apparently Euclid's own contribution. Much later, it was realized that more rigor had to be brought to bear on any axiomatic approach to geometry. Moritz Pasch (1843-1930) provided an important discussion of these issues in his book Vorlesungen *über* neuere Geometrie [Lectures on New Geometry], introducing axioms for the order of points on a line ["betweenness"] applied primarily to projective geometry. David Hilbert (1862-1943) developed a Euclidean geometry with more rigorous axioms, perhaps in a spirit closest to that of Euclid's "The Elements", in the book Grundlage der Geometrie [Foundations of Geometry] published first in 1899.

Postulates of Euclidean Geometry

A "geometry" is a system consisting of a set of noncontradictory postulates about the undefined terms "point", "line", and "plane" along with theorems deduced from these postulates.

Notations :

$A, B, C, O, P, Q, \cdots$ points,

l a straight line,

Π a plane,

AB a line segment with end points A and B,

$\overleftrightarrow{AB}$ a straight line passing through A and B,

$AB \cong CD$ BE is congruent to CD,

$\angle ABC$ an angle at a point B where two line segments AB and BC meet once.

(1) Postulate I : Given two points, there is a straight line passing through the two points.

Let A and B be in Π with $A \neq B$. Then we may define a straight line $\overleftrightarrow{AB}$(Figure1.1-1).

Figure1.1-1

(2) Postulate II : Given two line segments, one line segment is extended by another.

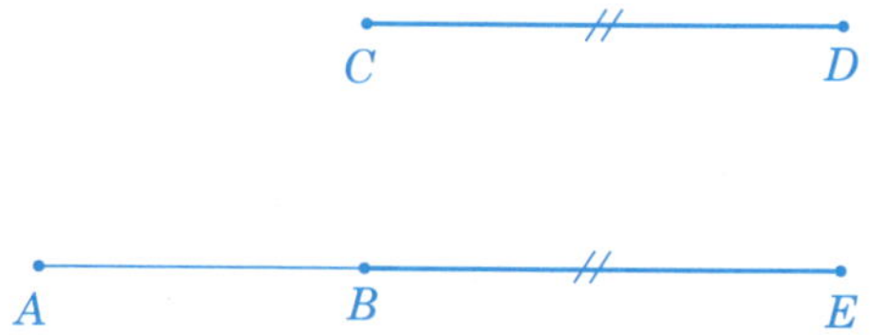

Figure1.1-2

Let AB and CD be two line segments. Then there is a point E in $\overleftrightarrow{AB}$ so that $BE \cong CD$ and B is between A and E(Figure1.1-2).

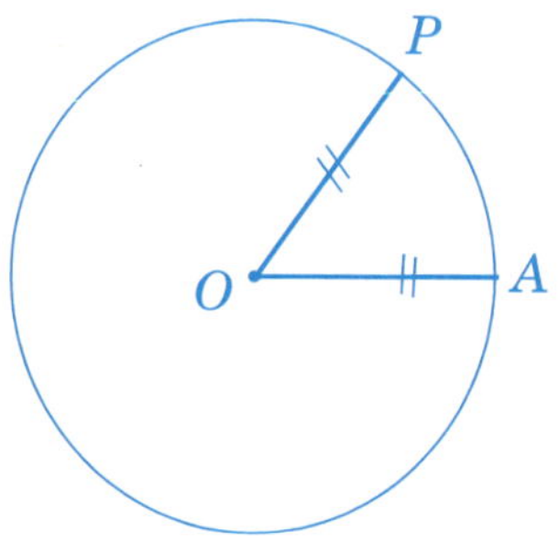

Figure1.1-3

(3) Postulate III : Given two points, there is a circle whose center is one point and whose radius is given by the line segment determined by the two points.

Let O and A be distinct two points. Then we may define a circle with center O and radius OA by a set $\{P \in \Pi \mid OP \cong OA\}$(Figure1.1-3).

(4) Postulate IV : All right angles are congruent.

Let $A, B, C \in \overleftrightarrow{AC}$ and E not in $\overleftrightarrow{AC}$. Then we may define a right angle at B when $\angle ABE \cong \angle CBE$(Figure1.1-4). For any two distinct right angles $\angle ABC$ and $\angle EFG$, we have $\angle ABC \cong \angle EFG$(Figure1.1-5).

(5) Postulate V : Given a point and a straight line not including the point, there is a unique straight line passing through the given point

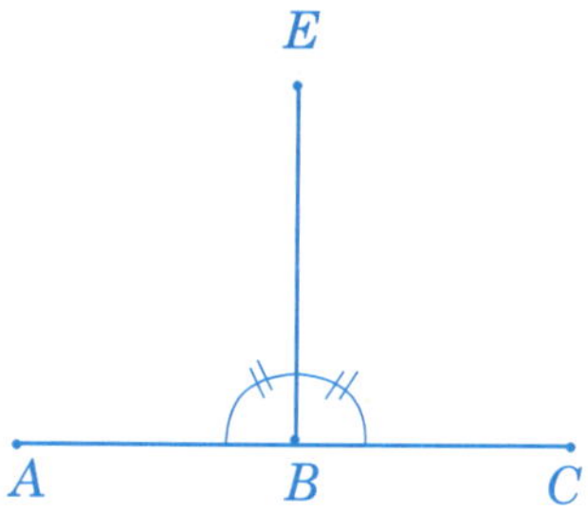

Figure1.1-4

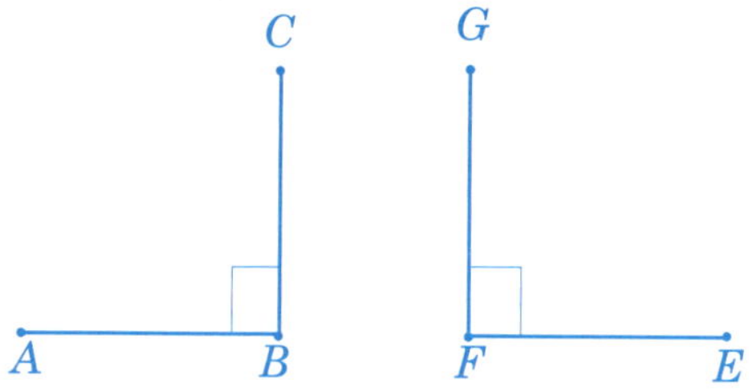

Figure1.1-5

never meeting the given the straight line

Let P be a point and P not on $\overleftrightarrow{AB}$. Then there is a unique straight line $\overleftrightarrow{CD}$ passing through P and indefinitely $\overleftrightarrow{AB} \cap \overleftrightarrow{CD} = \emptyset$(Figure1.1-6).

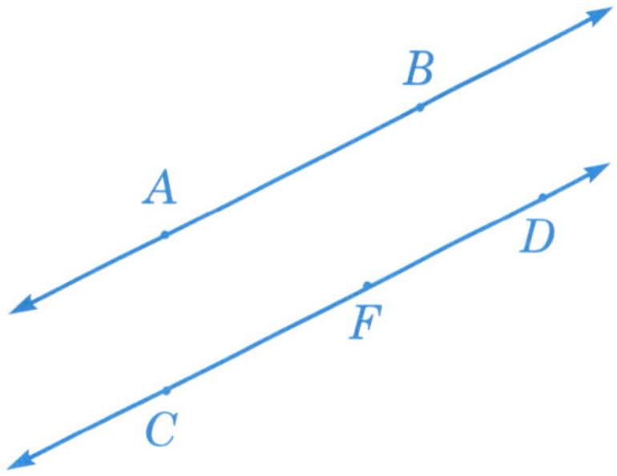

Figure1.1-6

Equivalents of Postulate V

For the five postulates in Euclid's *Elements*, the fifth postulate is originally stated as

Euclid's Postulate V : Let a straight line falling on two straight lines make the interior angles on the same side less than two right angles. Then the two straight lines meet ultimately on that side(Figure1.1-7).

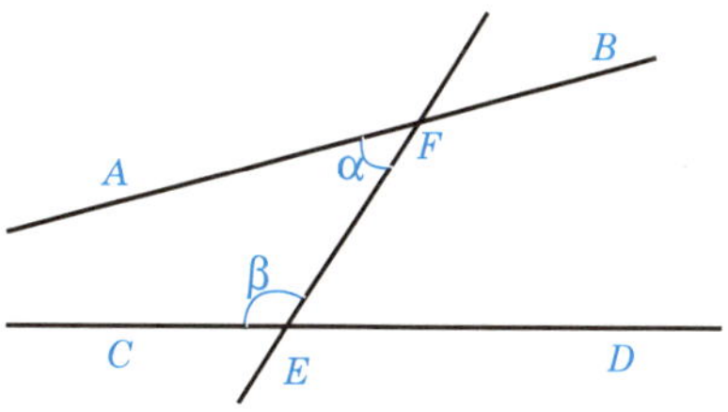

Figure1.1-7

Euclid formulated Postulate V in this way to avoid as much as possible dealing with lines of infinite length, preferring to use only finite line segments in geometry. Many geometers doubted the validity of the fifth postulate since the other four postulates held in a finite region of a plane but the fifth needed an infinite region of a plane. Thus they tried to prove this fifth postulate using the other four postulates. However, they failed to prove it but obtained other equivalents of the fifth postulate as follows.

(1) **Playfair's Axiom :** For a point not on a line, there is a unique line through the point parallel to the given line.

(2) The set of equidistant points from a given line is also a line.

(3) The sum of the interior angles of a triangle is two right angles.

(4) **Wallis's Postulate :** For any given triangle, there is a similar triangle of arbitrary magnitude (Figure 1.1-8).

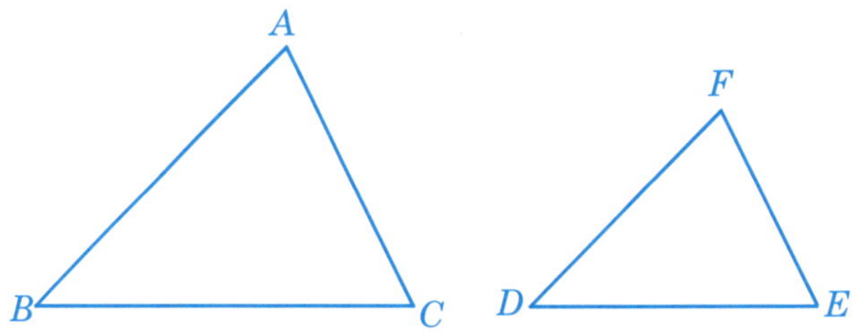

Figure 1.1-8

(5) The perpendicular bisectors of the sides of a triangle are concurrent (Figure 1.1-9).

(6) There is a point equidistant from any three noncollinear points(Figure 1.1-10).

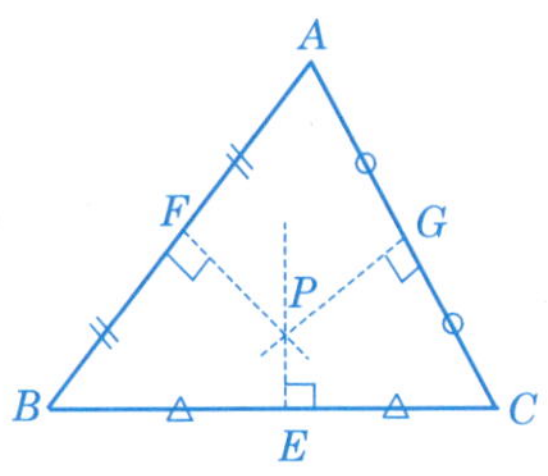

Figure 1.1-9

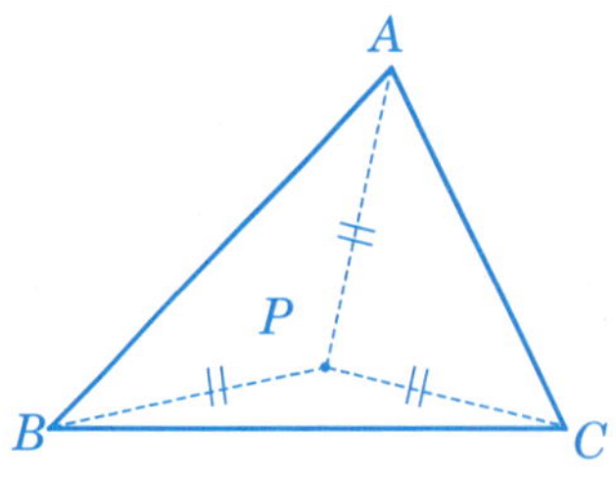

Figure 1.1-10

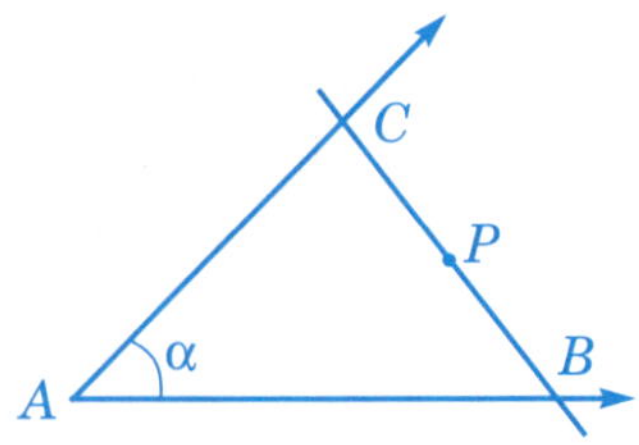

Figure 1.1-11

(7) **Clairaut's Axiom*** **:** There is a rectangle.

(8) There is an acute angle such that every point in the interior of the angle is on a line intersecting both rays of the angle away from the vertex(Figure 1.1-11).

---

*In his geometry text of 1741, Alexis Clairaut(1713-1765) argued that since we see rectangles all around us, the validity of Euclid's Postulate V should certainly be evident.

(9) Any pair of parallel lines has a common perpendicular(Figure1.1-12).

(10) In a right angled triangle the square on the side subtending the right angle is equal to the sum of squares on the sides containing the right angle (Figure 1.1-13).

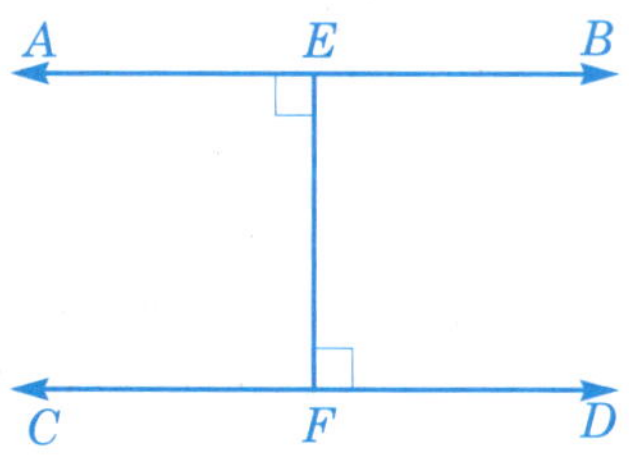

Figure1.1-12

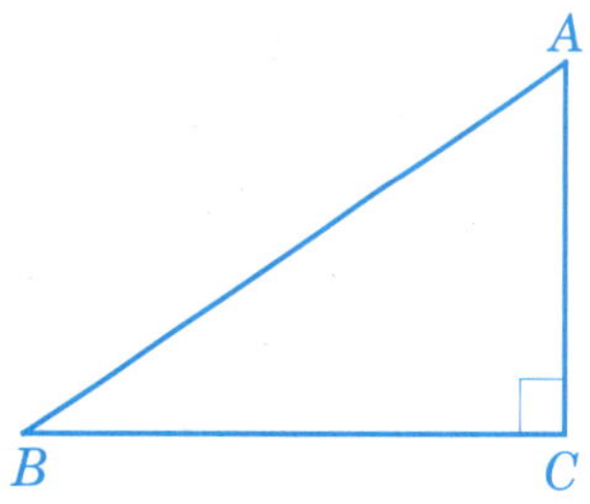

Figure1.1-13

Non-Euclidean Models

By the 1800's, many failed attempts had been made to find a substitute for the Parallel Postulate or to deduce this postulate from the other postulates. Gerolamo Sacchieri (1667-1733) of Italy studied certain quadrilaterals with two right angles in an unsuccessful attempt to prove the Parallel Postulate and Johann Lambert (1728-1777) of Germany made a related analysis for quadrilaterals with three right angles. Hence, the possible existence of non-Euclidean geometries had been seriously considered by a number of scientists by the 1800's. The German mathematician Carl Friedrich Gauss (1777-1855) had created a non-Euclidean geometry by 1816. However, he refrained from publishing his results, thinking the ideas were too controversial. Independently, the Russian mathematician Nikolay Ivanovich Lobachevsky (1793-1856) in a series of papers from 1826 to 1835 and the Hungarian army officer Janos Bolyai (1802-1860) by 1825 had worked out their models for non-Euclidean geometry. Bolyai, who had studied at the Royal College of Engineers in Vienna, published his results in 1832 in an appendix to a book written by his father, a mathematician, who was a fellow student of Gauss's in Göttingen, and later a high school teacher in Transylvania.

To develop the different geometries from Euclidean geometry we construct an elementary geometry called the **incidence geometry**. The incidence geometry is defined by a relation " a point P lies on a line l"(or "l is passing through P and Q") satisfying three axioms:

(1) For any two distinct points P and Q there exists a unique line l passing through P and Q.

(2) For any line l there exist at least two distinct points on l.

(3) There are three points such that no line is passing through all three of them.

There are two alternative geometries to Euclidean Geometry classified by the Parallel Postulate: *A straight line has only one parallel line.* The hyperbolic geometry established by Gauss, Lobachevsky and Bolyai as referred to above is the incidence geometry with **the hyperbolic axiom** (i.e., a straight line has more than one parallel line) while the elliptic geometries such as the projective plane are the incidence geometries with **the elliptic axiom** (i.e., a straight line has no parallel lines). Let l_0 be a line and Q be a point not on l_0. Drop a perpendicular segment PQ from P to l_0, and let l be the line through P perpendicular to PQ(Figure 1.1-14). We denote the Euclidean plane by $\mathbf{E}^2$ to distinguish it

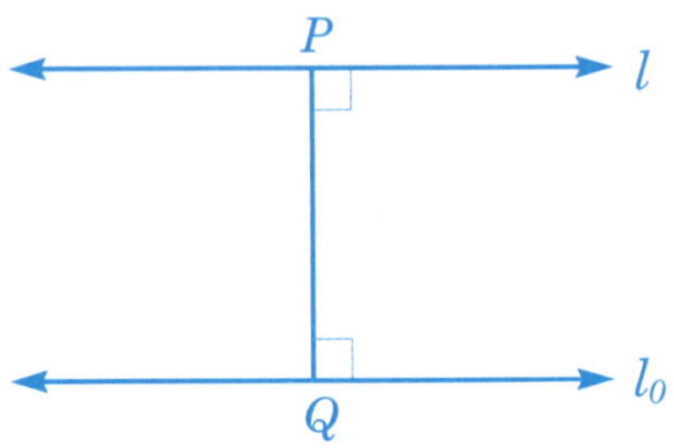

Figure1.1-14

from the non-Euclidean models which follow. In $\mathbf{E}^2$, the perpendicular distance between l and l_0 remains the same as the distance from P to Q as we move to the right along l and l_0. However, in hyperbolic geometry, the distance between rays increases as we move to the right (Greek "hyperballein" = "to exceed") and in elliptic geometry, the distance between the rays decreases and they eventually meet (Greek "elleipein" = "to fall short").

(1) Geometry on the sphere

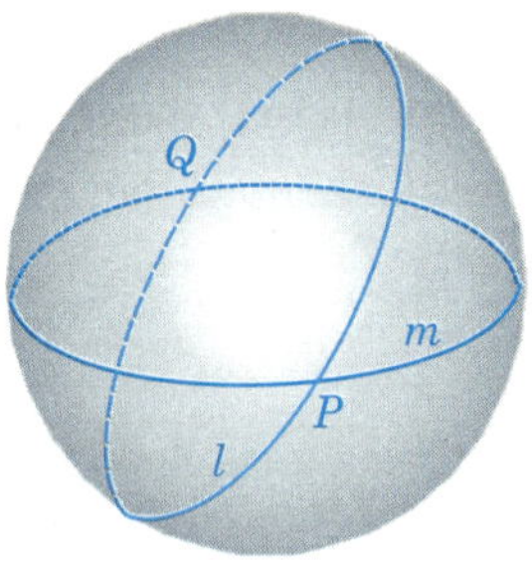

Figure1.1-15

A sphere in $\mathbb{R}^3$ gives some intuitive motivation for non-Euclidean geometry because the line l passing through a point P on the sphere $\mathbf{S^2}$ is a great circle of $\mathbf{S^2}$ passing through P and its *antipodal point** Q. Hence, any two lines meet at exactly two points of $\mathbf{S^2}$. But since infinitely many lines pass through any pair of antipodal points P and $Q = -P$, the spherical geometry does not satisfy the first axiom (1) of incidence geometry, even though the spherical geometry does satisfy axioms (2) and (3) of the incidence geometry(Figure1.1-15).

(2) Projective plane

The projective plane $\mathbf{P}^2$ may be defined as the set $\mathbf{P}^2 = \{(P, Q) | Q = -P,\ P, Q \in \mathbf{S^2}\}$. In other words, $\mathbf{P}^2$ is the set of all equivalence classes of

*Two points P and Q are said to be **antipodal** if $P = -Q$

antipodal points in $\mathbf{S}^2$. Note that each equivalence class of points corresponds to an Euclidean line passing through the origin. Since antipodal points of $\mathbf{S}^2$ are identified, $\mathbf{P}^2$ now satisfies axiom (1) of incidence geometry if the lines of $\mathbf{P}^2$ are defined as the great circles of $\mathbf{S}^2$ with all pairs of antipodal points identified. Alternatively, a projective line may be viewed as a plane through $O = (0, 0, 0)$ in $\mathbf{E}^3$ interpreted as a family of lines passing through the origin(Figure1.1-16). As well as satisfying the incidence axioms, $\mathbf{P}^2$ satisfies the elliptic axiom since any two projective lines meet at exactly one point of $\mathbf{P}^2$. Thus the projective plane is a model of elliptic geometry.

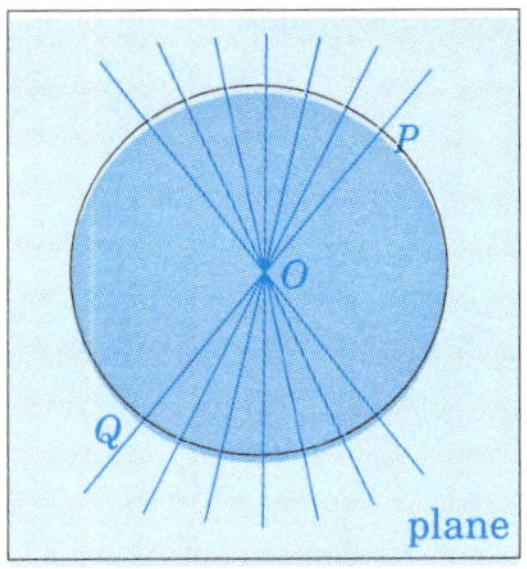

Figure1.1-16

(3) Beltrami-Klein Model

The Beltrami-Klein model (or briefly the Klein model) considers the region strictly inside an Euclidean circle γ. Note that the region does not include points on the circle itself. Thus, all points of this model are the points in the interior of γ. Moreover, we define a line in this model as any chord connecting two points on γ with the endpoints excluded. Note that the Klein model satisfies the axioms of the incidence geometry. Also we may see that it satisfies the hyperbolic axiom. Thus, it is one of models of the hyperbolic geometry(Figure 1.1-17).

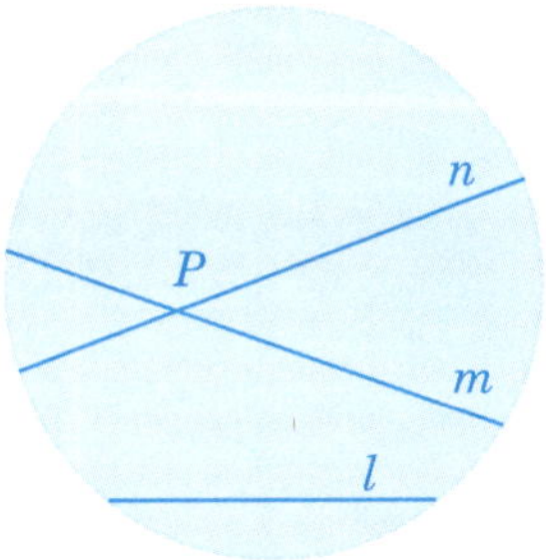

Figure1.1-17

(4) Poincaré Model

The Poincaré disk also represents all points in the interior of an Euclidean circle γ like the Klein model, but its lines are represented as arcs of the circle whose ends are perpendicular to the disk's boundary and diameters are also permitted as its lines. Note that it satisfies the axioms of the incidence geometry. Also we may see that it satisfies the hyperbolic axiom in this model and thus it is also a model of the hyperbolic geometry(Figure1.1-18).

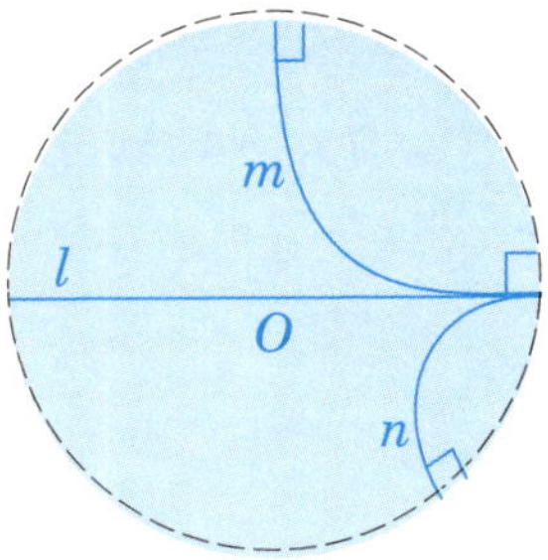

Figure1.1-18

(5) Minkowski Model

The Minkowski model is a two sheeted hyperboloid $\mathbf{H}^2$ represented by the equation

$$x^2 + y^2 - t^2 = -1$$

with coordinates x, y, t(t stands for time). Since the two sheets are dual, we just choose the upper hyperboloid for $t \geq 1$ and we call each point of the upper hyperboloid "a future directed timelike vector" from the origin O. A line on the hyperboloid is a hyperbola which is the section of $\mathbf{H}^2$ cut out by a plane passing through O. We may check that $\mathbf{H}^2$ satisfies the axioms of the incidence geometry. Moreover, it satisfies the hyperbolic axiom(Figure1.1-19).

In Figure1.1-19 we may also observe that this same plane cuts out $\mathbf{S}^2$ along a great circle. Thus, there is a one to one correspondence between the hyperbolas

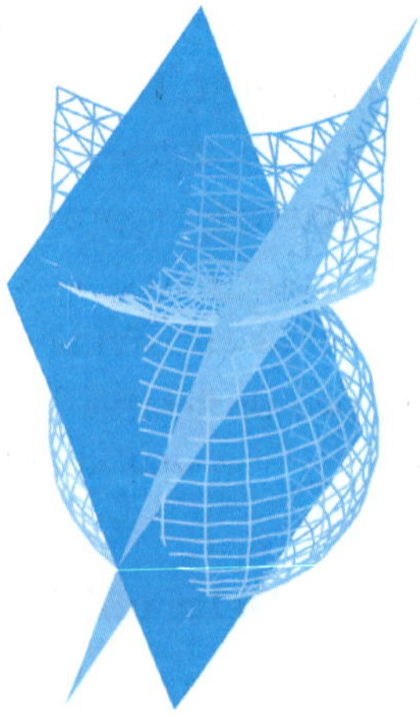

Figure 1.1-19

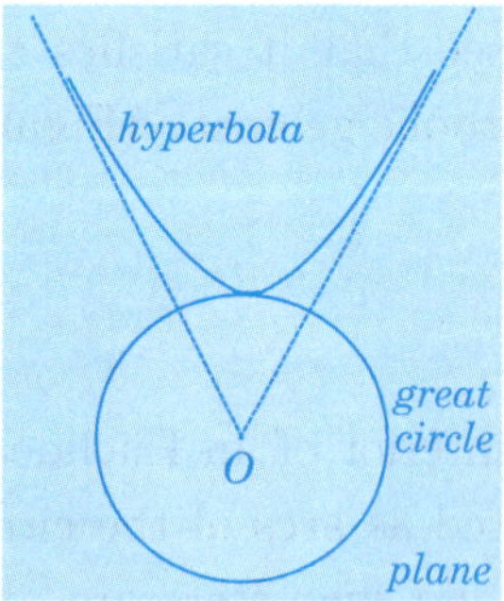

Figure 1.1-20

of $\mathbf{H}^2$ and the great circles of $\mathbf{S}^2$(Figure 1.1-20).

For further study, see the following references.

1. Marvin J. Greenberg, Euclidean and Non-Euclidean Geometry, Development and History, 3rd Edition, W. H. Freeman and Company, San Francisco, 1993.

2. Patrick J. Ryan, Euclidean and Non-Euclidean Geometry, An Analytic Approach, Cambridge University Press, New York, 1999.

1.2 Tangent Vectors and Vector Fields

While Nicole Oresme (c. 1323-1382), Bishop of Liseux and a teacher in the Parisian College of Navarre, is credited with the creation of coordinate geometry and the graphical representation of functions, the lawyer Pierre de Fermat (1601-1665) of Toulouse and scientist René Descartes (1596-1650) of Touraine were the first to systematically use coordinate geometry to study general curves and loci. A point in the plane could be completely determined by traveling a specified distance A from a fixed point O on a given base line (playing the role of the x-axis) to a point Z on this line, and then traveling along a line transversal to the given base line at Z for a distance E to reach J. Thus Fermat is using what are now called "oblique coordinates" for which the x and y axes are not necessarily parallel and no negative coordinates are used. Thus (A, E) corresponds to our (x, y). Similarly, Descartes independently used coordinates of points in the plane to represent a curve, also preferring positive x and y and not requiring the axes to be perpendicular. Fermat believed in the continuation of the Greek tradition in mathematics and regarded his use of coordinate geometry as a reformulation of much earlier work of Apollonius (c. 262-190 B.C.) on the conics, defined geometrically as intersections of planes with a circular cone, either right or oblique. But Descartes wished to break with the Greek tradition and consider a much wider variety of curves with uniform methods. Hence, Descartes is more often associated with the foundation of coordinate geometry and "analytic geometry" - the use of algebraic equations to systematically study curves and surfaces. We now denote the plane and 3-space with such coordinates (assuming perpendicular axes) by $\mathbb{R}^2$ and $\mathbb{R}^3$ respectively. John Wallis (1616-1703), a professor of geometry at Oxford, in his book on conic sections published in 1655 was the first to systematically allow for negative x and y values.

Tangent Vectors

In physics and engineering vectors are used widely to indicate forces, velocities, angular momenta, and many other quantities which have both magnitude and direction. A vector is often represented by an "arrow" whose length represents the magnitude of the vector and the arrow points in the direction of the vector.

Definition 1.2.1 A **tangent vector** v_P to $\mathbb{R}^3$ is defined by two points P, v of $\mathbb{R}^3$, where P is its point of application and v is its vector part.

Note that

$$v_P = w_Q \Longleftrightarrow P = Q \text{ and } v = w.$$

If $v = w$ and $P \neq Q$, we say the tangent vectors v_P and w_Q are **parallel**(figure1.2-21).

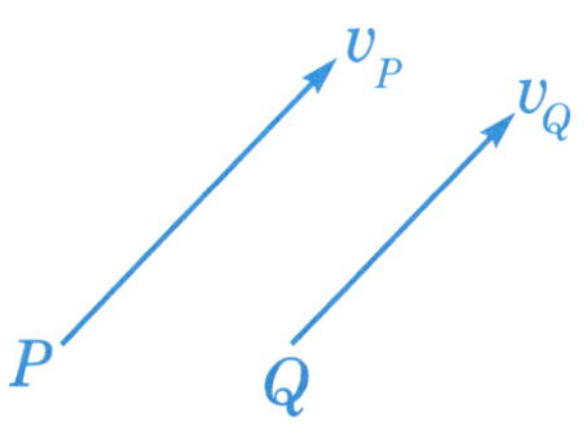

Figure1.2-21

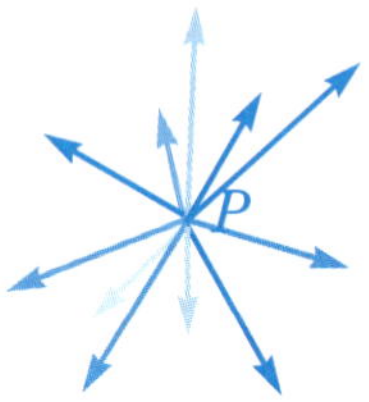

Figure 1.2-22

Definition 1.2.2 For $P \in \mathbb{R}^3$, the **tangent space** $T_P(\mathbb{R}^3)$ of $\mathbb{R}^3$ at P is defined as the set of all tangent vectors with application point P(figure 1.2-22).

Now, we define the **addition** and **scalar multiplication** of tangent vectors in $T_P(\mathbb{R}^3)$ by using the "parallelogram law" and stretching the arrow as follows: For any $u_P, v_P, w_P \in T_P(\mathbb{R}^3)$ and $k \in \mathbb{R}$,

(1) $v_P + w_P := (v + w)_P$

(2) $k(u_P) := (ku)_P$

where $v + w$ and ku are the addition and scalar multiplication of the vector space $\mathbb{R}^3$(figure 1.2-23).

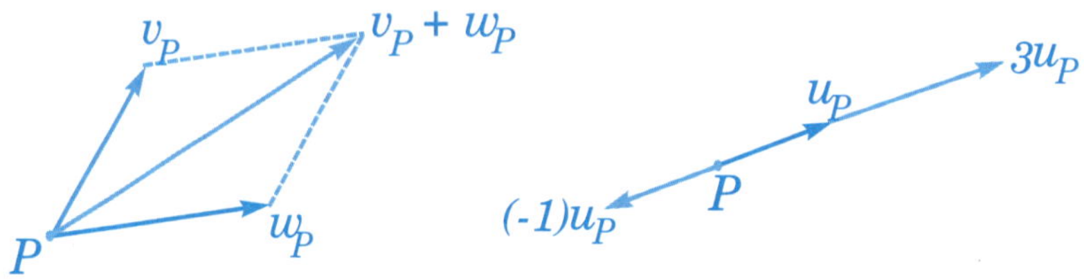

Figure 1.2-23

These two operations make this set $T_P(\mathbb{R}^3)$ a vector space for each $P \in \mathbb{R}^3$ and $T_P(\mathbb{R}^3)$ is isomorphic to $T_Q(\mathbb{R}^3)$ for $Q \in \mathbb{R}^3$ by the linear map* $v_P \mapsto v_Q$. Especially, since $\mathbb{R}^3 \equiv T_O(\mathbb{R}^3)$ for the origin $O \in \mathbb{R}^3$, $\mathbb{R}^3$ is isomorphic to $T_P(\mathbb{R}^3)$ for $P \in \mathbb{R}^3$ by $v \mapsto v_P$.

Definition 1.2.3 Let v_P be a tangent vector to $\mathbb{R}^3$ and f be a function† on $\mathbb{R}^3$. Then the **derivative of f in the direction** v_P is defined by

$$v_P[f] := \frac{d}{dt}\left(f(P + tv)\right)\Big|_{t=0}.$$

If $P = (p_1, p_2, p_3)$ and $v = (v_1, v_2.v_3)$, then $P + tv = (p_1 + tv_1, p_2 + tv_2, p_3 +$

*For two vector spaces V, W, a map $f : V \to W$ is is said to be **linear** if f satisfies for any $v, w \in V$ and $k \in \mathbb{R}$ (1) $f(v + w) = f(v) + f(w)$ and (2) $f(kv) = kf(v)$.

†Unless otherwise noted, the term "function" will mean "real-valued differentiable function"

$tv_3)$. By the chain rule,

$$\begin{aligned} v_P[f] &= \frac{d}{dt}(f(P+tv))\Big|_{t=0} \\ &= \frac{\partial f}{\partial x}(P+tv)\Big|_{t=0}\frac{d}{dt}(p_1+tv_1)\Big|_{t=0} \\ &+ \frac{\partial f}{\partial y}(P+tv)\Big|_{t=0}\frac{d}{dt}(p_2+tv_2)\Big|_{t=0} \\ &+ \frac{\partial f}{\partial z}(P+tv)\Big|_{t=0}\frac{d}{dt}(p_3+tv_3)\Big|_{t=0} \\ &= v_1\frac{\partial f}{\partial x}(P)+v_2\frac{\partial f}{\partial y}(P)+v_3\frac{\partial f}{\partial z}(P). \end{aligned}$$

Equivalently

$$v_P[f] = \sum_{i=1}^{3} v_i \frac{\partial f}{\partial x_i}(P)$$

where $x_1 = x,\ x_2 = y,\ x_3 = z$.

Hence we have the following theorem.

Theorem 1.2.4 *For $v_P, w_P \in T_P(\mathbb{R}^3)$, functions f, g on $\mathbb{R}^3$ and $s, k \in \mathbb{R}$,*

(1) $(sv_P + kw_P)[f] = sv_P[f] + kw_P[f]$

(2) $v_P[sf + kg] = sv_P[f] + kv_P[g]$

(3) $v_P[fg] = v_P[f]g(P) + f(P)v_P[g]$

Denote the **standard basis** for $\mathbb{R}^3$ by $\mathbf{e}_1 = (1,0,0), \mathbf{e}_2 = (0,1,0), \mathbf{e}_3 = (0,0,1)$, then every vector of $\mathbb{R}^3$ is spanned by $\{\mathbf{e}_1, \mathbf{e}_2, \mathbf{e}_3\}$.

Vector Fields

Definition 1.2.5 A **vector field** on $\mathbb{R}^3$ is a map

$$V : \mathbb{R}^3 \to \dot{\bigcup_{P\in\mathbb{R}^3}} T_P(\mathbb{R}^3)$$

defined by $Q \mapsto V(Q)$ where $V(Q)$ is a tangent vector in $T_Q(\mathbb{R}^3)$ and $\dot{\bigcup}_{P\in\mathbb{R}^3} T_P(\mathbb{R}^3)$ is the disjoint union of all tangent spaces $T_P(\mathbb{R}^3)$ for all $P \in \mathbb{R}^3$.

We also may define two operations, addition and scalar multiplication, for vector fields on $\mathbb{R}^3$ by using the addition and scalar multiplication of the tangent space as follows: For any vector field V, W and any function f on $\mathbb{R}^3$

(1) $(V+W)(P) := V(P) + W(P)$

(2) $(fV)(P) := f(P)V(P)$

for all $P \in \mathbb{R}^3$.

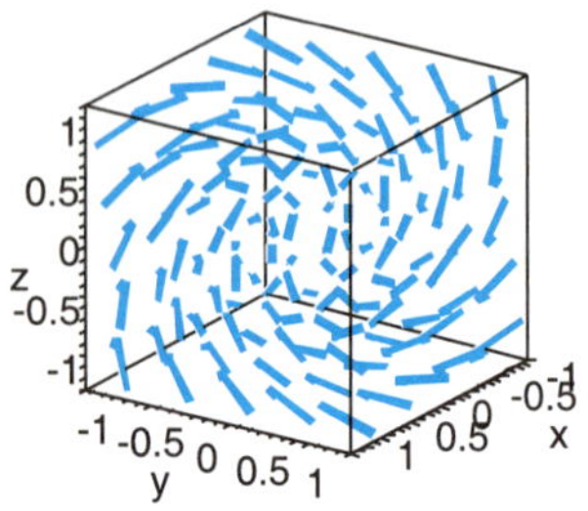

Figure 1.2-24

For example, $V = (y, z, x)$ is a vector field on $\mathbb{R}^3$(figure 1.2-24). For $P = (1, -2, 3) \in \mathbb{R}^3$, $V(P) = (y, z, x)_P = (-2, 3, 1)$ is a tangent vector to $\mathbb{R}^3$ at P(Figure 1.2-25).

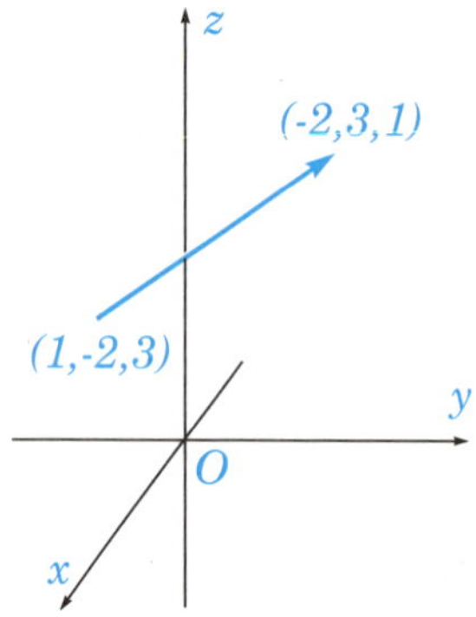

Figure 1.2-25

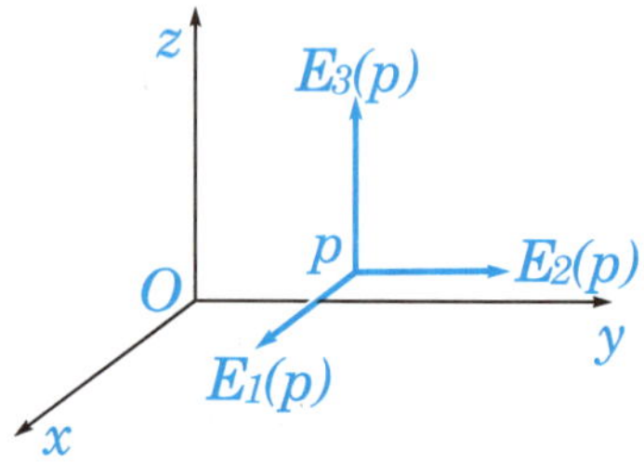

Figure 1.2-26

The **standard frame field** $\{E_1, E_2, E_3\}$ on $\mathbb{R}^3$ is defined by

$$E_1(P) := (\mathbf{e}_1)_P = (1, 0, 0)_P$$
$$E_2(P) := (\mathbf{e}_2)_P = (0, 1, 0)_P$$
$$E_3(P) := (\mathbf{e}_3)_P = (0, 0, 1)_P$$

for each $P \in \mathbb{R}^3$(figure 1.2-26).

Now, the derivative $V[f]$ of a function f with respect to a vector field V on $\mathbb{R}^3$ is defined by $V(P)[f]$ for each point P

Hence, for $i = 1, 2, 3$

$$E_i[f] = \frac{\partial f}{\partial x_i}.$$

Thus we may also obtain following theorem corresponding to Theorem 1.2.4.

Theorem 1.2.6 *For vector fields V, W on $\mathbb{R}^3$, functions f, g, h on $\mathbb{R}^3$, and $s, k \in \mathbb{R}$*

(1) $(fV + gW)[h] = fV[h] + gW[h]$

(2) $V[sf + kg] = sV[f] + kV[g]$

(3) $V[fg] = V[f]g + fV[g]$

Definition 1.2.7 Let $f : D \subset \mathbb{R}^3 \to \mathbb{R}$ be a differentiable function. The **gradient** of f is the vector field $grad\, f$ on $\mathbb{R}^3$ given by

$$grad\, f = \sum_{j=1}^{3} \frac{\partial f}{\partial x_j} E_j.$$

For further study, see the following reference.

Barret O'Neill, Elementary Differential Geometry, Academic Press Inc., New York, 1966.

1.3 Minkowski Space-times

In 1905, Albert Einstein (1879-1955) introduced Special Relativity, which was a very novel physical theory, since the older theory of aether was entirely discarded. Even though Hendrik Lorentz (1853-1928) and Jules Henri Poincaré (1854-1912) had considered Lorentz-FitzGerald contractions as mathematical constructs slightly earlier, Einstein was the first to employ these ideas in a full fledged physical theory. While Einstein was a university student at the ETH in Zurich, Switzerland, one of his professors was Herman Minkowski (1864-1909). Later while at Göttingen, Minkowski developed the 4-dimensional differential geometry behind Special Relativity, introducing the current terminologies "pseudo-Euclidean" and "world line" (for the trajectory of a particle in 4-dimensions in this model). Minkowski popularized Special Relativity and his own mathematical treatment at an Assembly of the German Natural Scientists and Physicians held at Cologne in 1908 in a famous lecture called "Space and Time". During this period, Einstein was apparently unsympathetic to Minkowski's viewpoint. But when Einstein adopted the formalism of differential geometry in his later theory of General Relativity published in 1916, he took a kindlier view.

Inner Products and Cross Products in $\mathbb{R}^3$

For two vectors $x = (x_1, x_2, x_3)$ and $y = (y_1, y_2, y_3)$ in $\mathbb{R}^3$ the inner product of x and y is defined by

$$< x, y >= x_1y_1 + x_2y_2 + x_3y_3.$$

The **length** of a vector $x = (x_1, x_2, x_3)$ in $\mathbb{R}^3$ is given by

$$|x| = \sqrt{< x, x >} = \sqrt{x_1^2 + x_2^2 + x_3^2}.$$

The inner product induces the Euclidean geometry on $\mathbb{R}^3$, and thus we denote $\mathbb{R}^3$ with this inner product by

$$\mathbf{E}^3 = (\mathbb{R}^3, < \ , \ >).$$

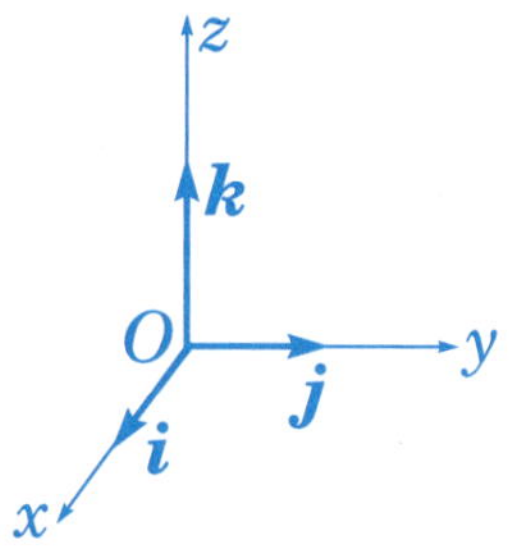

Figure1.3-27

Recall that x is a unit vector if $|x| = 1$ and x and y are orthogonal(or perpendicular) if $< x, y >= 0$. Hence, any three vectors $\{x, y, z\}$ which are unit orthogonal vectors in $\mathbf{E}^3$ are said to be an **orthonormal basis** for $\mathbf{E}^3$. The standard basis $\mathbf{i} = (1, 0, 0)$, $\mathbf{j} = (0, 1, 0)$, $\mathbf{k} = (0, 0, 1)$ for $\mathbf{E}^3$ is also orthonormal(Figure1.3-27).

This inner product on $\mathbf{E}^3$ satisfies the following properties.

(1) (positive definiteness): For any vector x, $< x, x >\geq 0$ and $< x, x >= 0 \Longleftrightarrow x = 0$.

(2) (linearity): For any three vectors x, y, z and real number k
$< kx, y >= k < x, y >$,
$< x, y + z >=< x, y > + < x, z >$

(3) (commutativity): $< x, y >=< y, x >$

The **cross product** for two vectors $a = (a_1, a_2, a_3), b = (b_1, b_2, b_3) \in \mathbb{R}^3$ is defined by $a \times b = (a_2b_3 - a_3b_2, a_3b_1 - a_1b_3, a_1b_2 - a_2b_1)$.

This definition seems more complicated than the inner product but has the matrix representation:

$$\begin{aligned} a \times b &= (a_2b_3 - a_3b_2)\mathbf{i} - (a_1b_3 - a_3b_1)\mathbf{j} + (a_1b_2 - a_2b_1)\mathbf{k} \\ &= \begin{vmatrix} a_2 & a_3 \\ b_2 & b_3 \end{vmatrix} \mathbf{i} - \begin{vmatrix} a_1 & a_3 \\ b_1 & b_3 \end{vmatrix} \mathbf{j} + \begin{vmatrix} a_1 & a_2 \\ b_1 & b_2 \end{vmatrix} \mathbf{k} \end{aligned}$$

Hence,

$$a \times b = \begin{vmatrix} \mathbf{i} & \mathbf{j} & \mathbf{k} \\ a_1 & a_2 & a_3 \\ b_1 & b_2 & b_3 \end{vmatrix}.$$

Using matrix properties, we may obtain the following theorem. Property (5) may be derived by successive applications of properties (3) and (4).

Theorem 1.3.1 *For $a, b, c, d \in \mathbf{E}^3$*

(1) $< a \times b, a >=< a \times b, b >= 0$

(2) $a \times b = -b \times a$

(3) $< a \times b, c >=< a, b \times c >$

(4) $(a \times b) \times c =< a, c > b- < b, c > a$

(5) $< a \times b, c \times d >=< a, c >< b, d > - < b, c >< a, d >$

Further, we may compute the length of $a \times b$ in $\mathbf{E}^3$ as follows.

Theorem 1.3.2

$$|a \times b| = |a||b| \sin \theta.$$

Proof.

$$
\begin{aligned}
|a \times b|^2 =& (a \times b) \cdot (a \times b) \\
=& (a_2b_3 - a_3b_2)^2 + (a_3b_1 - a_1b_3)^2 + (a_1b_2 - a_2b_1)^2 \\
=& a_2^2b_3^2 - 2a_2a_3b_2b_3 + a_3^2b_2^2 + a_3^2b_1^2 - 2a_1a_3b_1b_3 \\
& + a_1^2b_3^2 + a_1^2b_2^2 - 2a_1a_2b_1b_2 + a_2^2b_1^2 \\
=& (a_1^2 + a_2^2 + a_3^2)(b_1^2 + b_2^2 + b_3^2) - (a_1b_1 + a_2b_2 + a_3b_3)^2 \\
=& |a|^2|b|^2 - (a \cdot b)^2 \\
=& |a|^2|b|^2 - |a|^2|b|^2 \cos^2 \theta \\
=& |a|^2|b|^2 \sin^2 \theta
\end{aligned}
$$

□

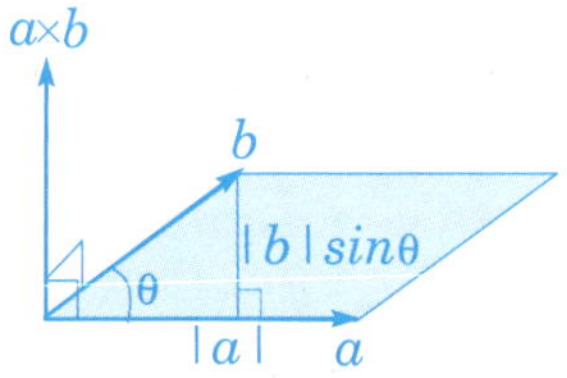

Figure1.3-28

Thus, the vector $a \times b$ is perpendicular to the 2-plane spanned by a and b, and its length $|a \times b|$ is the area of the parallelogram given by a and b(Figure 1.3-28). This formula in Theorem 1.3.2 is often used in calculus to demonstrate that two nonzero vectors a, b are parallel if and only if $a \times b = 0$.

The Lorentz Metric

Now we introduce a second bilinear form on $\mathbb{R}^3$ defined by

$$g(x, y) = x_1y_1 + x_2y_2 - x_3y_3$$

for two vectors $x = (x_1, x_2, x_3)$ and $y = (y_1, y_2, y_3)$ in $\mathbb{R}^3$. Since the meaning will be clear from context, we will still denote the Lorentz metric g by $<\ ,\ >$ as in the Euclidean case. The Lorentz bilinear form satisfies properties (2) (linearity) and (3) (commutativity) of the Euclidean inner product, but property (1) (positive definiteness) fails. Especially

$$< x, x >= (x_1)^2 + (x_2)^2 - (x_3)^2$$

may be negative or zero with the vector x nonzero. However, the Lorentz metric satisfies the following important property which serves as a partial substitute for the positive definiteness of the Euclidean inner product :

(4) (nondegeneracy): if $< x, y >= 0$ for all y in $\mathbb{R}^3$, then $x = 0$.

This may be easily established by noting that $< x, \mathbf{i} >= x_1, < x, \mathbf{j} >= x_2$ and $< x, \mathbf{k} >= -x_3$.

An analogous bilinear form g on $\mathbb{R}^4$ is used in Einstein's Special Relativity to study a 4-dimensional cosmos $(\mathbb{R}^4, g)$ called **Minkowski space-time** (where units are chosen such that the speed of light $c = 1$).

Definition 1.3.3 A nonzero vector x in $\mathbb{R}^3$ is said to be

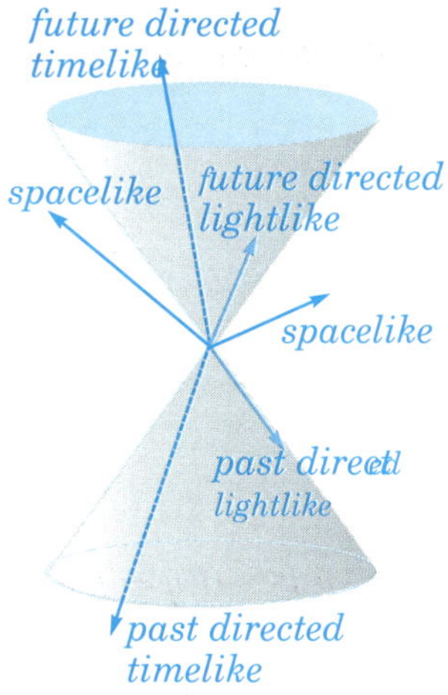

Figure1.3-29

i. **spacelike** if $< x, x >> 0$,

ii. **timelike** if $< x, x >< 0$,

iii. **lightlike**(or **null**) if $< x, x >= 0$,

iv. **nonspacelike** if $< x, x >\leq 0$.

Further, we may classify all nonspacelike vectors into **future-directed** vectors if $< x, \mathbf{k} >< 0$ and **past-directed** vectors if $< x, \mathbf{k} >> 0$(Figure1.3-29).

From now on, we call g defined as above the **Lorentz metric** and we denote the 3 dimensional Minkowski space by $\mathbf{L}^3 = (\mathbb{R}^3, g)$.

Somewhat as in $\mathbf{E}^3$ the length of a vector x in $\mathbf{L}^3$ is defined by

$$|x| = \sqrt{|< x, x >|} = \sqrt{|x_1^2 + x_2^2 - x_3^2|}.$$

Moreover, a vector v is said to be **unit** if $|v| = 1$. Note that $|x|^2 = -< x, x >$ for x timelike.

Example 1.3.4 For $x = (2, 1, -3)$, $y = (1, -2, -2)$ in $\mathbb{R}^3$ compute $< x, y >$ and $|x|$ in $\mathbf{E}^3$ and $\mathbf{L}^3$ respectively.

Solution. In $\mathbf{E}^3$, $< x, y >= 2 \cdot 1 + 1 \cdot (-2) + (-3) \cdot (-2) = 6$,
$|x| = \sqrt{2^2 + 1^2 + (-3)^2} = \sqrt{14}$.
In $\mathbf{L}^3$, $< x, y >= 2 \cdot 1 + 1 \cdot (-2) - (-3) \cdot (-2) = -6$,
$|x| = \sqrt{|2^2 + 1^2 - (-3)^2|} = \sqrt{4}$ □

In $\mathbf{L}^3$, a spacelike vector x satisfies $< x, x >> 0$, but it is not true that $< x, x >= 0 \Longrightarrow x = 0$.

The Reverse Triangle Inequality

Theorem 1.3.5 *$x = (x_1, x_2, t_1)$ is future directed timelike if and only if $t_1 > 0$ and $x_1^2 + x_2^2 < t_1^2$.*

Proof. Supposing x is timelike

$$< x, x >= x_1^2 + x_2^2 - t_1^2 < 0.$$

Hence, $x_1^2 + x_2^2 < t_1^2$. Moreover, $< x, \mathbf{k} >= -t_1 < 0$ implies $t_1 > 0$. □

Theorem 1.3.6 *Let $x, y \in \mathbf{L}^3$ be nonzero future directed timelike vectors and $z \in \mathbf{L}^3$ be a future directed lightlike vector. Then*

(1) $< x, y >< 0$.

(2) $x + y$ *is future directed.*

(3) $< x, z >< 0$ *and* $x + z$ *is future directed.*

(4) If $< x, w >= 0$ *for nonzero* $w \in \mathbf{L}^3$, *then* w *is spacelike.*

(5) $- < x, y >\geq |x||y|$ *and equality holds iff for some real* $k > 0$ $y = kx$.

(6) $|x + y| \geq |x| + |y|$ *and equality holds iff for some real* $k > 0$ $y = kx$.

Proof.

(1) Since x, y are future directed and since

$$t_1 > 0,\ x_1^2 + x_2^2 < t_1^2$$

$$t_2 > 0,\ y_1^2 + y_2^2 < t_2^2$$

we have

$$(x_1^2 + x_2^2)(y_1^2 + y_2^2) < t_1^2 t_2^2$$

and

$$(x_1 y_1 + x_2 y_2)^2 \leq (x_1^2 + x_2^2)(y_1^2 + y_2^2).$$

Since

$$x_1 y_1 + x_2 y_2 \leq |x_1 y_1 + x_2 y_2| < t_1 t_2$$

we obtain

$$< x, y >= x_1 y_1 + x_2 y_2 - t_1 t_2 < 0.$$

(2) Since $< x + y,\ x + y >=< x, x > +2 < x, y > + < y, y >$, $< x, x >< 0$, $< y, y >< 0$ and from (1), we obtain $< x + y,\ x + y >< 0$. Moreover, $t_1 > 0, t_2 > 0$ imply $t_1 + t_2 > 0$.

(3) Now let x be future directed timelike and z be future directed lightlike. Then

$$t_1 > 0,\ x_1^2 + x_2^2 < t_1^2$$

$$t_2 > 0,\ z_1^2 + z_2^2 = t_2^2$$

so that as in (1) and (2) above,

$$(x_1^2 + x_2^2)(z_1^2 + z_2^2) < t_1^2 t_2^2$$

and

$$< x, z >= x_1 z_1 + x_2 z_2 - t_1 t_2 < 0.$$

Also,

$$< x + z, x + z >=< x, x > +2 < x, z >< 0$$

since $< x, x >< 0$, $< z, z >= 0$ and $< x, z >< 0$. As $t_1 + t_2 > 0$, $x + z$ is future directed.

(4) Suppose $x \in \mathbf{L}^3$ is future directed timelike and $< x, w >= 0$ with $w \in \mathbf{L}^3$ nonzero. If w were nonspacelike, replacing w with $-w$ if necessary, we may suppose w is future directed. But then $< x, w >< 0$ from (1) or (3), in contradiction.

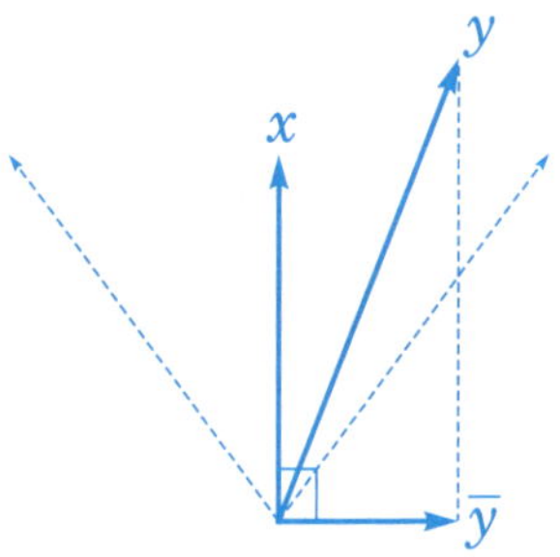

Figure1.3-30

(5) Given x, y future directed timelike, form

$$\vec{y} = y - \frac{< y, x >}{< x, x >} x$$

and note that $< x, \vec{y} >= 0$. Hence by (4), $\vec{y} = 0$ or $\vec{y}$ is spacelike. Putting $k = \dfrac{< y, x >}{< x, x >} > 0$, we may write the above decomposition as $y = kx + \vec{y}$(Figure 1.3-30). Since $< x, \vec{y} >= 0$, $\vec{y}$ is zero or spacelike. Thus we obtain

$$< y, y >= k^2 < x, x > + < \vec{y}, \vec{y} >$$

so also

$$k^2 = \frac{< y, y > - < \vec{y}, \vec{y} >}{< x, x >}.$$

Returning to $k = \dfrac{< y, x >}{< x, x >}$, we have

$$\begin{aligned} < x, y >^2 &= k^2 (< x, x >)^2 \\ &= \left(\frac{< y, y > - < \vec{y}, \vec{y} >}{< x, x >} \right) (< x, x >)^2 \\ &= (< y, y > - < \vec{y}, \vec{y} >) < x, x > . \end{aligned}$$

As $\vec{y}$ is spacelike or zero,

$$(< x, y >)^2 \geq < y, y >< x, x >= (- < y, y >)(- < x, x >).$$

Hence, $| < x, y > | \geq |x||y|$ which implies that

$$- < x, y >\geq |x||y|$$

since $< x, y >< 0$. From the derivation, equality holds in (5) iff $\vec{y} = 0$, which holds iff $y = kx$.

(6) Now consider

$$(|x|+|y|)^2 = |x|^2 + 2|x||y| + |y|^2 \\ \leq |x|^2 - 2 < x, y > + |y|^2.$$

from (5). Since x, y and $x+y$ are future directed timelike, we further obtain

$$(|x|+|y|)^2 \leq - < x, x > -2 < x, y > - < y, y > \\ = - < x+y,\ x+y > \\ = |x+y|^2.$$

Thus $|x+y| \geq |x|+|y|$. From the derivation, equality holds iff $- < x, y >= |x||y|$, which by (5) holds iff $y = kx$. □

The hyperbolic cosine $\cosh\theta$ satisfies $\cosh\theta \geq 1$ and $\cosh\theta$ is strictly increasing for $\theta \geq 0$ (cf. Section 1.4 for the definition). Since $\dfrac{- < x, y >}{|x||y|} \geq 1$ from (5), by the Intermediate Value Theorem given x, y future directed timelike, there exists a unique $\theta \geq 0$ so that

$$\frac{- < x, y >}{|x||y|} = \cosh\theta.$$

The corresponding real number θ is said to be the **hyperbolic angle** between x and y. Thus,

$$< x, y >= -|x||y|\cosh\theta.$$

Remark 1.3.7 Theorem 1.3.6 (5) does not hold when x, y are spacelike vectors in $\mathbf{L}^3$ in general but it works in L^2. That is, $< x, y >\geq |x||y|$ for two spacelike vectors $x,\ y$ in $\mathbf{L^2}$ and equality holds iff for some real $k > 0$ $y = kx$. Therefore, there is also a hyperbolic angle θ such that

$$< x, y >= |x||y|\cosh\theta.$$

Remark 1.3.8 The inequality in Theorem 1.3.6 (6) is called the **reverse triangle inequality**. This inequality may be related to the Twin Paradox in Special Relativity, which has generated more controversy than any other topic in relativity theory in the words of a basic physics text.

One of two twin brothers starts at O to travel to a star, reaching $A(1/2, 1)$ after 1 year. Soon, he starts back to rejoin his twin brother who has remained at O awaiting him. We may represent the starting point at time zero as $O(0,0)$ and the same place two years later as $B(0,2)$ in $\mathbf{L}^2$ (Figure1.3-31). For $x = \vec{OA}$ and $y = \vec{AB}$, we have $|x+y| = 2$, $|x| = \sqrt{3/4}$, $y = \sqrt{3/4}$ and $|x|+|y| = \sqrt{3}$. Thus, $|x|+|y| < |x+y|$ and the traveling twin brother will be younger than his brother who remained on earth at O.

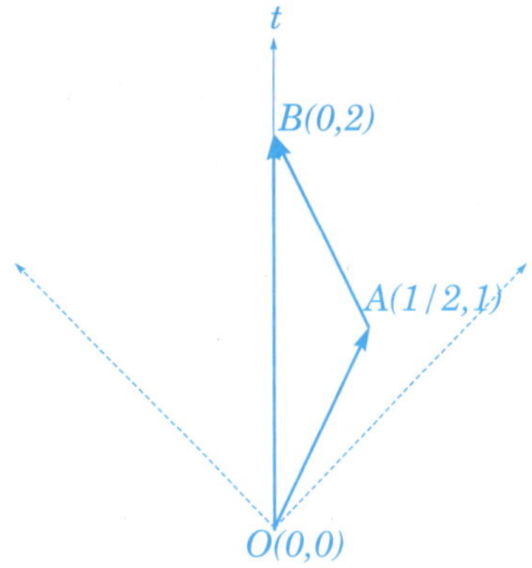

Figure1.3-31

Orthonormal Bases in $\mathbf{L}^3$

Motivated by the Euclidean definition, a basis $\{e_1, e_2, e_3\}$ for $\mathbf{L}^3$ is said to be **orthonormal** if $|e_1| = |e_2| = |e_3| = 1$ and $< e_i, e_j >= 0$ for all i not equal to j. As for $\mathbf{E}^3$, $\{\mathbf{i}, \mathbf{j}, \mathbf{k}\}$ is the standard basis for $\mathbf{L}^3$. Note that $\{\mathbf{i}, \mathbf{j}, \mathbf{k}\}$ is an orthonormal basis for $\mathbf{L}^3$. Especially, $\mathbf{i}, \mathbf{j}$ are spacelike and $\mathbf{k}$ is timelike.

Theorem 1.3.9 *Every orthonormal basis has one timelike vector and two spacelike vectors.*

Proof. Let $\{e_1, e_2, e_3\}$ be an orthonormal basis for $\mathbf{L}^3$. Then for any vector v

$$v = \sum_{i=1}^{3} v_i e_i.$$

Suppose that $\{e_1, e_2, e_3\}$ are all spacelike vectors. Taking the inner product on both sides, we have for v nonzero that

$$< v, v >= \sum_{i=1}^{3} v_i^2 < e_i, e_i >> 0,$$

which means that v is always spacelike. This contradicts that $\mathbf{k}$ is timelike. Thus, one of $\{e_1, e_2, e_3\}$, say e_3, must be timelike(it could not be lightlike since $|e_i| = 1,\ i = 1, 2, 3.$). But then as $< e_1, e_2 >=< e_1, e_3 >= 0$, the vectors e_1 and e_2 are spacelike by Theorem 1.3.6 (4). □

Sometimes in Relativity, a **pseudo-orthonormal basis** $\{n_1, n_2, e_1\}$ with n_1, n_2 lightlike, e_1 spacelike, $< n_1, e_1 >=< n_2, e_1 >= 0, |e_1| = 1$, and $< n_1, n_2 >= -1$ is useful. Such a basis may be constructed from an orthonormal basis $\{e_1, e_2, e_3\}$ with e_3 timelike by putting $n_1 = \dfrac{e_3 + e_2}{\sqrt{2}}$ and $n_2 = \dfrac{e_3 - e_2}{\sqrt{2}}$.

The Lorentz Cross Product

For nonzero orthogonal vectors $a, b, c \in \mathbf{E}^3$ (i.e., $< a, b >=< a, c >=< b, c >= 0$), the Euclidean cross product $(a \times b) \times c = 0$ as $a \times b$ must be parallel to c. Consider now Minkowski 3-space $\mathbf{L}^3 = (\mathbb{R}^3, g)$ and the Euclidean cross product. Let $u = \mathbf{i} + \mathbf{j} + \mathbf{k}$, $v = \mathbf{i} - \mathbf{j}$ and $w = \mathbf{i} + \mathbf{j} + 2\mathbf{k}$. Note that $g(u, v) = g(u, w) = g(v, w) = 0$ so that $\{u, v, w\}$ is a Lorentz orthogonal basis for $\mathbf{L}^3$. But $u \times v = \mathbf{i} + \mathbf{j} - 2\mathbf{k}$ so that $g(u \times v, u) = 4$. Thus, generally, $u \times v$ will not be Lorentz orthogonal to u and v. Also $(u \times v) \times w = 4\mathbf{i} - 4\mathbf{j}$ is nonzero.

Hence, a modified Lorentz cross product $u \times_g v$ is needed with $g(u \times_g v, u) = g(u \times_g v, v) = 0$. Fortunately, given a nondegenerate bilinear form b for $\mathbb{R}^3$, a cross product $u \times_b v$ may be constructed for $(\mathbb{R}^3, b)$ which satisfies $b(u \times_b v, u) = b(u \times_b v, v) = 0$ and many other properties analogous to those of the Euclidean cross product for $\mathbf{E}^3$. Take any Euclidean orthonormal basis $\{e_1, e_2, e_3\}$ for $\mathbb{R}^3$ and calculate $\det(B) = \det(B(e_i, e_j))$. Fixing $u, v \in \mathbb{R}^3$, the map $\mathbb{R}^3 \to \mathbb{R}$ given by $z \mapsto \sqrt{|\det(B)|}\det(z, u, v)$ is linear. Hence, because b is nondegenerate, there exists a unique $w \in \mathbb{R}^3$ with

$$b(w, z) = \sqrt{|\det(B)|}\det(z, u, v)$$

for all $z \in \mathbb{R}^3$. Put $u \times_b v := w$.

Now we particularize to the case that the bilinear form is the Lorentz metric $g =< \, , >$ for $\mathbb{R}^3$. Selecting $\{\mathbf{i}, \mathbf{j}, \mathbf{k}\}$ for the Euclidean orthonormal basis, we have

$$B = \begin{pmatrix} 1 & 0 & 0 \\ 0 & 1 & 0 \\ 0 & 0 & -1 \end{pmatrix}$$

whence $\det(B) = -1$ and $\sqrt{|\det(B)|} = +1$. Thus if we fix $a, b \in \mathbf{L}^3$, $w = a \times_g b$ must satisfies

$$g(w, z) = \det(z, a, b)$$

for all $z \in \mathbf{L}^3$. Especially taking $z = \mathbf{i}, \mathbf{j}$ and $\mathbf{k}$, we find that for $w = A\mathbf{i} + B\mathbf{j} + C\mathbf{k}$

$$A =< w, \mathbf{i} >= \det(\mathbf{i}, a, b) = a_2 b_3 - a_3 b_2$$

$$B =< w, \mathbf{j} >= \det(\mathbf{j}, a, b) = -(a_1 b_3 - a_3 b_1)$$

$$-C =< w, \mathbf{k} >= \det(\mathbf{k}, a, b) = a_1 b_2 - a_2 b_1,$$

where we will frequently use $< \, , >$ for the Lorentz metric g below. Thus $A = a_2 b_3 - a_3 b_2$, $B = -(a_1 b_3 - a_3 b_1)$ and $C = -(a_1 b_2 - a_2 b_1)$ so that in a determinant formulation

$$a \times_g b = \begin{vmatrix} \mathbf{i} & \mathbf{j} & -\mathbf{k} \\ a_1 & a_2 & a_3 \\ b_1 & b_2 & b_3 \end{vmatrix}.$$

Especially, if $a \times b = A\mathbf{i} + B\mathbf{j} + C\mathbf{k}$ with the Euclidean cross product, then $a \times_g b = A\mathbf{i} + B\mathbf{j} - C\mathbf{k}$. Also, from the properties of the determinant, $b \times_g a = -a \times_g b$. As another by product of the defining equation $g(w, z) = \det(z, a, b)$ for the Lorentz cross product, we have

$$g(a, b \times_g c) = g(b \times_g c, a) = \det(a, b, c)$$

$$= \begin{vmatrix} a_1 & a_2 & a_3 \\ b_1 & b_2 & b_3 \\ c_1 & c_2 & c_3 \end{vmatrix}$$

just as for the Euclidean theory. While the direction of the Euclidean cross product of nonzero, nonparallel vectors may always be determined by the right hand rule(Figure 1.3-28), the Lorentz cross product sometimes obeys a right hand rule and sometimes a left hand rule. For $a = \mathbf{i}$ spacelike and $b = \mathbf{i} + \mathbf{j} + 2\mathbf{k}$ future directed timelike, $a \times_g b = -2\mathbf{j} - \mathbf{k}$ has a direction relative to a and b consistent with the usual right hand rule(Figure1.3-32). But for $a = 2\mathbf{i} + \mathbf{k}$ and $b = \mathbf{j}$ spacelike vectors, $a \times_g b = -\mathbf{i} - 2\mathbf{k}$ has a direction relative to a and b with a left hand rule(Figure1.3-33). By contrast with property (1) of Theorem 1.3.10 below, the Lorentz cross product of two nonzero, nonorthogonal spacelike vectors may be null. Take $a = \mathbf{i} + \mathbf{j} + \mathbf{k}$ and $b = \mathbf{i} + 2\mathbf{j} + \mathbf{k}$, then $a \times_g b$ is null.

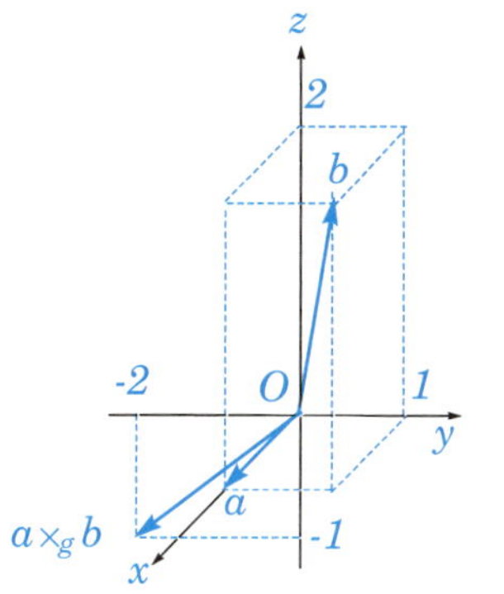

Figure1.3-32

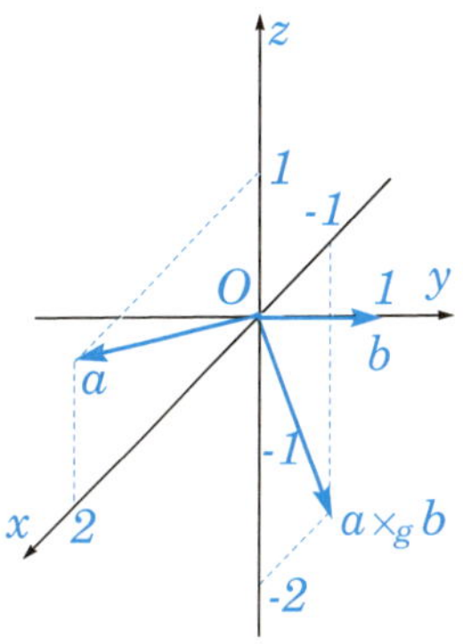

Figure1.3-33

Theorem 1.3.10 *For a, b, c in $\mathbf{L}^3$,*

(1) $< a \times_g b, a >=< a \times_g b, b >= 0$. Hence, if a is timelike, $a \times_g b$ is spacelike or zero.

(2) $b \times_g a = -a \times_g b$.

(3) $< a \times_g b, c >=< a, b \times_g c >$.

(4) $(a \times_g b) \times_g c = -(< a, c > b- < b, c > a)$.

(5) Nonzero a, b in $\mathbf{L}^3$ are parallel iff $a \times_g b = 0$.

Proof. We will write $a \times b$ for $a \times_g b$ in the proof, as well as $< \, , \, >$ for the Lorentz metric g.

(1) Since (2) has already been noted, $< a \times b, b >= - < b \times a, b >= 0$ will follow from $< a \times b, a >= 0$. But $< a \times b, a >=< a, a \times b >= a_1(a_2b_3 - a_3b_2) - a_2(a_1b_3 - a_3b_1) - a_3(-(a_1b_2 - a_2b_1)) = 0$. The second assertion now follows from Theorem 1.3.6 (4), replacing a with $-a$ if a is past directed.

(3) Recalling the above formula for $< a, b \times c >$, we have $< a, b \times c >= \det(a, b, c) = \det(c, a, b) =< c, a \times b >=< a \times b, c >$.

(4) Since we are calculating in low dimensions, one may simply calculate $(a \times b) \times c$ using the determinant formula for the Lorentz cross product twice in succession, then verify that a direct calculation of $- < a, c > b+ < b, c > a$ yields the same result. A somewhat more conceptional argument may also be given, valid for general nondegenerate bilinear forms, where this formula is studied for a Lorentz orthonormal basis and an appeal is made to Theorem 1.3.11 below.

(5) As for the Euclidean cross product, determinant properties imply that $a \times b = 0$ if $b = \lambda a$, $\lambda \in \mathbb{R}$. Now suppose $a \times b = 0$ with a, b nonzero. Then $< a \times b, z >= \det(z, a, b) = 0$ for all $z \in \mathbf{L}^3$. Taking $z = \mathbf{k}, \mathbf{i}$ and $\mathbf{j}$ successively,

(i) $a_1b_2 = a_2b_1$

(ii) $a_1b_3 = a_3b_1$

(iii) $a_2b_3 = a_3b_2$

As $a \neq 0$ is assumed, by symmetry we may suppose $a_1 \neq 0$. Put $\lambda = \dfrac{b_1}{a_1}$. Then using (i) and (ii), $\lambda(a_1, a_2, a_3) = \left(b_1, \dfrac{a_2b_1}{a_1}, \dfrac{a_3b_1}{a_1}\right) = \left(b_1, \dfrac{a_1b_2}{a_1}, \dfrac{a_1b_3}{a_1}\right) = (b_1, b_2, b_3)$. Hence, $b = \lambda a$ as desired. □

Theorem 1.3.11 *Let $\{e_1, e_2, e_3\}$ be an orthonormal basis for $\mathbf{L}^3$. Then replacing e_3 with $-e_3$ if necessary,*

$$e_i \times_g e_j =< e_k, e_k > e_k,$$

where $i \neq j, j \neq k, i \neq k,\ \ k = 1, 2, 3.$

With these results, further parallels between the Euclidean and Lorentz cross product may be obtained, including a partial analogue to the basic result Theorem 1.3.2.

Theorem 1.3.12 *For $a, b, c, d \in \mathbf{L}^3$,*

(1) $< a \times_g b, c \times_g d >= -(< a, c >< b, d > - < b, c >< a, d >)$ Hence a, b are nonzero, nonnull vectors in $\mathbf{L}^3$ which are orthogonal, then $a \times_g b$ is nonzero and nonnull.

(2) if a and b are future directed timelike and $\theta \geq 0$ is the hyperbolic angle between a and b, then

$$|a \times_g b| = |a||b| \sinh \theta.$$

Proof. (1) Let $v = a \times_g b$. Then

$$\begin{aligned} &< a \times b, c \times d >=< v, c \times d >=< v \times c, d > \\ &\quad =< (a \times b) \times c, d >=< - < a, c > b+ < b, c > a, d > \\ &\quad = -(< a, c >< b, d > - < b, c >< a, d >). \end{aligned}$$

Especially, if $< a, b >= 0$, then $< a \times b, a \times b >= - < a, a >< b, b >$ from which the second claim follows.

(2) Given a, b future directed timelike, recall from the proof of Theorem 1.3.6 that $< a, b >= -|a||b| \cosh \theta$. Also, since a is timelike and $a \times b$ is orthogonal to a, we have $a \times b$ is spacelike or zero. If $a \times b = 0$, then a and b are parallel by Theorem 1.3.10 (5), $\theta = 0$, and both sides of the desired equality are zero.

Thus, assume $a \times b$ is spacelike. Then

$$\begin{aligned}|a \times b|^2 &=< a \times b, a \times b > \\ &= -(< a,a >< b,b > -(< a,b >)^2) \\ &= -(< a,a >< b,b > -(|a||b| \cosh \theta)^2) \\ &= -[< a,a >< b,b > -((- < a,a >)(- < b,b >) \cosh^2 \theta)] \\ &=< a,a >< b,b > (\cosh^2 \theta - 1) \\ &= (- < a,a >)(- < b,b >) \sinh^2 \theta.\end{aligned}$$

Hence $|a \times b| = |a||b| \sinh \theta$, as $\theta \geq 0$. □

The correspondences between the Euclidean and Lorentz cross product may be viewed on a more abstract algebraic basis as follows. Consider the linear transformation $T : \mathbb{R}^3 \to \mathbb{R}^3$ given by $T((a_1, a_2, a_3)) = (a_1, a_2, -a_3)$. Then T is an isomorphism(i.e., $1-1$ and onto) and because $a \times b = A\mathbf{i} + B\mathbf{j} + C\mathbf{k}$ iff $a \times_g b = A\mathbf{i} + B\mathbf{j} - C\mathbf{k}$ as noted above, $T(a \times b) = a \times_g b$ and $T(a \times_g b) = a \times b$. Further the Euclidean and Lorentz inner products are related by

$$g(u,v) =< u, T(v) >$$

for all $u, v \in \mathbb{R}^3$. Hence, the isomorphism T may be used to transfer properties of the Euclidean cross product to the Lorentz cross product. Consider (1) of Theorem 1.3.10 as an example. Taking $a, b \in \mathbf{L}^3 = \mathbb{R}^3$,

$$\begin{aligned}g(a \times_g b, a) &= g(a, a \times_g b) \\ &=< a, T(a \times_g b) > \\ &=< a, a \times b > \\ &= 0.\end{aligned}$$

In Minkowski space-time $\mathbf{L}^3$, we usually use the coordinate x, y, t, for example, $x = (x_1, x_2, t_1)$, $y = (y_1, y_2, t_2)$. We also call t the **time axis** instead of the z axis.

While the differential and integral calculus were developed over several centuries with the systematic work of Newton and Leibniz occurring in the 1600's, the vector analysis methods used in this section are much more recent. While searching for 3-dimensional complex numbers, William R. Hamilton of Dublin, Ireland discovered quaternions with their noncommutative (but still associative multiplication) in 1843. James Clerk Maxwell, in formulating his equations of electromagnetism in 1864, employed a quaternionic formulation. Dissatisfied with the complexities of this technique in which vector analysis implicitly lurked through the separate treatments of the scalar and vector parts of quaternions,

Josiah Willard Gibbs of Yale University and Oliver Heaviside in England independently developed 3-dimensional vector analysis in the 1880s. (Apart from Hamilton's work, Gibbs and Heaviside were also acquainted with the alternative formalism of the German high school teacher Herman Günther Grassmann, first published in 1844.) Heaviside was the first to formulate the Maxwell's equations in the vector analysis form used today. In the 1880's, certain aspects of vector analysis were quite startling; not only is the vector cross product noncommutative, but it is also nonassociative since $(a \times b) \times c$ and $a \times (b \times c)$ are generally different vectors.

For further study, see the following references. The first reference deals primarily with $\mathbf{L}^2$.

1. Beem, J. K., P. E. Ehrlich and K. L. Easley, Global Lorentzian Geometry, Second Edition, Marcel Dekker Inc. New York, 1996.

2. Birman, G. and K. Nomizu, *Trigonometry in Lorentzian Geometry*, Amer. Math. Monthly 91(1984), 543-549.

3. Birman, R. and K. Nomizu, The Gauss-Bonnet theorem for two-dimensional space-time, Mich. Math. J. 31(1984) 77-81.

4. Patrick J. Ryan, Euclidean and Non-Euclidean Geometry, An Analytic Approach, Cambridge University Press, New York, 1999.

1.4 Hyperbolic Trigonometry

In Section 1.1, Johann Lambert's studies of quadrilaterals with 3 right angles in his efforts to prove the Parallel Postulate were mentioned. These investigations led Lambert in 1768 to consider the trigonometric functions of an imaginary angle iA, thus, in effect, introducing the hyperbolic functions drawing on earlier results of Euler and the Bernoullis.

Hyperbolic Functions

Using the Maclaurin series for e^{iy}, we have

$$\begin{aligned} e^{iy} &= 1 + iy - \frac{y^2}{2!} - \frac{iy^3}{3!} + \frac{y^4}{4!} - \cdots \\ &= \left(1 - \frac{y^2}{2!} + \frac{y^4}{4!} - \cdots\right) + i\left(y - \frac{y^3}{3!} + \frac{y^5}{5!} - \cdots\right) \\ &= \cos y + i \sin y. \end{aligned}$$

where $i = \sqrt{-1}$. From this formula, we obtain

$$\cos y = \frac{e^{iy} + e^{-iy}}{2}, \ i \sin y = \frac{e^{iy} - e^{-iy}}{2}.$$

We now introduce the hyperbolic functions

$$\cosh x = \frac{e^x + e^{-x}}{2}, \ \sinh x = \frac{e^x - e^{-x}}{2}.$$

Thus, it is interesting to observe the relation between the hyperbolic functions of real radial angles and the trigonometric functions of imaginary angles as follows.

$$\cos(iy) = \cosh y, \ \sin(iy) = i \sinh y.$$

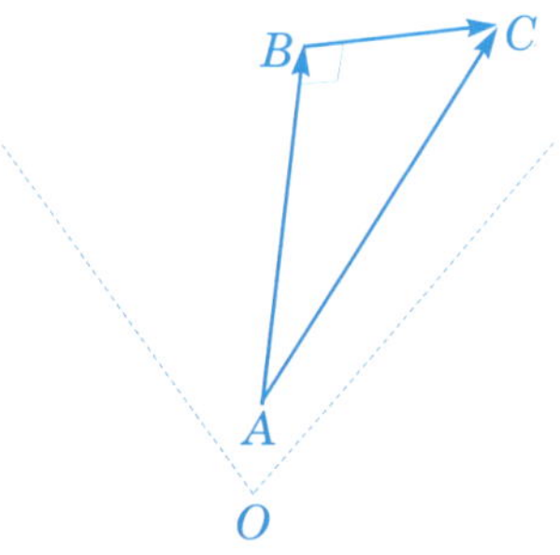

Figure1.4-34

Moreover, we may also compare the geometric roles of triangles with a right angle.

Let A, B, C be three non-collinear points in $\mathbf{L}^2$, with $\overrightarrow{AB}$ and $\overrightarrow{AC}$ future directed timelike vectors and $\overrightarrow{BC}$ a spacelike vector with $< \overrightarrow{AB}, \overrightarrow{BC} >= 0$(Figure 1.4-34). Then recalling the discussion after Theorem 1.3.6 just before Remark 1.3.7,

$$\begin{aligned} \cosh\theta &= -\frac{< \overrightarrow{AB}, \overrightarrow{AC} >}{|\overrightarrow{AB}||\overrightarrow{AC}|} = -\frac{< \overrightarrow{AB}, \overrightarrow{AB} + \overrightarrow{BC} >}{|\overrightarrow{AB}||\overrightarrow{AC}|} \\ &= -\frac{< \overrightarrow{AB}, \overrightarrow{AB} >}{|\overrightarrow{AB}||\overrightarrow{AC}|} = \frac{|\overrightarrow{AB}|^2}{|\overrightarrow{AB}||\overrightarrow{AC}|} = \frac{|\overrightarrow{AB}|}{|\overrightarrow{AC}|}, \end{aligned}$$

and for $\theta \geq 0$

$$\sinh\theta = \sqrt{\cosh^2\theta - 1} = \sqrt{\frac{|\overrightarrow{AB}|^2}{|\overrightarrow{AC}|^2} - 1} = \sqrt{\frac{|\overrightarrow{AB}|^2 - |\overrightarrow{AC}|^2}{|\overrightarrow{AC}|^2}}$$
$$= \frac{\sqrt{- < \overrightarrow{AB}, \overrightarrow{AB} > + < \overrightarrow{AC}, \overrightarrow{AC} >}}{|\overrightarrow{AC}|}$$
$$= \frac{\sqrt{- < \overrightarrow{AB}, \overrightarrow{AB} > + < \overrightarrow{AB} + \overrightarrow{BC},\ \overrightarrow{AB} + \overrightarrow{BC} >}}{|\overrightarrow{AC}|}$$
$$= \frac{\sqrt{< \overrightarrow{BC},\ \overrightarrow{BC} >}}{|\overrightarrow{AC}|} = \frac{|\overrightarrow{BC}|}{|\overrightarrow{AC}|}.$$

Hence, the hyperbolic functions on $\mathbf{L}^2$ and the trigonometric functions on on $\mathbf{E}^2$ have the same geometric interpretation.

The following identities for the hyperbolic functions are easily derived from their definitions in terms of exponential functions:

(1) $\sinh(-\alpha) = -\sinh\alpha$, $\cosh(-\alpha) = \cosh\alpha$

(2) $\cosh^2\alpha - \sinh^2\alpha = 1$

(3) $\sinh(\alpha \pm \beta) = \sinh\alpha\cosh\beta \pm \cosh\alpha\sinh\beta$

(4) $\cosh(\alpha \pm \beta) = \cosh\alpha\cosh\beta \pm \sinh\alpha\sinh\beta$

(5) $\sinh 2\alpha = 2\sinh\alpha\cosh\alpha$

Rotation on $\mathbf{L}^2$

Now, we consider a rotation transformation $F_\theta : \mathbf{L}^2 \longrightarrow \mathbf{L}^2$ corresponding in a Minkowski space to rotation about the origin in $\mathbf{E}^2$. Given any x in $\mathbf{L}^2$, put $x = x_1\mathbf{i} + x_2\mathbf{j}$ where $\mathbf{i} = (1, 0)$, $\mathbf{j} = (0, 1)$. Then if $F_\theta(\mathbf{i})$ and $F_\theta(\mathbf{j})$ are given and we set $F_\theta(x_1\mathbf{i} + x_2\mathbf{j}) = x_1F_\theta(\mathbf{i}) + x_2F_\theta(\mathbf{j})$, F_θ so defined will be a linear transformation. To define $F_\theta(\mathbf{i})$ and $F_\theta(\mathbf{j})$, we employ the **unit hyperbolic circle** $\mathbf{H}^1 = \{x \in \mathbf{L}^2 |\ |x| = 1\}$ as in Figure 1.4-35. Then $\mathbf{H}^1 = \mathbf{H}^1_t \cup \mathbf{H}^1_s$, where the vectors in $\mathbf{H}^1_t$ are timelike and in $\mathbf{H}^1_s$ are spacelike. Moreover, as the hyperbolic angle θ increases, on $\mathbf{H}^1_t$ the unit timelike vector moves from left to right and on $\mathbf{H}^1_s$ the unit spacelike vector moves from below to above on the hyperbolic circle, so called the **boost transformation**(Figure 1.4-36). Also, remark that each pair of the same color of vectors in Figure 1.4-36 is perpendicular.

This suggests defining $F_\theta(\mathbf{i}) = (\cosh\theta,\ \sinh\theta)$ and $F_\theta(\mathbf{j}) = (\sinh\theta,\ \cosh\theta)$. By linearity,

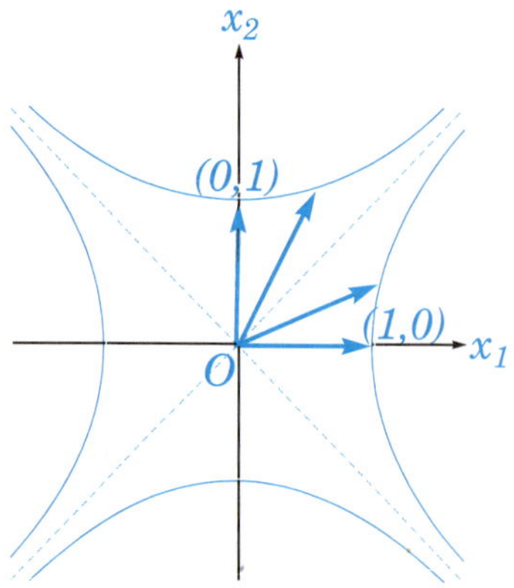

Figure1.4-35

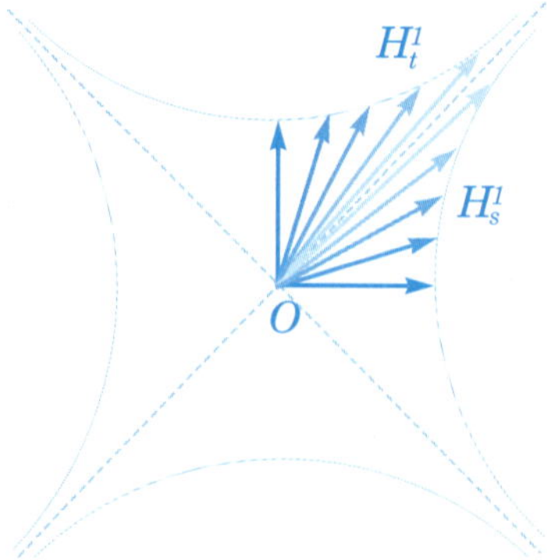

Figure1.4-36

$$\begin{aligned} F_\theta(x) &= x_1(\cosh\theta,\ \sinh\theta) + x_2(\sinh\theta, \cosh\theta) \\ &= (x_1\cosh\theta + x_2\sinh\theta,\ x_1\sinh\theta + x_2\cosh\theta) \\ &= \begin{pmatrix} \cosh\theta & \sinh\theta \\ \sinh\theta & \cosh\theta \end{pmatrix} \begin{pmatrix} x_1 \\ x_2 \end{pmatrix} \\ &= \begin{pmatrix} \cosh\theta & \sinh\theta \\ \sinh\theta & \cosh\theta \end{pmatrix} x. \end{aligned}$$

Recalling the Minkowski inner product $< x, y >= x_1y_1 - x_2y_2$ for $\mathbf{L}^2$, it may be checked that the hyperbolic identity $\cosh^2\theta - \sinh^2\theta = 1$ implies that $< F_\theta(x), F_\theta(y) >=< x, y >$. Thus the linear transformation $F_\theta : \mathbf{L}^2 \to \mathbf{L}^2$ is an isometry and the matrix $\begin{pmatrix} \cosh\theta & \sinh\theta \\ \sinh\theta & \cosh\theta \end{pmatrix}$ is said to be the **hyperbolic rotation transformation** for the hyperbolic angle θ.

Using matrix multiplication and the hyperbolic identities given above, for any $\alpha, \beta \in \mathbb{R}$ the corresponding hyperbolic rotations F_α, F_β satisfy

$$F_\alpha F_\beta = F_\beta F_\alpha = F_{\alpha+\beta}$$

and

$$(F_\alpha)^{-1} = F_{-\alpha}.$$

Now let $x, y \in H_t^1$ be future directed unit timelike vectors. Choose unique $\alpha, \beta \in \mathbb{R}$ so that $x = F_\alpha(\mathbf{j})$ and $y = F_\beta(\mathbf{j})$. Then if $\theta = \beta - \alpha$, we have $y = F_\theta(x)$

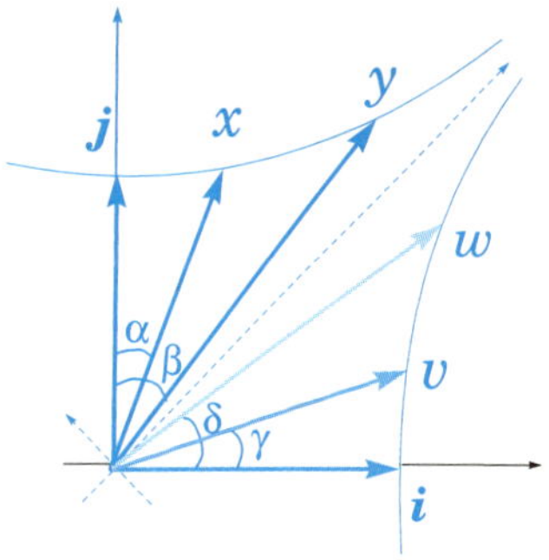

Figure1.4-37

since

$$\begin{aligned} y = F_\beta(\mathbf{j}) &= F_\beta((F_\alpha)^{-1}F_\alpha)(\mathbf{j}) \\ &= (F_\beta F_{-\alpha})(F_\alpha(\mathbf{j})) = F_{\beta-\alpha}(x) \\ &= F_\theta(x). \end{aligned}$$

Similarly, for unit spacelike vectors $v, w \in H_s^1$, choose unique $\gamma, \delta \in \mathbb{R}$ so that $v = F_\gamma(\mathbf{i})$ and $w = F_\delta(\mathbf{i})$. Then if $\theta = \delta - \gamma$, we have $w = F_\theta(v)$.

Hence, $\theta = \beta - \alpha$(or $\theta = \delta - \gamma$) might be considered to be the hyperbolic angle between x and y(v and w) as in Figure 1.4-37. On the other hand, in Remark 1.3.7 in Section 1.3 the hyperbolic angle was defined for future directed timelike vectors in $\mathbf{L}^3$ in terms of the Lorentz inner product for $\mathbf{L}^3$. An analogous formulation is valid for $(\mathbf{L}^2, < \ , \ >)$; for with $\theta = \beta - \alpha$ and $|x| = |y| = 1$ as above,

$$\begin{aligned} < x, y > &=< F_\alpha(\mathbf{j}), F_\beta(\mathbf{j}) >=< (\sinh\alpha, \cosh\alpha), (\sinh\beta, \cosh\beta) > \\ &= \sinh\alpha\sinh\beta - \cosh\alpha\cosh\beta \\ &= -\cosh(\alpha - \beta) = -\cosh(\beta - \alpha) \\ &= -\cosh\theta. \end{aligned}$$

By the same manner, $< v, w >= \cosh\theta$ for $v, w \in H_s^1$.

With these calculation at hand, for two future directed timelike vectors $x, y \in \mathbf{L}^2$, define $\theta \in \mathbb{R}$ to be the **oriented hyperbolic angle** from x to y if $F_\theta(x) = y$. Because $(F_\theta)^{-1} = F_{-\theta}$, the oriented hyperbolic angle from y to x is $-\theta$. The unoriented hyperbolic angle between x and y is defined by $|\theta|$. For general future directed timelike vectors a and b, define their hyperbolic angle θ to be the same as for $x = \dfrac{a}{|a|}$ and $y = \dfrac{b}{|b|}$. Then

$$\cosh\theta = -\frac{< a, b >}{|a||b|}.$$

For the spacelike vectors c and d in $\mathbf{L}^2$ with their hyperbolic angle θ, we may also obtain

$$\cosh\theta = \frac{< c, d >}{|c||d|}.$$

Pure Triangles in $\mathbf{L}^2$

For three non-collinear points A, B, C on $\mathbf{L}^2$, if $\overrightarrow{AB}, \overrightarrow{BC}$ are future directed timelike vectors, $\triangle ABC$ is said to be a **pure timelike triangle**. Similarly, if $\overrightarrow{AB}, \overrightarrow{BC}$ are spacelike vectors, $\triangle ABC$ is said to be a **pure spacelike triangle**.

Let $\triangle ABC$ be a pure timelike triangle. Then $\overrightarrow{AC} = \overrightarrow{AB} + \overrightarrow{BC}$ is also a future directed timelike vector and B is said to be a **middle vertex**. As in

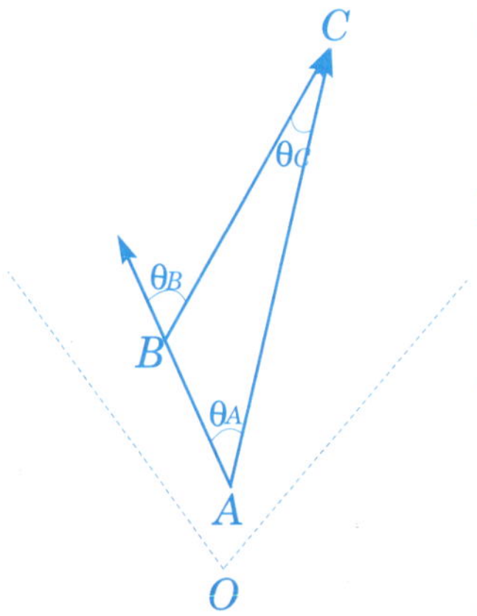

Figure1.4-38

Figure 1.4-38 denote the hyperbolic angles between $\overrightarrow{AB}$ and $\overrightarrow{AC}$ at A, between $\overrightarrow{AB}$ and $\overrightarrow{BC}$ at B, and between $\overrightarrow{AC}$ and $\overrightarrow{BC}$ at C by $\theta_A, \theta_B, \theta_C$ respectively.

Remark 1.4.1 The hyperbolic angle θ_B at the middle vertex B is the exterior angle of the Euclidean triangle, because it is measured in the increasing direction from the left to the right between two future directed timelike lines. Note that $\theta_B = \theta_A + \theta_C$.

For if $x = \dfrac{\overrightarrow{AB}}{|\overrightarrow{AB}|}$, $y = \dfrac{\overrightarrow{AC}}{|\overrightarrow{AC}|}$, $z = \dfrac{\overrightarrow{BC}}{|\overrightarrow{BC}|}$ and u is the hyperbolic angle from x to y and v is the hyperbolic angle from $-y$ to $-z$, then $F_u(x) = y$, $F_v(y) = z$ and by composition $F_vF_u = F_{u+v}$ carries $\overrightarrow{AB}$ to $\overrightarrow{BC}$. Hence $F_{u+v}(x) = z$ and $u + v$ is the hyperbolic angle at B, which translates into the desired identity.

Let $x = (x_1, x_2)$, $y = (y_1, y_2)$ be the two point on $\mathbf{L}^2$ such that $\overrightarrow{Ox}$ and $\overrightarrow{Oy}$ are future directed timelike vectors at the origin O. From more general concepts for arbitrary space-times, the area element in $\mathbf{L}^2$ is known to be the same as the area element in Euclidean space $\mathbf{E}^2$. Also, given general vectors $x, y \in \mathbf{L}^2$ their cross product may be defined as the Lorentz cross product of $x = (x_1, 0, x_2)$ and $y = (y_1, 0, y_2)$ in $\mathbf{L}^3$; hence $x \times_g y = (0, x_1y_2 - x_2y_1) \in \mathbf{L}^2$. Thus from the properties of the Euclidean cross product recalled in Section 1.3, the area of $\triangle Oxy = \frac{1}{2}|x \times_g y| = \frac{1}{2}|x_1y_2 - x_2y_1|$.

We now observe that hyperbolic rotations preserve area.

Theorem 1.4.2 *A hyperbolic rotation of a pure timelike triangle preserves its area.*

Proof.

$$F_\theta(x) = \begin{pmatrix} \cosh\theta & \sinh\theta \\ \sinh\theta & \cosh\theta \end{pmatrix} \begin{pmatrix} x_1 \\ x_2 \end{pmatrix} = \begin{pmatrix} x_1\cosh\theta + x_2\sinh\theta \\ x_1\sinh\theta + x_2\cosh\theta \end{pmatrix}$$

$$F_\theta(y) = \begin{pmatrix} \cosh\theta & \sinh\theta \\ \sinh\theta & \cosh\theta \end{pmatrix} \begin{pmatrix} y_1 \\ y_2 \end{pmatrix} = \begin{pmatrix} y_1\cosh\theta + y_2\sinh\theta \\ y_1\sinh\theta + y_2\cosh\theta \end{pmatrix}$$

But

$$
\begin{aligned}
&\text{area of } \triangle OF_\theta(x)F_\theta(y)\\
&= \frac{1}{2}|(x_1\cosh\theta + x_2\sinh\theta)(y_1\sinh\theta + y_2\cosh\theta)\\
&- (x_1\sinh\theta + x_2\cosh\theta)(y_1\cosh\theta + y_2\sinh\theta)|\\
&= \frac{1}{2}|x_1y_1\sinh\theta\cosh\theta + x_1y_2\cosh^2\theta + x_2y_1\sinh^2\theta + x_2y_2\sinh\theta\cosh\theta\\
&- x_1y_1\sinh\theta\cosh\theta - x_1y_2\sinh^2\theta - x_2y_1\cosh^2\theta - x_2y_2\sinh\theta\cosh\theta|\\
&= \frac{1}{2}|x_1y_2(\cosh^2\theta - \sinh^2\theta) - x_2y_1(\cosh^2\theta - \sinh^2\theta)|\\
&= \frac{1}{2}|x_1y_2 - x_2y_1|\\
&= \text{area of } \triangle Oxy.
\end{aligned}
$$

□

Note that Theorem 1.4.2 also holds for the pure spacelike triangles. Also corresponding to the triangle in Figure1.4-38, the following hyperbolic analogue of a well known area formula for triangles in $\mathbf{L}^2$ may be derived.

Proposition 1.4.3 *The are of the pure timelike triangle $\triangle ABC$ is given by*

$$\textit{Area } \triangle ABC = \frac{1}{2}(|\overrightarrow{AB}||\overrightarrow{AC}|\sinh\theta_A)$$

where θ_A is the unoriented hyperbolic angle at A.

Proof. We may assume that $A = (0,0)$, $B = |\overrightarrow{AB}|(\sinh u, \cosh u)$ and $C = |\overrightarrow{AC}|(\sinh v, \cosh v)$ with $\theta_A = v - u$, recalling the discussion of oriented hyperbolic angle above. Then

$$
\begin{aligned}
\text{Area } \triangle ABC &= \frac{1}{2}|x_1y_2 - x_2y_1|\\
&= \frac{1}{2}|\overrightarrow{AB}||\overrightarrow{AC}|\sinh u\cosh v - \cosh u\sinh v|\\
&= \frac{1}{2}|\overrightarrow{AB}||\overrightarrow{AC}|\sinh(u-v)|\\
&= \frac{1}{2}|\overrightarrow{AB}||\overrightarrow{AC}\sinh|u-v|.
\end{aligned}
$$

□

Triangles and Hyperbolic Circles

For a point $P \in \mathbf{L}^2$ and a real number $r > 0$ the hyperbolic circle $\mathbf{H}^1_P(r) = \{x \in L^2| \ |\overrightarrow{Px}| = r\}$ is divided into two parts if the vector $\overrightarrow{Px}$ is timelike or

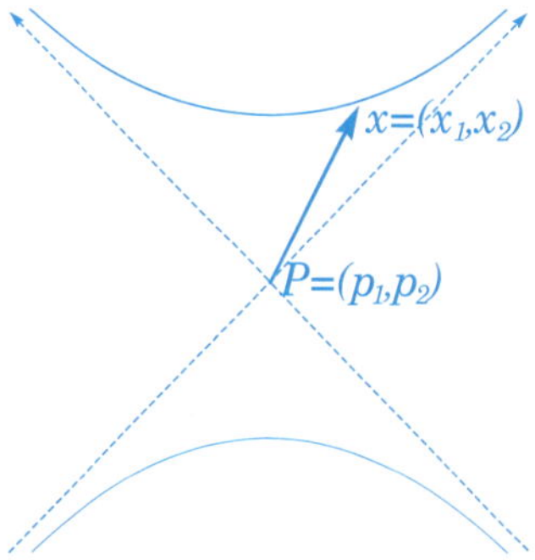

Figure1.4-39

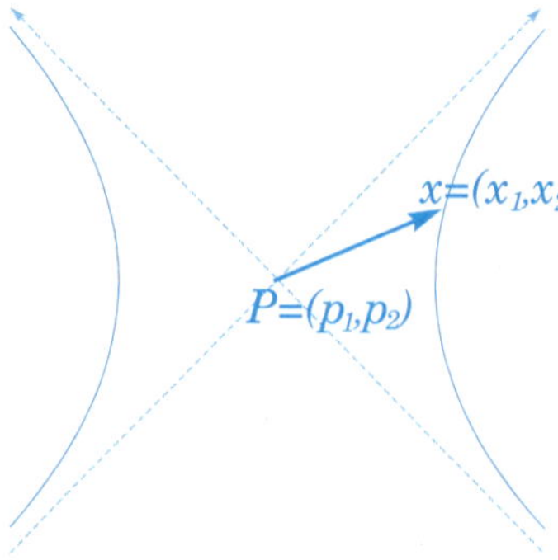

Figure1.4-40

spacelike as is the unit hyperbolic circle $\mathbf{H}^1$. If $P=(p_1,p_2)$, $x=(x_1,x_2)$, we have $r=|\overrightarrow{Px}|=\sqrt{|-(x_1-p_1)^2+(x_2-p_2)^2|}$.

(a) If $\overrightarrow{Px}$ is timelike, then $-(x_1-p_1)^2+(x_2-p_2)^2>0$. Hence,

$$-(x_1-p_1)^2+(x_2-p_2)^2=r^2(\text{Figure}\,1.4\text{-}39).$$

(b) if $\overrightarrow{Px}$ is spacelike, then $-(x_1-p_1)^2+(x_2-p_2)^2<0$. Hence,

$$+(x_1-p_1)^2-(x_2-p_2)^2=r^2(\text{Figure}\,1.4\text{-}40).$$

Theorem 1.4.4 *Let x be on a hyperbolic circle $\mathbf{H}^1_P(r)$ of center P and radius r with $\overrightarrow{Px}$ timelike. Then the timelike vector $\overrightarrow{Px}$ is perpendicular in the Lorentz inner product to the vector tangent to $\mathbf{H}^1_P(r)$ at x.*

Proof. Consider the parametric curve* $x(\theta)=(r\sinh\theta+p_1,\ r\cosh\theta+p_2)$ which passes through x for some hyperbolic angle θ. We have $\overrightarrow{Px}(\theta)=(r\sinh\theta,$ $r\cosh\theta)$. Hence, $\overrightarrow{Px}'(\theta)=\dfrac{d\overrightarrow{Px}(\theta)}{d\theta}=(r\cosh\theta,\ r\sinh\theta)$. Thus, we have $<\overrightarrow{Px}(\theta),\ \overrightarrow{Px}'(\theta)>=-r^2\sinh\theta\cosh\theta+r^2\sinh\theta\cosh\theta=0$. □

Similarly, for a point x on a hyperbolic circle of center P with radius r and $\overrightarrow{Px}$ spacelike, the vector $\overrightarrow{Px}$ is perpendicular to the vector tangent to $\mathbf{H}^1_P(r)$ at x.

Theorem 1.4.5 *Let A,B,C be three successive points on $\mathbf{H}^1_P(r)$. Then the triangle inscribed along $\mathbf{H}^1_P(r)$ with vertices A,B,C is pure spacelike(Figure 1.4-41).*

*The parametric curve will be studied more in Section 2.1.

Proof. We may assume that $P = O$ and suppose that $A = (r\sinh\theta_A, r\cosh\theta_A)$, $B = (r\sinh\theta_B,\ r\cosh\theta_B)$, $C = (r\sinh\theta_C,\ r\cosh\theta_C)$ with $\theta_A < \theta_B < \theta_C$. Then

$$\begin{aligned} < \overrightarrow{AB}, \overrightarrow{AB} > =&< (r\sinh\theta_B - r\sinh\theta_A,\ r\cosh\theta_B - r\cosh\theta_A), \\ &(r\sinh\theta_B - r\sinh\theta_A,\ r\cosh\theta_B - r\cosh\theta_A) > \\ &= -r^2(\cosh\theta_B - \cosh\theta_A)^2 + r^2(\sinh\theta_B - \sinh\theta_A)^2 \\ &= r^2(-\cosh^2\theta_B + 2\cosh\theta_A\cosh\theta_B - \cosh^2\theta_A \\ &+ \sinh^2\theta_B - 2\sinh\theta_A\sinh\theta_B + \sinh^2\theta_A) \\ &= -r^2(2 + 2(\sinh\theta_A\sinh\theta_B - \cosh\theta_A\cosh\theta_B)) \\ &= -2r^2(1 - \cosh(\theta_B - \theta_A)) > 0,\ \text{since } \theta_B - \theta_A > 0. \end{aligned}$$

Hence, $\overrightarrow{AB}$ is spacelike. Similarly, $\overrightarrow{BC}$, $\overrightarrow{AC}$ are also spacelike. □

In the same manner, it may be shown that if $\overrightarrow{Px}$ is spacelike for $x \in \mathbf{H}^1_P(r)$, the triangle inscribed along $\mathbf{H}^1_P(r)$ is pure timelike(Figure1.4-41).

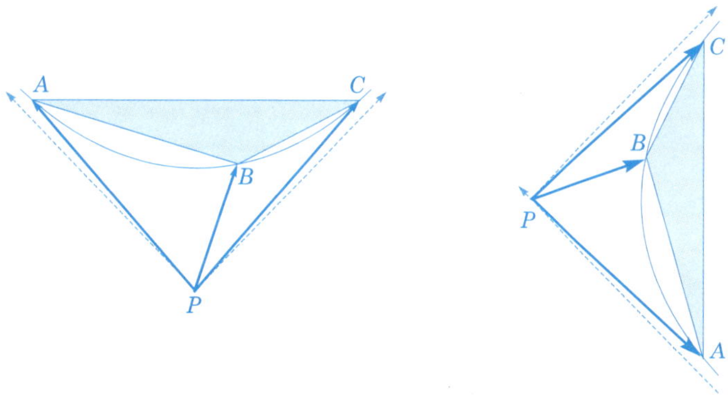

Figure1.4-41

Hyperbolic Sine and Cosine Laws

For three non-collinear points $A,\ B,\ C$ on $\mathbf{L^2}$, if $\overrightarrow{AB}, \overrightarrow{BC}$ are future directed timelike vectors, $\triangle ABC$ is said to be a **pure timelike triangle**. Similarly, if $\overrightarrow{AB}, \overrightarrow{BC}$ are spacelike vectors, $\triangle ABC$ is said to be a **pure spacelike triangle**.

Let $\triangle ABC$ be a pure triangle inscribed along $\mathbf{H}^1_t$ and set $a = |\overrightarrow{BC}|,\ b = |\overrightarrow{CA}|,\ c = |\overrightarrow{AB}|$. By applying a hyperbolic rotation to $\triangle ABC$, we may assume that the point B is a point at which the t-axis meets $\mathbf{H}^1_t$.As in Figure1.4-42 the point D is given when B is projected orthogonally to the segment AC. Then $< \overrightarrow{BD},\ \overrightarrow{AC} >= 0$ and $\overrightarrow{BD}$ is timelike. Let $\alpha,\ \beta,\ \gamma$ be the hyperbolic angles at the vertices $A,\ B,\ C$. Then $\sinh\alpha = \dfrac{|\overrightarrow{BD}|}{|\overrightarrow{AB}|} = \dfrac{|\overrightarrow{BD}|}{c}$. Since $\sinh\gamma = \dfrac{|\overrightarrow{BD}|}{|\overrightarrow{BC}|} =$

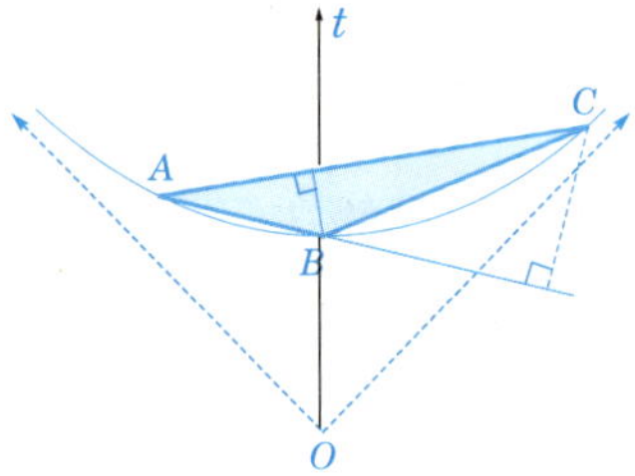

Figure 1.4-42

$\dfrac{|\overrightarrow{BD}|}{a}$, we have $\dfrac{\sinh\alpha}{a} = \dfrac{\sinh\gamma}{c}$.

Let the point E be also given when C is projected orthogonally to the segment AB. Then $< \overrightarrow{CE}, \overrightarrow{AB} >= 0$ and $\overrightarrow{CE}$ is timelike. Hence, $\sinh\beta = \dfrac{|\overrightarrow{CE}|}{a}$. Since $\sinh\alpha = \dfrac{|\overrightarrow{CE}|}{b}$, we have $\dfrac{\sinh\beta}{b} = \dfrac{\sinh\alpha}{a}$. Therefore,

$$\frac{\sinh\alpha}{a} = \frac{\sinh\beta}{b} = \frac{\sinh\gamma}{c}$$

holds for pure timelike triangles.

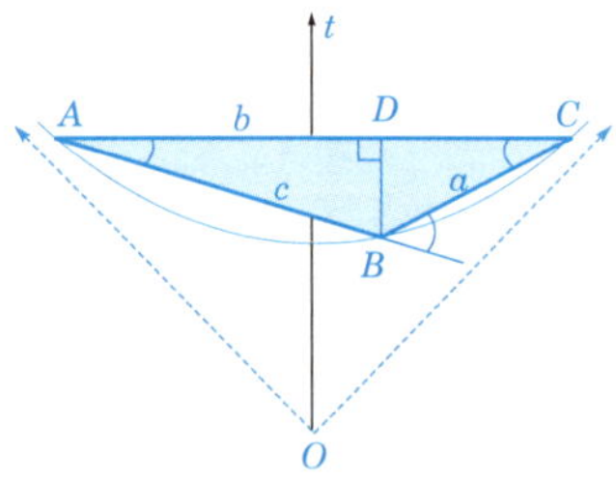

Figure 1.4-43

Theorem 1.4.6 *For a pure spacelike triangle $\triangle ABC$ inscribed along $\mathbf{H}_t^1$,*

$$a^2 = b^2 + c^2 - 2bc\cosh\alpha$$

$$b^2 = a^2 + c^2 + 2ac\cosh\beta$$

$$c^2 = a^2 + b^2 - 2ab\cosh\gamma$$

where $a = |\overrightarrow{BC}|$, $b = |\overrightarrow{AC}|, c = |\overrightarrow{AB}|$.

Proof. Without loss of generality, we rotate $\triangle ABC$ along $\mathbf{H}_t^1$ so that $< \overrightarrow{BD}, \overrightarrow{AC} >= 0$ as in Figure 1.4-43. We may observe that

$$b = |\overrightarrow{AD}| + |\overrightarrow{CD}| = c\cosh\alpha + a\cosh\gamma \tag{4.1}$$

As in Figure 1.4-44 we take a point E so that $< \overrightarrow{CE}, \overrightarrow{AB} >= 0$. Since $|\overrightarrow{EB}| = a\cosh\beta$, we have $c + |\overrightarrow{EB}| = b\cosh\alpha$. Hence,

$$c = b\cosh\alpha - a\cosh\beta \tag{4.2}$$

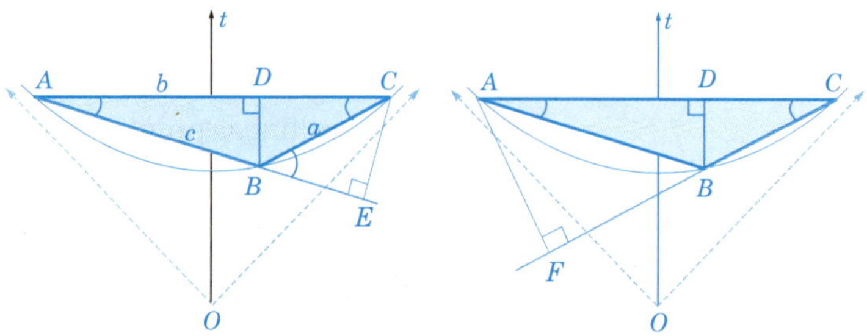

Figure 1.4-44

Also as in Figure 1.4-44 we take a point F so that $< \overrightarrow{AF}, \overrightarrow{BC} >= 0$. Then $|\overrightarrow{BF}| = c\cosh\beta$ and $a + |\overrightarrow{BF}| = b\cosh\gamma$. Hence,

$$a = b\cosh\gamma - c\cosh\beta \tag{4.3}$$

Multiply b, c, a to the equations (4.1), (4.2) (4.3) respectively. Then we obtain

$$b^2 = bc\cosh\alpha + ab\cosh\gamma \tag{4.4}$$

$$c^2 = bc\cosh\alpha - ac\cosh\beta \tag{4.5}$$

$$a^2 = ab\cosh\gamma - ac\cosh\beta \tag{4.6}$$

Adding (4.4) to (4.5) and comparing with (4.6), we obtain $a^2 = b^2 + c^2 - 2bc\cosh\alpha$.

The other two formulas follow similarly. □

Corollary 1.4.7 *For a pure timelike triangle $\triangle ABC$ inscribed along $\mathbf{H}^1_s$, we may also obtain*

$$a^2 = b^2 + c^2 - 2bc\cosh\alpha$$

$$b^2 = a^2 + c^2 + 2ac\cosh\beta$$

$$c^2 = a^2 + b^2 - 2ab\cosh\gamma$$

where $a = |\overrightarrow{BC}|,\ b = |\overrightarrow{AC}|,\ c = |\overrightarrow{AB}|$.

For further study, see the following reference:

1. Birman G. and K. Nomizu, *Trigonometry in Lorentzian Geometry*, Amer. Math. Monthly 91(1984), 543-549.

2. Birman, R. and K. Nomizu, The Gauss-Bonnet theorem for two-dimensional space-time, Mich. Math. J. 31(1984) 77-81.

3. do Carmo, M., Differential Geometry of Curves and Surfaces, Prentice-Hall Inc, 1976.

4. Millman, R. and G. Parker, Elements of Differential Geometry, Prentice-Hall Inc, 1977.

5. O'Neill, B., Elementary Differential Geometry, Academic Press Inc, 1966.

Chapter 2

Theory of Curves and Vector Calculus

2.1 Reparametrization by Arc Length

In $\mathbb{R}^3$, a particle moves along a curve. Thus physical applications provided a motivation for the study of such curves. The general theory of space curves, the first major development in 3-dimensional differential geometry, was launched by Alexis-Claude Clairaut(1713-1765) in a publication in 1731 written in 1729 when Clairaut was sixteen. Gaspard Monge(1746-1818), one of the organizers of Napoleon's engineering college, the Ecole Polytechnique in Paris, France, founded a school of differential geometry as well as doing research on curves and surfaces while a professor at that institution during 1795-1809. While the radius of curvature treated in Section 2.2 goes back to Leonhard Euler(1707-1783) and Clairaut formulated the idea that a space curve has two different curvatures, Monge's student Michel-Ange Lancret(1774-1807) was the first to formulate an explicit analytic definition of this second curvature quantity, which is called the torsion. The fundamental formulas linking the rate of change of the tangent, normal and binormal vector fields along a curve in $\mathbf{E}^3$ to its curvature and torsion were published independently by Joseph Alfred Serret(1819-1885) in 1851 and Fredric-Jean Frenet(1816-1900) in 1852. As in Chapter 1, we will often use $<\ ,\ >$ to denote either the Euclidean metric on $\mathbf{E}^2$ or $\mathbf{E}^3$ or to denote the Lorentz metric on $\mathbf{L}^2$ or $\mathbf{L}^3$.

Length of a Regular Curve

Let I be an open interval of $\mathbb{R}$. Consider a function $\gamma : I \to \mathbb{R}^3$ given by $\gamma(t) = (x_1(t), x_2(t), x_3(t))$ for each t as a parameter in I where x_1, x_2, x_3 are called its **coordinate functions**. Then γ is differentiable in I if its coordinate functions are differentiable in I. Especially, γ is said to be a **curve** if $\gamma : I \to \mathbb{R}^3$ is differentiable. The velocity vector of γ at t is the tangent vector $\gamma'(t)$ given by

$$\gamma'(t) = \left(\frac{dx_1(t)}{dt}, \frac{dx_2(t)}{dt}, \frac{dx_3(t)}{dt} \right)_{\gamma(t)}.$$

Now a differentiable function $\gamma : I \to \mathbb{R}^3$ is said to be a **regular curve** if $\gamma'(t) \neq 0$, for all $t \in I$. In this book, we will use only regular curves and hence from now on we call "regular curves" just "curves" without ambiguity. In $\mathbf{L}^3$, γ is said to be a **timelike(lightlike or spacelike) curve** if its velocity vector

$\gamma'(t)$ for all $t \in I$ is a timelike(lightlike or spacelike) vector respectively. Further, if $\gamma'(t)$ is always either a timelike or a lightlike vector for all $t \in I$, then γ is said to be a **nonspacelike curve**.

If a curve γ is represented by coordinate functions x, y, z, that is, $\gamma(t) = (x(t), y(t), z(t))$ in $\mathbf{E}^3$, the speed of γ at t is

$$v(t) = |\gamma'(t)| = \sqrt{x'(t)^2 + y'(t)^2 + z'(t)^2}.$$

In $\mathbf{L}^3$ the speed of γ is

$$v(t) = |\gamma'(t)| = \sqrt{|x'(t)^2 + y'(t)^2 - z'(t)^2|}.$$

For a closed subinterval $[c, d]$ of I, the length the arc $\gamma(t)$ restricted to $[c, d]$ is defined by $\int_c^d v(t)dt$ in $\mathbf{E}^3$ and $\mathbf{L}^3$.

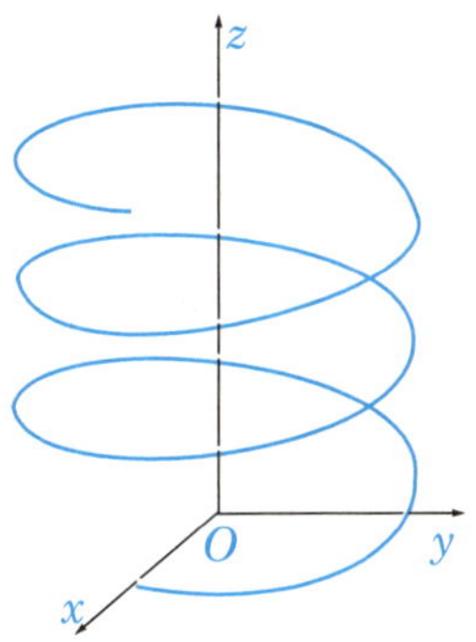

Figure2.1-1

Example 2.1.1 Let $\gamma(t) = (a\cos 6\pi t, a\sin 6\pi t, bt)$, $a \neq 0$, $b > 0$, $t \in [0, 1]$ be an arc. Compute the length of the arc in $\mathbf{E}^3$ and $\mathbf{L}^3$, with $I = \mathbb{R}$(Figure2.1-1).

Solution. Since $\gamma'(t) = 6\pi(-a\sin 6\pi t, a\cos 6\pi t, b)_{\gamma(t)}$, $v(t) = |\gamma'(t)| = 6\pi\sqrt{a^2 + b^2}$ in $\mathbf{E}^3$. Thus, $\int_0^1 v(t)dt = 6\pi\sqrt{a^2 + b^2}$. In $\mathbf{L}^3$ $v(t) = |\gamma'(t)| = 6\pi\sqrt{|a^2 - b^2|}$. Thus, $\int_0^1 v(t)dt = 6\pi\sqrt{|a^2 - b^2|}$. □

Note that $\gamma(t) = (a\cos 6\pi t, a\sin 6\pi t, bt)$, $a \neq 0$, $b > 0$, $t \in [0, 1]$ is a timelike (lightlike or spacelike) curve in $\mathbf{L}^3$ if and only if $a^2 < b^2$ ($a^2 = b^2$ or $a^2 > b^2$).

Parametrization by Arc Length

Definition 2.1.2 Let I and J be open intervals of $\mathbb{R}$ and $\gamma : I \to \mathbb{R}^3$ be a given curve. For a differentiable function $h : J \to I$ with $|h'(s)| > 0$ on J, the composite function β defined by $\beta(s) = \gamma(h(s))$ is a curve called the **reparametrization** of γ by h.

Theorem 2.1.3 *A future directed timelike curve $\gamma : I \to \mathbf{L}^3$ has a unit speed future directed timelike reparametrization.*

Proof. Take a point $a \in I$ and consider a length function

$$s(t) = \int_a^t |\gamma'(t)|dt.$$

Since $s'(t) = |\gamma'(t)|$ and $\gamma'(t) \neq 0$, we have $s'(t) > 0$. Hence, the length function s is an strictly increasing function and thus s has an inverse function $t = t(s)$

with domain $J = s(I)$.From elementary calculus, $\dfrac{dt}{ds}$ at $s = s(t)$ equals $1\Big/\dfrac{ds}{dt}$ at $t = t(s)$. Then the reparametrizarion of γ by s given by $\beta(s) = \gamma(t(s))$ satisfies $\dfrac{d\beta}{ds} = \dfrac{d\gamma}{dt}\dfrac{dt}{ds} = \dfrac{d\gamma}{dt}\Big/\dfrac{ds}{dt} = \dfrac{\gamma'(t)}{s'(t)} = \dfrac{\gamma'(t)}{|\gamma'(t)|}$. Now, since $< \beta'(s), \beta'(s) > = < \dfrac{\gamma'(t)}{|\gamma'(t)|}, \dfrac{\gamma'(t)}{|\gamma'(t)|} > = \dfrac{< \gamma'(t), \gamma'(t) >}{|\gamma'(t)|^2} = -1$ and since $\gamma'(t)$ is future directed, β is a future directed unit speed timelike curve. □

The proof of Theorem 2.1.3 holds for a regular curve in $\mathbf{E}^3$ and a similar proof also works for a spacelike curve in $\mathbf{L}^3$.

Corollary 2.1.4 *A spacelike curve $\gamma : I \to \mathbf{L}^3$ has a unit speed spacelike reparametrization.*

Remark 2.1.5 In Theorem 2.1.3 the curve γ in $\mathbf{E}^3$ or $\mathbf{L}^3$ is reparametrized in terms of arc length from a given point. In this case γ is said to be *parametrized by arc length.*

Example 2.1.6 Find a reparametrization β by arc length of the helix γ given by $\gamma(t) = (a\cos t, a\sin t, bt)$ in both $\mathbf{E}^3$ and $\mathbf{L}^3$.

Solution. In $\mathbf{E}^3$ we have $v(t) = \sqrt{a^2 + b^2}$ as in Example2.1-1, so

$$s(t) = \int_0^t \sqrt{a^2 + b^2}dt = \sqrt{a^2 + b^2}t,$$

$$t(s) = \frac{s}{\sqrt{a^2 + b^2}}, \text{ and}$$

$$\beta(s) = \left(a\cos\frac{s}{\sqrt{a^2 + b^2}}, a\sin\frac{s}{\sqrt{a^2 + b^2}}, \frac{bs}{\sqrt{a^2 + b^2}}\right).$$

In $\mathbf{L}^3$ we have $v(t) = \sqrt{|a^2 - b^2|}$,

$$s(t) = \int_0^t \sqrt{|a^2 - b^2|}dt = \sqrt{|a^2 - b^2|}t,$$

$$t(s) = \frac{s}{\sqrt{|a^2 - b^2|}}, \text{ and}$$

$$\beta(s) = \left(a\cos\frac{s}{\sqrt{|a^2 - b^2|}}, a\sin\frac{s}{\sqrt{|a^2 - b^2|}}, \frac{bs}{\sqrt{|a^2 - b^2|}}\right).$$

If γ is timelike, then $\beta(s) = \left(a\cos\dfrac{s}{\sqrt{b^2 - a^2}}, a\sin\dfrac{s}{\sqrt{b^2 - a^2}}, \dfrac{bs}{\sqrt{b^2 - a^2}}\right)$.

If γ is spacelike, then $\beta(s) = \left(a\cos\dfrac{s}{\sqrt{a^2 - b^2}}, a\sin\dfrac{s}{\sqrt{a^2 - b^2}}, \dfrac{bs}{\sqrt{a^2 - b^2}}\right)$. □

Note that a regular curve $\gamma : I \rightarrow L^3$ is future directed timelike (or spacelike) if and only if a reparametrization β of γ by arc length is future directed timelike (or spacelike).

The Frenet-Serret Apparatus

Definition 2.1.7 For a unit speed curve $\beta(s)$ in $\mathbf{E}^3$ the **Frenet-Serret apparatus** is

$$\{T(s), N(s), B(s), \kappa(s), \tau(s)\}$$

where

(1) $T(s) = \beta'(s)$ is called the **principal tangent vector field** on β.

(2) $\kappa(s) = |T'(s)|$ is called the **curvature function** of β.

(3) $N(s) = \dfrac{T'(s)}{\kappa(s)}$ is called the **principal normal vector field** on β.

(4) $B(s) = T(s) \times N(s)$ is called the **principal binormal vector field** on β.

(5) $\tau(s) = -< B'(s), N(s) >$ is called the **torsion function** of β.

Here is assumed that the curve $\beta(s)$ satisfies $\kappa(s) > 0$ on its domain in (1)-(4). For a straight line in $\mathbf{L}^3$ parametrized by arc length, while $\kappa(s) = 0$ may be calculated since $T(s)$ is just the (constant) unit direction of the given line and $T'(s) = 0$, $N(s)$ and $B(s)$ are not defined.

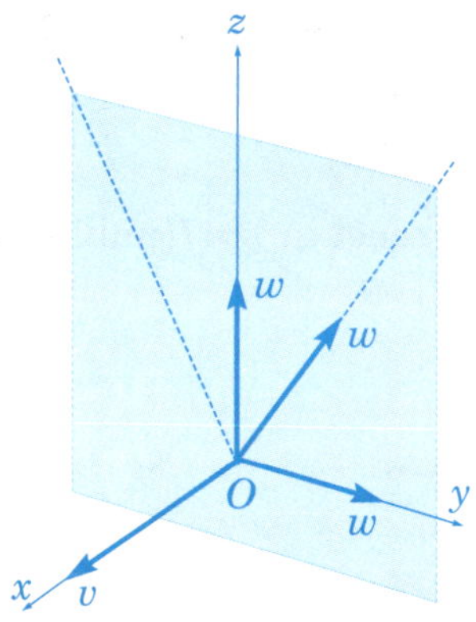

Figure 2.1-2

For the Frenet-Serret frame of a unit speed curve in $\mathbf{L}^3$, the same definitions (1)-(5) as in the Euclidean case will be employed, except the Lorentz cross product in $\mathbf{L}^3$ is used in (4) in defining $B(s) = T(s) \times_g N(s)$ instead of the Euclidean cross product. Note that Theorem 1.3.12-(1) implies that $B(s)$ is always nonzero and nonnull, since $< T(s), N(s) >= 0$. A timelike tangent vector v in $\mathbf{L}^3$ has the property that if $< v, w >= 0$ and w is nonzero, then w must be spacelike (cf. Theorem 1.3.6-(4)). However, if v is spacelike and $< v, w >= 0$ with w nonzero, then w may be timelike, spacelike or lightlike(Figure 2.1-2). Hence the Frenet-Serret theory is more complicated for unit spacelike curves in $\mathbf{L}^3$ than unit timelike curves in $\mathbf{L}^3$. For $\beta(s)$ a spacelike helix as in Example 2.1.6 with $a^2 > b^2$, we have $T(s) = \beta'(s)$ and $N(s)$ are unit spacelike. But for $\beta(s) = \left(\dfrac{\sinh s}{\sqrt{2}}, \dfrac{s}{\sqrt{2}}, \dfrac{\cosh s}{\sqrt{2}}\right)$, we have $T(s) = \beta'(s)$ unit spacelike but $N(s)$ unit timelike. (If $T'(s_0)$ were lightlike for some s_0 in I, then $\kappa(s_0) = |T'(s_0)| = 0$. Hence $N(s_0)$ and the Frenet-Serret frame can not be constructed. Thus this case does not need to be considered in the context of Theorem 2.1.9.)

Remark 2.1.8 If $\beta(s)$ is a future directed unit speed timelike curve in $\mathbf{L}^3$,

$T(s)$ is the principal unit timelike tangent vector field on β,

$N(s)$ is the principal spacelike normal vector field on β,

$B(s) = T(s) \times_g N(s)$ is the principal spacelike binormal vector field on β(Figure2.1-3).

Moreover, if $\beta(s)$ is a unit speed spacelike curve in L^3 with $N(s)$ spacelike,

$T(s)$ is the principal spacelike tangent vector field on β,

$N(s)$ is the principal spacelike normal vector field on β,

$B(s) = T(s) \times_g N(s)$ is the principal timelike binormal vector field on β.

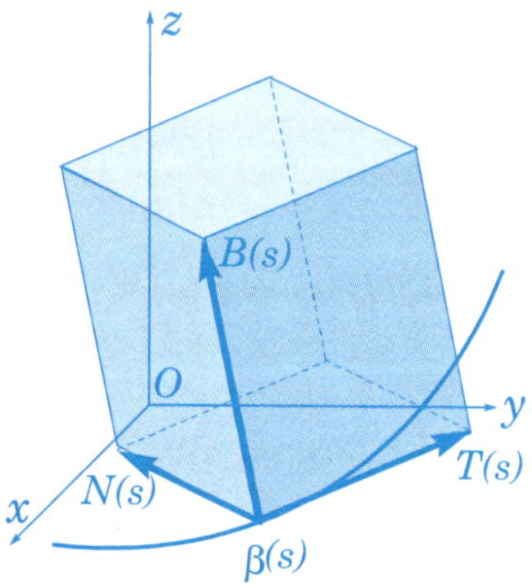

Figure2.1-3

A plane perpendicular to $B(s)$ is called the **osculating plane** of $\beta(s)$. A plane perpendicular to $T(s)$ is called the **normal plane** of $\beta(s)$. A plane perpendicular to $N(s)$ is called the **rectifying plane** of $\beta(s)$.

The following theorem was independently found by Frenet in 1847(published in 1852) and Serret in 1851 in the Euclidean case.

Theorem 2.1.9 (**Frenet-Serret**) *(a) For a unit speed Euclidean curve $\beta(s)$ in $\mathbf{E}^3$*

$T'(s) = \kappa(s)N(s),$

$N'(s) = -\kappa(s)T(s) + \tau(s)B(s)$

$B'(s) = -\tau(s)N(s).$

(b) For a unit timelike curve $\beta(s)$ in $\mathbf{L}^3$

$T'(s) = \kappa(s)N(s),$

$N'(s) = \kappa(s)T(s) + \tau(s)B(s),$

$B'(s) = -\tau(s)N(s).$

(c) For a unit speed spacelike curve $\beta(s)$ in $\mathbf{L}^3$ with $N(s)$ unit spacelike

$T'(s) = \kappa(s)N(s),$

$N'(s) = -\kappa(s)T(s) - \tau(s)B(s),$

$B'(s) = -\tau(s)N(s).$

(d) For a unit speed spacelike curve $\beta(s)$ in $\mathbf{L}^3$ with $N(s)$ unit timelike

$T'(s) = \kappa(s)N(s),$

$N'(s) = \kappa(s)T(s) + \tau(s)B(s),$

$B'(s) = \tau(s)N(s).$

Proof. In all four cases, $T'(s) = \kappa(s)N(s)$ in the Frenet-Serret apparatus. By differentiating $< T(s), B(s) >= 0$, $< T'(s), B(s) > + < T(s), B'(s) >= 0$. Since $T'(s) = \kappa(s)N(s)$, $\kappa(s) < N(s), B(s) > + < T(s), B'(s) >= 0$. Since $< N(s), B(s) >= 0$, we have $< T(s), B'(S) >= 0$. By differentiating $< B(s), B(s) >= \pm 1$, $< B'(s), B(s) >= 0$. Hence, $B'(s)$ is perpendicular to $T(s)$ and $B(s)$. Thus, $B'(s)$ is parallel to $N(s)$. i.e., $B'(s) = a(s)N(s)$, or $< B'(s), N(s) >= a(s) < N(s), N(s) >$. Since $< B'(s), N(s) >= -\tau(s)$, $a(s) = -\tau(s)$ in the cases (a), (b) and (c) for which $< N(s), N(s) >= 1$, and thus we have $B'(s) = -\tau(s)N(s)$. But since $a(s) = \tau(s)$, we have $B'(s) = \tau(s)N(s)$ in case (d) for which $< N(s), N(s) >= -1$.

Now, for $\beta(s)$ Euclidean, since $T(s)$ is unit and since $B(s) = T(s) \times N(s)$, we have $B(s) \times T(s) = (T(s) \times N(s)) \times T(s) =< T(s), T(s) > N(s) - < N(s), T(s) > T(s) = N(s)$. By differentiating $N(s) = B(s) \times T(s)$,

$$\begin{aligned} N'(s) &= B'(s) \times T(s) + B(s) \times T'(s) \\ &= -\tau(s)N(s) \times T(s) + \kappa(s)B(s) \times N(s) \\ &= \tau(s)B(s) - \kappa(s)T(s). \end{aligned}$$

Now, for $\beta(s)$ timelike, since $T(s)$ is unit timelike and since $B(s) = T(s) \times_g N(s)$, by Theorem 1.3.10 (4) we have $B(s) \times_g T(s) = (T(s) \times_g N(s)) \times_g T(s) = -(< T(s), T(s) > N(s) - < N(s), T(s) > T(s) >) = N(s)$. By differentiating

$N(s) = B(s) \times_g T(s)$,

$$\begin{aligned} N'(s) &= B'(s) \times_g T(s) + B(s) \times_g T'(s) \\ &= -\tau(s)N(s) \times_g T(s) + \kappa(s)B(s) \times_g N(s) \\ &= \tau(s)B(s) + \kappa(s)T(s) \end{aligned}$$

since $B(s) \times_g N(s) = T(s)$ using Theorem 1.3.10 (4).

Finally, for $\beta(s)$ unit spacelike with $N(s)$ unit spacelike, since $T(s)$ is unit spacelike, by Theorem 1.3.10 (4) we have $N(s) = -B(s) \times_g T(s)$ in cases (c) and (d). Using the same computation as in the timelike case, we have in case (c)

$$\begin{aligned} N'(s) &= -B'(s) \times_g T(s) - B(s) \times_g T'(s) \\ &= \tau(s)N(s) \times_g T(s) - \kappa(s)B(s) \times_g N(s) \\ &= -\tau(s)B(s) - \kappa(s)T(s) \end{aligned}$$

since $B(s) \times_g N(s) = T(s)$ using Theorem 1.3.10 (4). Hence, $N'(s) = -\kappa(s)T(s) - \tau(s)B(s)$ in case (c).

In case (d), since $N(s)$ is unit timelike, we have $B(s) \times_g N(s) = -T(s)$ using Theorem 1.3.10 (4). Thus, $N'(s) = \kappa(s)T(s) - \tau(s)B(s)$. □

In the case that $\beta(s)$ is unit speed Euclidean, the equations of Theorem 2.1.9 (a) may be represented as follows:

$$\begin{pmatrix} T'(s) \\ N'(s) \\ B'(s) \end{pmatrix} = \begin{pmatrix} 0 & \kappa(s) & 0 \\ -\kappa(s) & 0 & \tau(s) \\ 0 & -\tau(s) & 0 \end{pmatrix} \begin{pmatrix} T(s) \\ N(s) \\ B(s) \end{pmatrix}.$$

In the case that $\beta(s)$ is unit timelike, the equations of Theorem 2.1.9 (b) may be represented as follows:

$$\begin{pmatrix} T'(s) \\ N'(s) \\ B'(s) \end{pmatrix} = \begin{pmatrix} 0 & \kappa(s) & 0 \\ \kappa(s) & 0 & \tau(s) \\ 0 & -\tau(s) & 0 \end{pmatrix} \begin{pmatrix} T(s) \\ N(s) \\ B(s) \end{pmatrix}.$$

But, for $\beta(s)$ spacelike and $N(s)$ unit spacelike, the equations of Theorem 2.1.9 may be represented as follows:

$$\begin{pmatrix} T'(s) \\ N'(s) \\ B'(s) \end{pmatrix} = \begin{pmatrix} 0 & \kappa(s) & 0 \\ -\kappa(s) & 0 & -\tau(s) \\ 0 & -\tau(s) & 0 \end{pmatrix} \begin{pmatrix} T(s) \\ N(s) \\ B(s) \end{pmatrix}.$$

Finally, for $\beta(s)$ spacelike and $N(s)$ unit timelike, the equations of Theorem 2.1.9 may be represented as follows:

$$\begin{pmatrix} T'(s) \\ N'(s) \\ B'(s) \end{pmatrix} = \begin{pmatrix} 0 & \kappa(s) & 0 \\ \kappa(s) & 0 & \tau(s) \\ 0 & \tau(s) & 0 \end{pmatrix} \begin{pmatrix} T(s) \\ N(s) \\ B(s) \end{pmatrix}.$$

Thus, in Euclidean case the associated matrix for the derivative of $\{T(s), N(s), B(s)\}$ is skew symmetric(i.e., $A^t = -A$). But the others are neither symmetric(i.e., $A^t = A$) nor skew symmetric except for case (d) which is symmetric

Example 2.1.10 Find the Frenet-Serret apparatus $\{T(s), N(s), B(s), \kappa(s), \tau(s)\}$ on the non-lightlike curve γ in $\mathbf{L}^3$ given by $\gamma(t) = (a\cos t, a\sin t, bt)$.

Solution. The reparametrization of $\gamma(t)$ by arc length is

$$\begin{aligned} \beta(s) &= \left(a\cos\frac{s}{\sqrt{|a^2-b^2|}}, a\sin\frac{s}{\sqrt{|a^2-b^2|}}, \frac{bs}{\sqrt{|a^2-b^2|}} \right) \\ &= \left(a\cos\frac{s}{c}, a\sin\frac{s}{c}, \frac{bs}{c} \right). \end{aligned}$$

where $c = \sqrt{|a^2-b^2|}$. Hence, $T(s) = \beta'(s) = \left(-\frac{a}{c}\sin\frac{s}{c}, \frac{a}{c}\cos\frac{s}{c}, \frac{b}{c} \right)$,

$$\begin{aligned} T'(s) &= \beta''(s) = \left(-\frac{a}{c^2}\cos\frac{s}{c}, -\frac{a}{c^2}\sin\frac{s}{c}, 0 \right), \\ \kappa(s) &= |T'(s)| = \frac{a}{|a^2-b^2|} = \frac{a}{c^2}, \\ N(s) &= \frac{T'(s)}{\kappa(s)} = \left(-\cos\frac{s}{c}, -\sin\frac{s}{c}, 0 \right), \\ B(s) &= \left(\frac{b}{c}\sin\frac{s}{c}, -\frac{b}{c}\cos\frac{s}{c}, -\frac{a}{c} \right), \\ B'(s) &= \left(\frac{b}{c^2}\cos\frac{s}{c}, \frac{b}{c^2}\sin\frac{s}{c}, 0 \right), \\ \tau(s) &= - < B'(s), N(s) > = \frac{b}{c^2} = \frac{b}{|a^2-b^2|}. \end{aligned}$$

Here, if $\gamma(t)$ is a timelike (or spacelike) curve, $c = \sqrt{b^2-a^2}$ (or $c = \sqrt{a^2-b^2}$) since in the spacelike case $N(s)$ is unit spacelike. □

If $\gamma = (a\cos t, a\sin t, bt)$ is a curve in $\mathbf{E}^3$, then $\kappa(s) = \dfrac{a}{c^2}$ and $\tau(s) = \dfrac{b}{c^2}$ where $c = \sqrt{a^2+b^2}$ much as in the solution of Example 2.1.10.

Example 2.1.11 Find the Frenet-Serret apparatus $\{T(s), N(s), B(s), \kappa(s), \tau(s)\}$ for $\beta(s) = \left(\dfrac{\sinh s}{\sqrt{2}}, \dfrac{s}{\sqrt{2}}, \dfrac{\cosh s}{\sqrt{2}}\right)$.

Solution. Note that $T(s) = \beta'(s) = \left(\dfrac{\cosh s}{\sqrt{2}}, \dfrac{1}{\sqrt{2}}, \dfrac{\sinh s}{\sqrt{2}}\right)$ satisfies $< T(s), T(s) >= 1$ so that $\beta(s)$ is indeed parametrized by arc length. $T'(s) = \left(\dfrac{\sinh s}{\sqrt{2}}, 0, \dfrac{\cosh s}{\sqrt{2}}\right)$ is timelike with $< T'(s), T'(s) >= -\dfrac{1}{2}$, whence $\kappa(s) = \dfrac{1}{\sqrt{2}}$ and $N(s) = (\sinh s, 0, \cosh s)$. Using the Lorentz cross product yields

$$B(s) = \left(\frac{\cosh s}{\sqrt{2}}, -\frac{1}{\sqrt{2}}, \frac{\sinh s}{\sqrt{2}}\right).$$

Hence $B'(s) = \left(\dfrac{\sinh s}{\sqrt{2}}, 0, \dfrac{\cosh s}{\sqrt{2}}\right)$ and $\tau(s) = - < B'(s), N(s) >= \dfrac{1}{\sqrt{2}}$. □

It may be verified that the Frenet-Serret equations in case (d) of Theorem 2.1.9 are satisfied by this example.

The case of the Frenet-Serret frame for $\beta(s)$ lightlike is more complicated and will not be discussed. A treatment may be found in the graduate text of Duggal and Bejancu given in the reference list at the end of this section.

The Osculating Circle

In this section, we will consider regular curves $\gamma(t) = (x(t), y(t))$ in $\mathbf{E}^2$ and $\mathbf{L}^2$, other than straight line segments. Here $v(t) = |\gamma'(t)|$ is given by $\sqrt{x'(t)^2 + y'(t)^2}$ in $\mathbf{E}^2$ (resp., $\sqrt{|x'(t)^2 - y'(t)^2|}$ in $\mathbf{L}^2$) and the arc length $s'(t) = \displaystyle\int_{t=a}^{b} v(t)dt$ as before. If $\beta(s)$ is a unit speed curve parametrized by arc length in either $\mathbf{E}^2$ or $\mathbf{L}^2$, using the Euclidean and Lorentz norms respectively, we may define $T(s) = \beta'(s)$ and $\kappa(s) = |T'(s)|$. Assuming $\kappa(s) > 0$, we may further define $N(s) = \dfrac{T'(s)}{\kappa(s)}$ as in the last section, and $< T(s), N(s) >= 0$.

In the historical formulation of the theory of the osculating circle for a plane curve in $\mathbf{E}^2$, a circle is a drawn through 3 noncollinear points, employing the basic geometric fact going back to Greek mathematics that the problem of passing a circle through 3 given noncollinear points in $\mathbf{E}^2$ has a unique solution. Recall from Section 1.4, Figures1.4-39 and 1.4-40, that circles in $\mathbf{L}^2$ are branches of a hyperbola. Hence for $\mathbf{L}^2$, a construction procedure analogous to that of classical differential geometry for curves in $\mathbf{E}^2$ raises the question of whether a single branch of a hyperbola may be drawn through 3 given noncollinear points. Since translation preserves the Lorentz metric on $\mathbf{L}^2$ (i.e., translation is an isometry), one of the three point may be taken to be $(0,0)$. Thus the problem reduces to studying whether a singe branch of a hyperbola of the form

$$y^2 - x^2 + Ax + By = 0$$

passes through two other given points of $\mathbf{L}^2$. What emerges, in studying various examples, is that the location of the two given points with respect to the light cone through the origin ($y = \pm x$) is decisive. For example, consider $(-1,1)$ on the light cone and $(2,3)$ inside the light cone. Then the hyperbola through $(0,0)$, $(-1,1)$ and $(2,3)$ degenerates into a pair of lines $y = -x$ and $y = x+1$ with two points on one line and one point on the other line. Given a case like $(2,3)$ and $(-2,3)$ of two points outside the light cone and in the upper half plane $y > 0$, the three points $(0,0)$, $(2,3)$ and $(-2,3)$ lie on the hyperbola $y^2 - x^2 - \frac{5}{3}y = 0$, but $(0,0)$ lies on one branch and $(2,3)$, $(2,-3)$ lie other other branch. However, in the intended application below to a unit spacelike curve or to a unit timelike curve in $\mathbf{L}^2$ with positive curvature, a single branch of a unique hyperbola may be drawn through the three given noncollinear points. For $(2,-3)$ and $(2,3)$ which lie inside the light cone and are connected by a future directed timelike curve passing through $(2,-3)$, $(0,0)$ and $(2,3)$ successively, the right branch of the hyperbola

$$y^2 - x^2 - \frac{5}{2}x = 0$$

contains all three points. For "spacelike" points $(-2,1)$, $(3,1)$ outside the light cone, the upper branch of the hyperbola

$$y^2 - x^2 + x + 5y = 0$$

contains $(-2,1)$, $(0,0)$ and $(3,1)$.

In the rest of this subsection, arguments in the style of 17th and 18th century differential geometry will be given. Thus issues of convergence and continuity will be ignored. Hence let $\beta : I \to \mathbb{R}^2$ be a regular curve in either $\mathbf{E}^2$ or $\mathbf{L}^2$ parametrized by arc length and satisfying $\kappa(s) > 0$. Temporarily fixing s in I, take s_1, s_2, s_3 near s so that the three points $\beta(s_1)$, $\beta(s_2)$, $\beta(s_3)$ are not collinear. By the above discussion, these three points determine a unique Euclidean (or hyperbolic) circle with center $P(s_1, s_2, s_3)$ in $\mathbf{E}^2$ (or $\mathbf{L}^2$). Now fix $P(s_1, s_2, s_3)$ and let s vary. Also let

$$f(s) = |\beta(s) - P(s_1, s_2, s_3)|^2.$$

Then, since $f(s)$ is the square of the radius of this circle, $f(s_1) = f(s_2) = f(s_3)$. By Rolle's Theorem, there exist $t_1 \in (s_1, s_2)$, $t_2 \in (s_2, s_3)$ so that

$$f'(t_1) = 0, \quad f'(t_2) = 0.$$

We apply Rolle's Theorem again and thus there exists $u \in (t_1, t_2)$ so that

$$f''(u) = 0.$$

Differentiating $f(s)$, we have $f'(s) = 2 < \beta'(s), \beta(s) - P(s_1, s_2, s_3) >$ if $\beta(s) - P(s_1, s_2, s_3)$ is Euclidean or spacelike. Or, $f'(s) = -2 < \beta'(s), \beta(s) - P(s_1, s_2, s_3) >$ if $\beta(s) - P(s_1, s_2, s_3)$ is timelike.

Differentiating $f'(s)$ again, $f''(s) = 2[< \beta''(s), \beta(s) - P(s_1, s_2, s_3) > + < \beta'(s), \beta'(s) >]$ if $\beta(s) - P(s_1, s_2, s_3)$ is Euclidean or spacelike, or, $f''(s) = -2[< \beta''(s), \beta(s) - P(s_1, s_2, s_3) > + < \beta'(s), \beta'(s) >]$ if $\beta(s) - P(s_1, s_2, s_3)$ is timelike.

Now, let $s_1,\ s_2,\ s_3 \longrightarrow s$ and assume that $P(s_1, s_2, s_3) \longrightarrow P(s)$ with $P(s) \neq \beta(s)$. Then $t_1,\ t_2 \longrightarrow s$ and $u \longrightarrow s$. Since $f'(t_1) = f'(t_2) = f''(u) = 0$, we have $0 = 2 < \beta'(s), \beta(s) - P(s) >$ if $\beta(s) - P(s)$ is Euclidean or spacelike, and $0 = -2 < \beta'(s), \beta(s) - P(s) >$ if $\beta(s) - P(s)$ is timelike. Further, $0 = 2[< \beta''(s), \beta(s) - P(s) > + < \beta'(s), \beta'(s) >]$ if $\beta(s) - P(s)$ is Euclidean or spacelike, and $0 = -2[< \beta''(s), \beta(s) - P(s) > + < \beta'(s), \beta'(s) >]$ if $\beta(s) - P(s)$ is timelike.

Since $< \beta'(s), \beta(s) - P(s) >= 0$, $\beta'(s)$ is timelike (or spacelike) if $\beta(s) - P(s)$ is spacelike (or timelike). Hence, $< \beta''(s), \beta(s) - P(s) >= -1$ if $\beta(s) - P(s)$ is Euclidean or timelike, and $< \beta''(s), \beta(s) - P(s) >= 1$ if $\beta(s) - P(s)$ is spacelike.

By differentiating the equation $< \beta'(s), \beta'(s) >= \pm 1$, we have

$$2 < \beta'(s), \beta''(s) >= 0.$$

Hence, $\beta''(s)$ is proportional to $\beta(s) - P(s)$. Since $\beta''(s) = T'(s) = \kappa(s)N(s)$, $N(s)$ is also proportional to $\beta(s) - P(s)$, and $N(s)$ and $\beta''(s)$ point in the same direction..

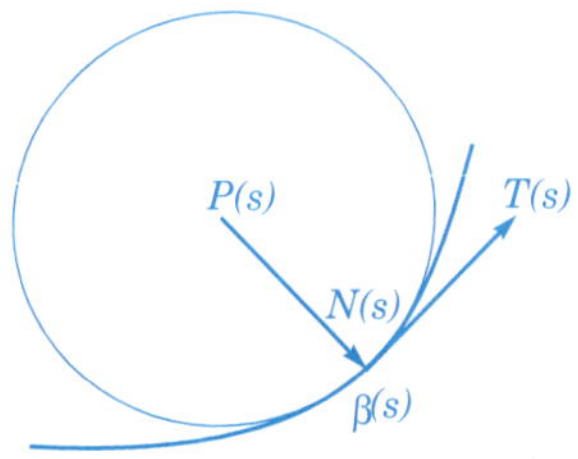

Figure 2.1-4

Thus, if $\beta(s) - P(s)$ is Euclidean, $N(s)$ has the opposite direction to $\beta(s) - P(s)$(Figure 2.1-4). If $\beta(s) - P(s)$ is timelike, $N(s)$ has the same direction as $\beta(s) - P(s)$(Figure 2.1-5). And if $\beta(s) - P(s)$ is spacelike, $N(s)$ has the same direction as $\beta(s) - P(s)$(Figure 2.1-6).

[For instance, suppose $\beta(s) - P(s)$ is spacelike and express $\beta''(s) = \lambda(\beta(s) - P(s))$. Then $1 =< \beta''(s), \beta(s) - P(s) >= \lambda|\beta(s) - P(s)|^2$ from which

$$\lambda = \frac{1}{|\beta(s) - P(s)|^2} > 0$$

and $N(s)$ and $\beta(s) - P(s)$ point in the same direction.]

Therefore, $N(s)$ always turns towards the Euclidean or hyperbolic circle inscribed the bending curve at $\beta(s)$.

In the Euclidean case, Leibniz in 1686 called this circle with center $P(s)$ the **osculating circle**. Also, $\kappa(s) = 1/|\beta(s) - P(s)|$ in all cases, so that the radius of the osculating circle is the reciprocal of the curvature. [For example,

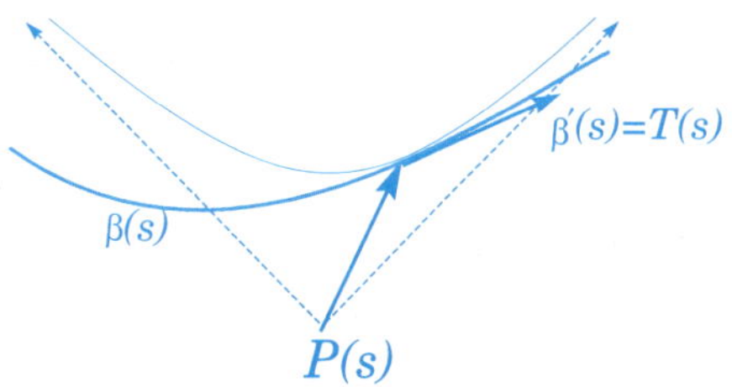

Figure 2.1-5

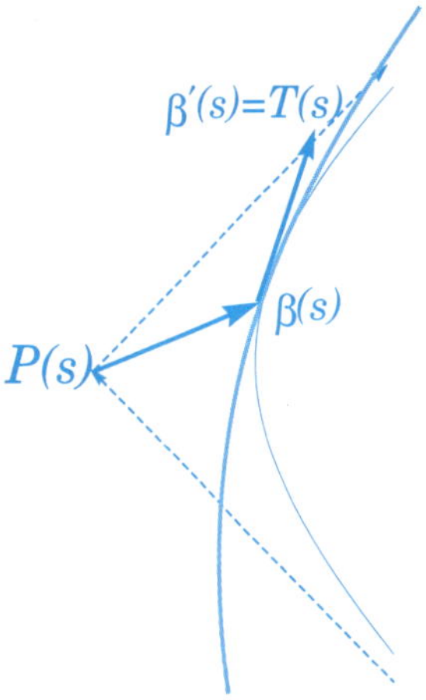

Figure 2.1-6

if $\beta(s) - P(s)$ and $N(s)$ are spacelike, then from the previous calculation

$$\beta''(s) = \frac{\beta(s) - P(s)}{|\beta(s) - P(s)|^2} = \kappa(s)N(s)$$

so that $\dfrac{|\beta(s) - P(s)|}{|\beta(s) - P(s)|^2} = \kappa(s)|N(s)| = \kappa(s)$ from which the above equation follows.] Newton in his *Geometria Analytica* published in 1736 noted that the osculating circle is the circle in $\mathbf{E}^2$ which makes the closest contact with the given curve β at the point $\beta(s)$.

The Geometric Meaning of Torsion

Let $\beta(s)$ be a unit speed curve in $\mathbf{E}^3$ or a non-lightlike unit speed curve in $\mathbf{L}^3$. If we approximate $\beta(s)$ by its Taylor series near $\beta(0)$, then we have

$$\beta(s) \approx \beta(0) + s\beta'(0) + \frac{s^2}{2}\beta''(0) + \frac{s^3}{6}\beta'''(0).$$

Since $\beta'(0) = T(0)$, $\beta''(0) = T'(0) = \kappa(0)N(0)$, and by Frenet-Serret Theorem for $\beta(s)$ Euclidean

$$\begin{aligned}\beta'''(0) =&\kappa'(0)N(0) + \kappa(0)N'(0)\\ =&\kappa'(0)N(0) + \kappa(0)(-\kappa(0)T(0) + \tau(0)B(0))\\ =& -\kappa(0)^2T(0) + \kappa'(0)N(0) + \kappa(0)\tau(0)B(0).\end{aligned}$$

Now, substituting the above derivatives into the approximation of $\beta(s)$, we obtain

$$\begin{aligned}\beta(s) \approx&\beta(0) + sT(0) + \frac{s^2}{2}\kappa(0)N(0)\\ &+\frac{s^3}{6}\left(-\kappa(0)^2T(0) + \kappa'(0)N(0) + \kappa(0)\tau(0)B(0)\right).\end{aligned}$$

Keeping only the terms in each component $T(0), N(0), B(0)$ containing the smallest power of s, we obtain the **Frenet approximation** for β Euclidean:

$$\beta(s) \approx \beta(0) + sT(0) + \frac{s^2}{2}\kappa(0)N(0) + \frac{s^3}{6}\kappa(0)\tau(0)B(0).$$

The same approximation formula is valid for $\beta(s)$ unit timelike, despite a sign change in one term of the formula for $N'(0)$ to $\kappa(0)T(0)$ since higher order terms in s for the component of $T(0)$ are being dropped
For $\beta(s)$ unit spacelike with $T'(s)$ timelike, using Theorem 2.1.9-(d), we obtain exactly the same Frenet approximation. But for $\beta(s)$ unit spacelike with $T'(s)$ spacelike, Theorem 2.1.9-(c) leads to the Frenet approximation

$$\beta(s) \approx \beta(0) + sT(0) + \frac{s^2}{2}\kappa(0)N(0) - \frac{s^3}{6}\kappa(0)\tau(0)B(0).$$

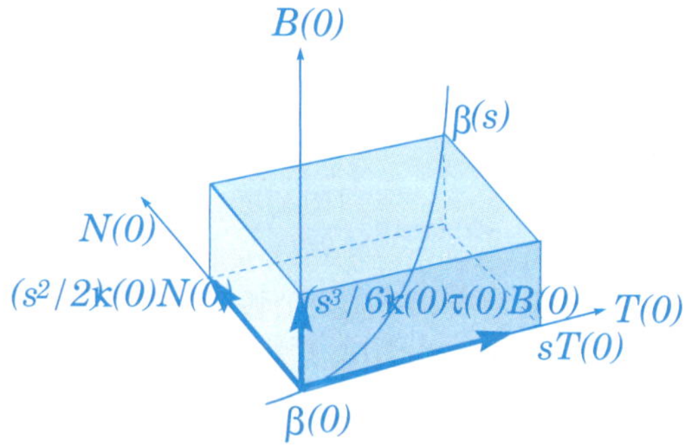

Figure2.1-7

Figure2.1-7 illustrates the geometric meaning of this expression in the Euclidean or timelike cases.

The linear approximation $\beta(0) + sT(0)$ of $\beta(s)$ near $\beta(0)$ corresponds to the tangent line at $\beta(0)$. The quadratic approximation $\beta(0) + sT(0) + \frac{s^2}{2}\kappa(0)N(0)$ of $\beta(s)$ near $\beta(0)$ is a parabola in the osculating plane of $\beta(s)$. Especially, $\kappa(0)\frac{s^2}{2}$ near $s = 0$ depends on $\kappa(0)$. Hence $\kappa(0) > 0$ intuitively controls the bending of the curve near $\beta(0)$. Finally, for $\beta(s)$ Euclidean or timelike, in the approximation $\beta(0) + sT(0) + \frac{s^2}{2}\kappa(0)N(0) + \frac{s^3}{6}\kappa(0)\tau(0)B(0)$, $\tau(0)$ controls the motion of $\beta(s)$ only in the direction of $B(0)$. Thus, if $\tau(0) \neq 0$, $\beta(s)$ near $s = 0$ twists out of the osculating plane. In other words, $\tau(0)$ measures how the osculating plane is turning along $\beta(s)$ near $s = 0$ as s increases. But for $\beta(s)$ spacelike with $T'(s)$ timelike the sign of $\tau(0)$ is opposite from the Euclidean and timelike cases.

For further study, see the following references.

1. Duggal, K. and Bejancu, A., Lightlike Submanifolds of Semi-Riemannian Manifolds and Applications, Kluwer Academic Publishers, 1996.

2. Millman, R. S. and G. D. Parker,Elements of Differential Geometry, Prentice-Hall, Englewood Cliffs, NJ, 1977.

3. McCleary, J., Geometry from a Differentiable Viewpoint, Cambridge University Press, 1994.

4. O'Neill, B., Elementary Differential Geometry, Academic Press, New York, 1966.

2.2 The Curvature and Torsion of Curves

In Section 2.1 we studied the Frenet-Serret apparatus for the reparametrization of a regular curve by arc length and introduced the curvature and torsion along the curve. Now, we classify curves by their curvature and torsion.

Straight Lines

In section 2.1 we observed that the curvature measures bending of the given curve. The following theorem shows the geometric meaning of the zero curvature for a given curve(Figure2.2-8).

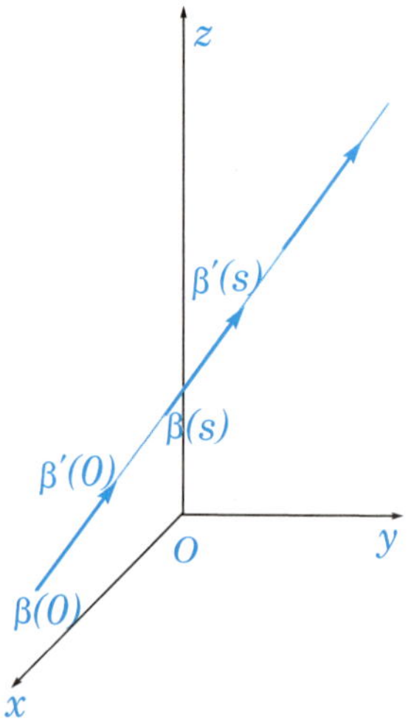

Figure2.2-8

Theorem 2.2.1 *Let $\beta(s)$ be a unit speed curve in $\mathbf{E}^3$ or a non-lightlike curve in $\mathbf{L}^3$. Then $\kappa(s) = 0$ on $[a, b]$ if and only if $\beta : [a, b] \to \mathbb{R}^3$ is a straight line.*

Proof. Since $\kappa(s) = |T'(s)|$, $T'(s) = 0$ on $[a, b]$ and thus $T(s) = A$ on (a, b) where A is a constant (non-lightlike) vector. Hence,

$$\int_a^s \beta'(r)dr = \int_a^s Adr.$$

That is, $\beta(s) - \beta(a) = (s - a)A$. Therefore,

$$\beta(s) = (s - a)A + \beta(a) = sA + (\beta(a) - aA)$$

on $[a, b]$.

Conversely, if $\beta(s) = sA + B$ on $[a, b]$ where A, B are constants, $T(s) = \beta'(s) = A$ and thus $T'(s) = 0$ on $[a, b]$. Therefore,

$$\kappa(s) = |T'(s)| = 0$$

on $[a, b]$. □

Plane curves

A curve is said to be **planar** or the **plane curve** if it lies in a single plane of $\mathbb{R}^3$. In Theorem 2.1.9 we observed that if $\tau(0) \neq 0$, $\beta(s)$ near $s = 0$ twists out of the osculating plane. However, if $\tau(s) = 0$ for $s \approx 0$, then $\beta(s)$ tends to remain on the osculating plane near $s = 0$ by the Frenet approximation of Section 2.1. The following theorem shows a more powerful implication of zero torsion. To obtain the proof, we need to discuss plane normals in $\mathbf{L}^3$ as well as the familiar case of $\mathbf{E}^3$. Thus let $\mathcal{P}$: $Ax + By + Cz = D$ be a plane in $\mathbb{R}$ and $P_0 = (x_0, y_0, z_0)$ be a point in $\mathcal{P}$. Then $P = (x, y, z)$ in $\mathbb{R}^3$ lies on $\mathcal{P}$ if and only if

$$A(x - x_0) + B(y - y_0) + C(z - z_0) = 0.$$

In $\mathbf{E}^3 = (\mathbb{R}^3, < , >)$ we form the vector $N = A\mathbf{i} + B\mathbf{j} + C\mathbf{k}$ and verify that this equation is equivalent to $< N, P_0P >= 0$.

In $\mathbf{L}^3 = (\mathbb{R}^3, g)$ we form the vector $N_1 = A\mathbf{i} + B\mathbf{j} - C\mathbf{k}$ and verify that the plane equation is equivalent to $< N_1, P_0P >= 0$. Hence, the plane $Ax + By + Cz = D$ has Lorentz normal $N_1 = A\mathbf{i} + B\mathbf{j} - C\mathbf{k}$.

Theorem 2.2.2 *Let $\beta(s)$ be a unit speed curve in $\mathbf{E}^3$ or a non-lightlike curve in $\mathbf{L}^3$ with $\kappa(s) > 0$. Then $\beta(s)$ is planar if and if $\tau(s) = 0$.*

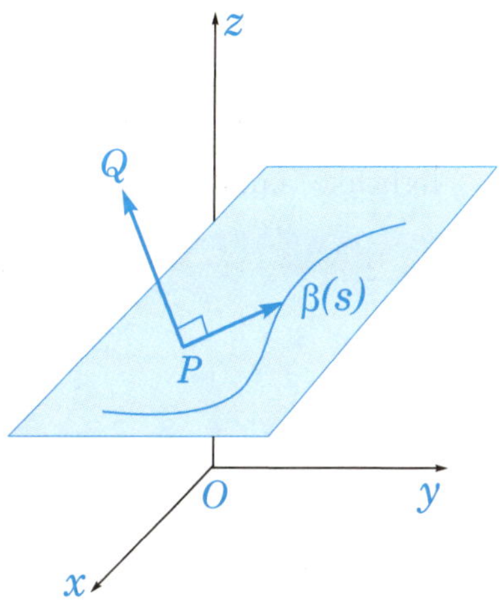

Figure2.2-9

Proof. If the unit speed (non-lightlike) curve $\beta(s)$ lies on a plane of $\mathbf{E}^3$ ($\mathbf{L}^3$), then for any fixed point P in the plane and vector Q orthogonal to the plane

$$< \beta(s) - P, Q >= 0$$

for all s(Figure2.2-9). By differentiating the above equation, we have

$$< \beta'(s), Q >= 0 \text{ and } < \beta''(s), Q >= 0$$

for all s. That is, $< T(s), Q >= 0$ and $< T'(s), Q >= 0$. Since $T'(s) = \kappa(s)N(s)$, Q is orthogonal to $T(s)$ and $N(s)$. Since $B(s)$ is also orthogonal to $T(s)$ and $N(s)$, we have Q and $B(s)$ are parallel. In $\mathbf{L}^3$, since $B(s)$ is nonnull, also Q is nonnull and we may conclude that

$$B(s) = \pm \frac{Q}{|Q|}$$

in either $\mathbf{E}^3$ or $\mathbf{L}^3$. Thus, $B'(s) = 0$. Therefore, $\tau(s) = - < B'(s), N(s) >= 0$.

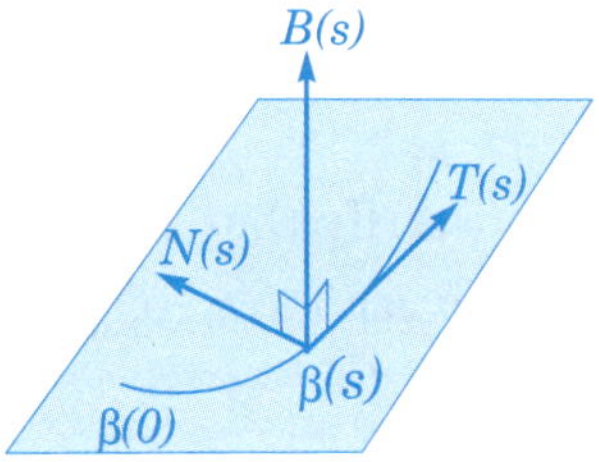

Figure2.2-10

Conversely, if $\tau(s) = 0$, then $B'(s) = \pm\tau(s)N(s) = 0$. Hence $B(s)$ is a constant vector in $\mathbb{R}^3$. Set

$$f(s) := <\beta(s) - \beta(0), B(s)>$$

for all s. Since $f'(s) =< \beta'(s), B(s) >=< T(s), B(s) >= 0$, $f(s)$=constant. Moreover, since $f(0) = 0$, we have $f(s) = 0$. That is, $< \beta(s) - \beta(0), B(s) >= 0$ for all s. Therefore, $\beta(s)$ lies on a plane orthogonal to the constant binormal vector $B(s)$(Figure2.2-10). □

Euclidean or Hyperbolic Circles

Example 2.2.3 Compute the curvature and torsion of the unit speed (non-lightlike) curve given by $\beta(s) = \left(a\cos\frac{s}{c}, a\sin\frac{s}{c}, \frac{bs}{c}\right)$ where $a > 0$ and

$$c = \begin{cases} (a^2+b^2)^{\frac{1}{2}} & \text{if } \beta(s) \text{ is Euclidean} \\ (b^2-a^2)^{\frac{1}{2}} & \text{if } \beta(s) \text{ is timelike} \\ (a^2-b^2)^{\frac{1}{2}} & \text{if } \beta(s) \text{ is spacelike} \end{cases}$$

Solution. Since $T(s) = \beta'(s) = \left(-\frac{a}{c}\sin\frac{s}{c}, \frac{a}{c}\cos\frac{s}{c}, \frac{b}{c}\right)$, we have $T'(s) = \beta''(s) = \left(-\frac{a}{c^2}\cos\frac{s}{c}, \frac{a}{c^2}\sin\frac{s}{c}, 0\right)$. Hence, $\kappa(s) = |T'(s)| = \frac{a}{c^2}$

$$= \begin{cases} \dfrac{a}{(a^2+b^2)^{\frac{1}{2}}} & \text{if } \beta(s) \text{ is Euclidean} \\ \dfrac{a}{(b^2-a^2)^{\frac{1}{2}}} & \text{if } \beta(s) \text{ is timelike} \\ \dfrac{a}{(a^2-b^2)^{\frac{1}{2}}} & \text{if } \beta(s) \text{ is spacelike} \end{cases}$$

which is positive. □

Note that if $b = 0$ then $\beta(s)$ is the Euclidean circle with curvature $\kappa(s) = 1$ in $\mathbf{E}^3$, but it is not the hyperbolic circle.

Theorem 2.2.4 *Let $\beta(s)$ be a unit speed curve in $\mathbf{E}^3$ or a non-lightlike curve in $\mathbf{L}^3$ with constant curvature $\kappa > 0$ and $\tau(s) = 0$. Then $\beta(s)$ is a part of a Euclidean or hyperbolic circle of radius $\frac{1}{\kappa}$.*

Proof. Since $\tau(s) = 0$, $\beta(s)$ is planar by Theorem 2.2.2. For $\beta(s)$ Euclidean, or spacelike with $N(s)$ spacelike, consider the curve given by $\gamma(s) = \beta(s) + \frac{1}{\kappa}N(s)$. Then

$$\begin{aligned}\gamma'(s) &= \beta'(s) + \frac{1}{\kappa}N'(s)\\ &= \beta'(s) + \frac{1}{\kappa}(-\kappa T(s))\\ &= T(s) - T(s)\\ &= 0.\end{aligned}$$

For $\beta(s)$ timelike with $N(s)$ spacelike, or spacelike with $N(s)$ timelike, consider the curve given by $\gamma(s) = \beta(s) - \frac{1}{\kappa}N(s)$. Then

$$\begin{aligned}\gamma'(s) &= \beta'(s) - \frac{1}{\kappa}N'(s)\\ &= \beta'(s) - \frac{1}{\kappa}(\kappa T(s))\\ &= T(s) - T(s)\\ &= 0.\end{aligned}$$

Hence, in all cases $\gamma(s) = c$ where c is constant. Thus $|\beta(s) - c| = |\pm(\beta(s) - \gamma(s))| = \frac{1}{\kappa}$. Therefore, the result follows by using either Euclidean or Lorentzian inner product. □

General Helix

Definition 2.2.5 A regular curve α in $\mathbf{E}^3$ ($\mathbf{L}^3$) is said to be a **(timelike) general helix** if there is a constant unit (future directed timelike) vector u such that $< T(s), u >= \cos\theta$ ($< T(s), u >= -\cosh\theta$) is nonzero constant for the unit (future directed timelike) tangent vector field $T(s)$ along α. The constant vector u is called the **axis of the (timelike) general helix** and θ is the **pitch** of α(Figure2.2-11).

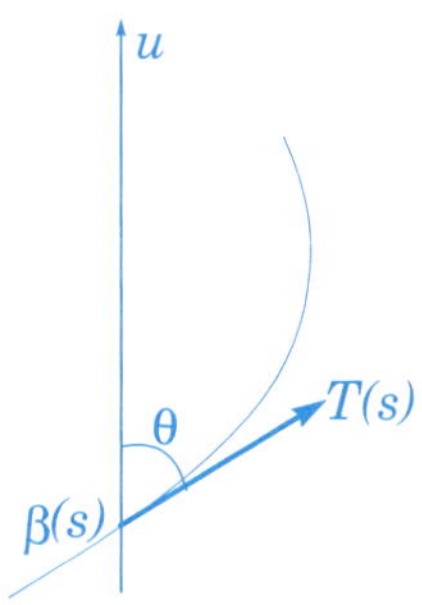

Figure2.2-11

Theorem 2.2.6 *A future directed unit speed timelike curve $\beta(s)$ with $\kappa(s) > 0$ is a timelike general helix if and only if $\frac{\tau(s)}{\kappa(s)}$ is a constant C with $|C| > 1$.*

Proof. Since $< T(s), u >$ is constant and $T(s)$, u are future directed timelike, we may write

$$< T(s), u >= -\cosh\theta$$

where θ is the fixed pitch of $\beta(s)$. Since $\frac{du}{ds} = 0$, we have

$$0 =< T(s), u >' =< T'(s), u >= \kappa(s) < N(s), u > .$$

Moreover, since $\kappa(s) > 0$, we have $< N(s), u >= 0$. Hence, u lies on the plane containing $T(s)$ and $B(s)$. Thus, recalling Figure 1.4-34, we have

$$u = \cosh\theta T(s) + \sinh\theta B(s).$$

Differentiating this equation, $0 = \cosh\theta T'(s) + \sinh\theta B'(s) = \kappa(s)\cosh\theta N(s) - \tau(s)\sinh\theta N(s)$. Therefore, since $\kappa(s) > 0$ precludes $\theta = 0$, we have $\sinh\theta \neq 0$ and we may form

$$\frac{\tau(s)}{\kappa(s)} = \frac{\cosh\theta}{\sinh\theta} = \coth\theta,$$

which is a constant C with $|C| > 1$.

Conversely, if $\dfrac{\tau(s)}{\kappa(s)} = C$ is constant with $|C| > 1$, by the Intermediate Value Theorem there exists θ such that

$$\frac{\tau(s)}{\kappa(s)} = \coth\theta.$$

Then $(\kappa(s)\cosh\theta - \tau(s)\sinh\theta)N(s) = 0$. By the Frenet-Serret Theorem we have $\dfrac{d}{ds}(\cosh\theta T(s) - \sinh\theta B(s)) = 0$. Hence, $\cosh\theta T(s) - \sinh\theta B(s) = u$ is constant and thus we have $< u, u >= \cosh^2\theta < T(s), T(s) > + \sinh^2\theta < B(s), B(s) >$. Since $T(s)$ is timelike and $B(s)$ is spacelike, we have $< T(s), T(s) >= -1$ and $< B(s), B(s) >= 1$. Thus, $< u, u >= -\cosh^2\theta + \sinh^2\theta = -1 < 0$. Further,

$$\begin{aligned} < T(s), u > &=< T(s), \cosh\theta T(s) + \sinh\theta B(s) > \\ &= \cosh\theta < T(s), T(s) > + \sinh\theta < T(s), B(s) > \\ &= -\cosh\theta \\ &= \text{constant}. \end{aligned}$$

□

Recall from Example 2.1.10 that the timelike curve $\gamma(t) = (a\cos t, a\sin t, bt)$, $a \neq 0$, $b > 0$, $b^2 > a^2$, has the curvature $\kappa(s) = \dfrac{a}{b^2 - a^2}$ and torsion $\tau(s) = \dfrac{b}{b^2 - a^2}$. Hence, $\dfrac{\tau(s)}{\kappa(s)} = \dfrac{b}{a}$ =constant with $|b/a| > 1$. Thus, $\gamma(t)$ is a timelike general helix according to Theorem 2.2.6.

Let us check that $u = \mathbf{k}$ is the axis, supposing also that $a > 0$. Recall that $c = \sqrt{b^2 - a^2}$ in Example 2.1.10. Given $\coth\theta = \dfrac{b}{a}$, then $a\cosh\theta = b\sinh\theta$ and using the basic hyperbolic identity $\cosh^2\theta - \sinh^2\theta = 1$, we may solve for $\cosh\theta = \dfrac{b}{c}$ and $\sinh\theta = \dfrac{a}{c}$. Using the formulas for $T(s)$ and $B(s)$ from Example 2.1.10,

$$u = \cosh\theta T(s) + \sinh\theta B(s) = \frac{b}{c}T(s) + \frac{a}{c}B(s) = (0, 0, 1).$$

Furthermore, the Euclidean analogue of Theorem 2.2.6 may be obtained similarly using the decomposition

$$u = \cos\theta T(s) + \sin\theta B(s).$$

Corollary 2.2.7 *A unit speed curve $\beta(s)$ in $\mathbf{E}^3$ with $\kappa(s) > 0$ is a general helix if and only if $\dfrac{\tau(s)}{\kappa(s)}$ is constant.*

Arbitrary Speed Curves

Let α be a regular curve in $\mathbf{E}^3$ or in $\mathbf{E}^3$. Then we observed that the reparameterization β of α by arc length is useful to obtain the Frenet-Serret apparatus. However, some arbitrary speed curves may not be explicitly parametrized by arc length as the following examples show.

Example 2.2.8 Let a curve be given by $\alpha(t) = (2\sin t, \cos t, 0)$. Then $\alpha'(t) = (2\cos t, -\sin t, 0)$ and $|\alpha'(t)| = \sqrt{4\cos^2 t + \sin^2 t} = \sqrt{4 - 3\sin^2 t} = 2\sqrt{1 - \frac{3}{4}\sin^2 t}$. But the indefinite integral of $\sqrt{1 - \frac{3}{4}\sin^2 t}$ can not be explicitly calculated. Therefore, the arc length $s(t) = \int_0^t |\alpha'(u)|du$ can not be explicitly calculated.

Example 2.2.9 Let a curve is given by $\alpha(t) = \left(t, \frac{t^2}{2}, 0\right)$. Then $\alpha'(t) = (1, t, 0)$. Since $|\alpha'(t)| = \sqrt{1+t^2}$, we have the length function

$$s(t) = \int_0^t \sqrt{1+u^2}du = \frac{1}{2}(t\sqrt{1+t^2} + \ln(t + \sqrt{1+t^2})).$$

But it is impossible to solve this equation explicitly for t as a function of s. Hence this curve can not be explicitly reparametrized by arc length parameter s.

Now we obtain the Frenet-Serret apparatus for an arbitrary speed curve $\alpha(t)$ by alternative computational techniques to circumvent these difficulties.

Theorem 2.2.10 *Let α be a regular curve with $\kappa > 0$ in $\mathbf{E}^3$, or a timelike curve or a spacelike curve in $\mathbf{L}^3$ with $N(t)$ unit spacelike(resp., a spacelike curve in $\mathbf{L}^3$ with $N(t)$ unit timelike). Then using the Euclidean cross product and inner product for $\mathbf{E}^3$ and the Lorentz cross product and inner product for $\mathbf{L}^3$, we have*

(1) $T(t) = \dfrac{\alpha'}{|\alpha'|}$

(2) $B(t) = \dfrac{\alpha' \times \alpha''}{|\alpha' \times \alpha''|}$

(3) $N(t) = B \times T$ *for $\alpha(t)$ Euclidean or timelike (resp., $N(t) = -B \times_g T$ for $\alpha(t)$ spacelike)*

(4) $\kappa(t) = \dfrac{|\alpha' \times \alpha''|}{|\alpha'|^3}$

(5) $\tau(t) = \dfrac{< \alpha' \times \alpha'', \alpha''' >}{|\alpha' \times \alpha''|^2}$ *for $\alpha(t)$ Euclidean, timelike or spacelike with $N(t)$ unit spacelike (resp., $\tau = -\dfrac{< \alpha' \times_g \alpha'', \alpha''' >}{|\alpha' \times_g \alpha''|^2}$ for $\alpha(t)$ spacelike with $N(t)$ unit timelike).*

Proof. (1) Recall that $s(t) = \int_a^t |\alpha'(u)|du$ and $\alpha(t) = \beta(s(t))$. Since $\alpha'(t) = \beta'(s)s'(t) = s'(t)T(s)$ and since $s'(t) = |\alpha'(t)|$, we have

$$T(s) = \frac{\alpha'(t)}{s'(t)} = \frac{\alpha'(t)}{|\alpha'(t)|}.$$

(4) $\alpha''(t) = s''(t)T(s) + s'(t)\dfrac{d}{dt}T(s) = s''(t)T(s) + s'(t)\dfrac{d}{ds}T(s)\dfrac{ds}{dt} = s''(t)T(s) + (s'(t))^2\dfrac{d}{ds}T(s) = s''(t)T(s) + (s'(t))^2\kappa(s)N(s)$. Since $T(s) \times T(s) = 0$, we have $\alpha'(t) \times \alpha''(t) = s'(t)T(s) \times (s''(t)T(s) + (s'(t))^2\kappa(s)N(s)) = \kappa(s)(s'(t))^3 B(s)$ and thus $|\alpha'(t) \times \alpha''(t)| = \kappa(s)(s'(t))^3$. Thus we have

$$\kappa(s) = \frac{|\alpha'(t) \times \alpha''(t)|}{|\alpha'(t)|^3}.$$

(2) Since $\kappa > 0$ and $s'(t) = |\alpha'(t)| > 0$, $B(s) = \dfrac{\alpha'(t) \times \alpha''(t)}{\kappa(s)(s'(t))^3} = \dfrac{\alpha'(t) \times \alpha''(t)}{|\alpha'(t) \times \alpha''(t)|}$.

(3) From Theorem 1.3.1-(4) and Theorem 1.3.10-(4), for $\alpha(t)$ Euclidean or timelike we have $N(s) = B(s) \times T(s)$ but for $\alpha(t)$ spacelike we have $N(s) = -B(s) \times_g T(s)$.

(5) By differentiating $\alpha''(t) = s''(t)T(s) + (s'(t))^2\kappa(s)N(s)$, we have for $\alpha(t)$ Euclidean or timelike

$$\begin{aligned}\alpha'''(t) =& s'''(t)T(s) + s''(t)s'(t)T'(s) + \frac{d}{dt}((s'(t))^2\kappa(s))N(s)\\ &+ (s'(t))^3\kappa(s)N'(s)\\ =& s'''(t)T(s) + s'(t)s''(t)\kappa(s)N(s) + \frac{d}{dt}((s'(t))^2\kappa(s))N(s)\\ &- (s'(t))^3\kappa^2(s)T(s) + (s'(t))^3\kappa(s)\tau(s)B(s)\\ =&(s'''(t) - (s'(t))^3\kappa^2(s))T(s)\\ &+ \left(s'(t)s''(t)\kappa(s) + \frac{d}{dt}((s'(t))^2\kappa(s))\right)N(s)\\ &+ (s'(t))^3\kappa(s)\tau(s)B(s).\end{aligned}$$

Hence,

$$\begin{aligned}&< \alpha'(t) \times \alpha''(t), \alpha'''(t) >=< \kappa(s)(s'(t))^3 B(s), \alpha'''(t) >\\ &= \tau(s)(\kappa(s)(s'(t))^3)^2 = \tau(s)|\alpha'(t) \times \alpha''(t)|^2.\end{aligned}$$

Thus, we have

$$\tau = \frac{< \alpha' \times \alpha'', \alpha''' >}{|\alpha' \times \alpha''|^2}.$$

In a same manner, this same formula may be derived for $\alpha(t)$ spacelike with $N(t)$ unit spacelike. Finally, for $\alpha(t)$ spacelike with $N(t)$ unit timelike, we obtain with Theorem 2.1.9-(c) that

$$\tau = -\frac{< \alpha' \times_g \alpha'', \alpha''' >}{|\alpha' \times_g \alpha''|^2}.$$

□

Remark 2.2.11 Alternate formulas (3) and (4) may be given in which these quantities are expressed in terms of $T'(t)$ despite the disadvantage of a quotient rule differentiation to calculate $T'(t)$ from (1). First, using the chain rule

$$\kappa(s) = |T'(s)| = \left|\frac{dT}{dt}\Big/\frac{ds}{dt}\right| = \frac{|T'(t)|}{ds/dt}$$

so that

$$(6) \quad \kappa(t) = \frac{|T'(t)|}{|\alpha'(t)|}.$$

Then $N(s) = \dfrac{T'(s)}{\kappa(s)} = \dfrac{T'(t)}{ds/dt}\Big/\dfrac{|T'(t)|}{|\alpha'(t)|}$ from which

$$(7) \quad N(t) = \frac{T'(t)}{|T'(t)|}.$$

Formula (6) expresses the curvature $\kappa(t)$ as a function of the given parameter. We also need to see how the torsion $\tau(s)$ relates to the given parameter t. Now Theorem 2.1.9 implies in all cases that $\tau(s) = \pm|B'(s)|$. Hence,

$$\tau(s) = \pm|B'(s)| = \pm\left|\frac{dB}{dt}\frac{dt}{ds}\right| = \pm\left|\frac{dB}{dt}\right| \Big/ \frac{ds}{dt} = \pm\frac{|B'(t)|}{|\alpha'(t)|}.$$

Hence we obtain

$$(8) \quad \tau(t) = \pm\frac{|B'(t)|}{|\alpha'(t)|}$$

and thus for the ratio $\tau(s)/\kappa(s)$ in Theorem 2.2.6 and Corollary 2.2.7, we have

$$\frac{\tau(s)}{\kappa(s)} = \pm\frac{|B'(t)|}{|T'(t)|} = \pm\frac{|B'(t)|}{|\alpha'(t)|}\Big/\frac{|T'(t)|}{|\alpha'(t)|} = \frac{\tau(t)}{\kappa(t)}.$$

Hence if the ratio of torsion to curvature is constant when calculated with the given parametrization $\alpha(t)$, it is also constant for the associated curve $\beta(s) = \alpha(s(t))$ reparametrized by arc length.

Example 2.2.12 Find the Frenet-Serret apparatus for a curve $\alpha(t) = (3t - t^3, 3t^2, 3t + t^3)$ in $\mathbf{E}^3$(Figure 2.2-12).

Solution. We may check that

$$\alpha'(t) = (3 - 3t^2, 6t, 3 + 3t^2) = 3(1 - t^2, 2t, 1 + t^2),$$
$$\alpha''(t) = (-6t, 6, 6t) = 6(-t, 1, t),$$
$$\alpha'''(t) = (-6, 0, 6) = 6(-1, 0, 1),$$
$$|\alpha'(t)| = 3\sqrt{(1 - t^2)^2 + 4t^2 + (1 + t^2)^2} = 3\sqrt{2}(1 + t^2),$$
$$\alpha'(t) \times \alpha''(t) = 18(-1 + t^2, -2t, 1 + t^2),$$
$$|\alpha'(t) \times \alpha''(t)| = 18\sqrt{2}(1 + t^2), \text{ and}$$
$$< \alpha'(t) \times \alpha''(t), \alpha'''(t) >= 216.$$

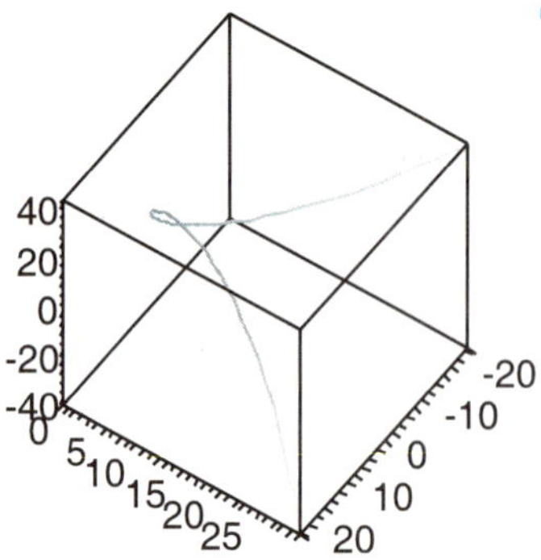

Figure 2.2-12

Hence, we have

$$T = \frac{\alpha'}{|\alpha'|} = \frac{(1 - t^2, 2t, 1 + t^2)}{\sqrt{2}(1 + t^2)},$$
$$B = \frac{\alpha' \times \alpha''}{|\alpha' \times \alpha''|} = \frac{(-1 + t^2, -2t, 1 + t^2)}{\sqrt{2}(1 + t^2)},$$
$$N = B \times T == \frac{(-2t, 1 - t^2, 0)}{1 + t^2},$$
$$\kappa = \frac{|\alpha' \times \alpha''|}{|\alpha'|^3} = \frac{1}{3(1 + t^2)^2}, \text{ and}$$
$$\tau = \frac{< \alpha' \times \alpha'', \alpha''' >}{|\alpha' \times \alpha''|^2} = \frac{1}{3(1 + t^2)^2}.$$ □

Since $\dfrac{\tau(t)}{\kappa(t)} = 1$, this curve is a generalized helix (with axis $u = \mathbf{k}$) by

Corollary 2.2.7 and Remark 2.2.11.

Note that the curve given by $\alpha(t) = (3t - t^3, 3t^2, 3t + t^3)$ is lightlike in $\mathbf{L}^3$.

For further study, see the following references.

1. Millman, R. S. and G. D. Parker,Elements of Differential Geometry, Prentice-Hall, Englewood Cliffs, NJ, 1977.
2. McCleary, J., Geometry from a Differentiable Viewpoint, Cambridge University Press, 1994.
3. O'Neill, B., Elementary Differential Geometry, Academic Press, New York, 1966.

2.3 Vector Calculus

Line Integrals of Vector Fields

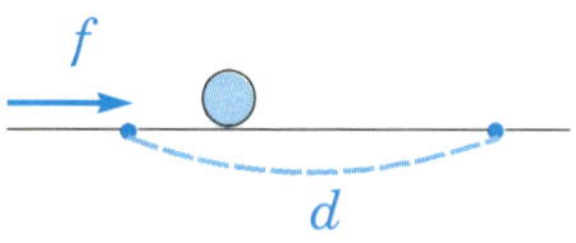

Figure2.3-13

If a particle moves along a straight line a distance d under a constant force f(Figure2.3-13), the amount of work done by this force is given by

$$f \cdot d.$$

Now, when a particle moves by a variable force field $F = f_1\mathbf{i} + f_2\mathbf{j} + f_3\mathbf{k}$ such as the gravitational vector field or the electric field along a smooth curve C, consider the amount of its work, starting with a smooth curve C in $\mathbf{E}^2$ and $F(x, y) = f_1\mathbf{i} + f_2\mathbf{j}$.

Theorem 2.3.1 *If F is a vector field along a smooth curve C in $\mathbf{E}^2$, the amount W of work done by F is a curve integral of $F \cdot T$ along C, i.e.,*

$$W = \int_C F \cdot T ds$$

where T is the unit tangent vector field of C.

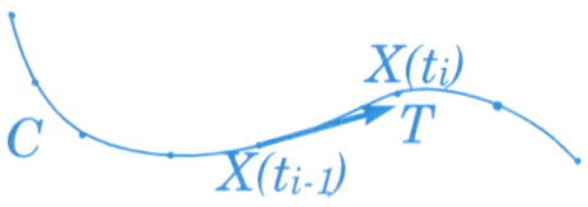

Figure2.3-14

Proof. Let a particle move by F along $C : X(t) = (x(t), y(t))$, $t \in [a, b]$. For any partition $P : a = t_0 < t_1 < \cdots < t_n = b$ of $[a, b]$,

$$X(t_0) = (x_0, y_0), X(t_1) = (x_1, y_1), \cdots, X(t_n) = (x_n, y_n)$$

are vertices of C for the partition P. Let (x_i, y_i) be any point between $X(t_{i-1})$ and $X(t_i)$. Then the work of the particle on the short interval between $X(t_{i-1})$ and $X(t_i)$ is approximated by the straight line movement in the direction $T(x_i, y_i)$ by $F(x_i, y_i) \cdot T(x_i, y_i)$ with distance Δs_i(Figure2.3-14). Hence, the amount ΔW_i of work of F on each interval is approximated by

$$\Delta W_i \approx \Big[F(x_i, y_i) \cdot T(x_i, y_i)\Big] \Delta s_i,$$

and the total amount W of work is given by

$$W = \lim_{|P| \to 0} \sum_{i=1}^{n} \Big[F(x_i, y_i) \cdot T(x_i, y_i)\Big] \Delta s_i = \int_C F \cdot T ds.$$

Therefore,

$$W = \int_C F \cdot T ds.$$

□

The **curve integral** or **line integral** of F along C in $\mathbf{E}^2$ is defined by

$$\int_C F = \int_C F \cdot T ds.$$

Now let

$$F(x, y, z) = f_1(x, y, z)\mathbf{i} + f_2(x, y, z)\mathbf{j} + f_3(x, y, z)\mathbf{k}$$

be a vector field along a smooth curve $C : X(t) = (x(t), y(t), z(t)),\ t \in [a, b]$ in $\mathbf{E^3}$. Then

$$\begin{aligned}\int_C F \cdot dX &= \int_C F \cdot T ds \\ &= \int_C \left[F \cdot \left(\frac{dX}{dt} \Big/ \left|\frac{dX}{dt}\right|\right)\right] \left|\frac{dX}{dt}\right| dt \\ &= \int_a^b F(X(t)) \cdot X'(t) dt.\end{aligned}$$

Thus using the curve integral of real valued functions

$$\begin{aligned}\int_C F \cdot dX &= \int_a^b (f_1(X(t)), f_2(X(t)), f_3(X(t))) \cdot X'(t) dt \\ &= \int_a^b f_1(X(t))x'(t)dt + f_2(X(t))y'(t)dt + f_3(X(t))z'(t)dt \\ &= \int_C f_1 dx + f_2 dy + f_3 dz.\end{aligned}$$

However, if F is a vector field along C in $\mathbf{L^3}$,

$$\begin{aligned}\int_C F \cdot dX &= \int_a^b (f_1(X(t)), f_2(X(t)), f_3(X(t))) \cdot X'(t) dt \\ &= \int_a^b f_1(X(t))x'(t)dt + f_2(X(t))y'(t)dt - f_3(X(t))z'(t)dt \\ &= \int_C f_1 dx + f_2 dy - f_3 dz.\end{aligned}$$

Example 2.3.2 When a particle is moving counterclockwise along the quadrant circle $X(t) = \cos t\mathbf{i} + \sin t\mathbf{j},\ 0 \le t \le \frac{\pi}{2}$ in $\mathbf{E}^2$, find the amount of work done by the force field $F(x, y) = x^2\mathbf{i} - xy\mathbf{j}$ on X.

Solution. Since $x = \cos t, y = \sin t, \quad 0 \le t \le \frac{\pi}{2}$,

$$\begin{aligned}F(X(t)) &= \cos^2 t\mathbf{i} - \cos t \sin t\mathbf{j} \\ X'(t) &= -\sin t\mathbf{i} + \cos t\mathbf{j}.\end{aligned}$$

Hence,

$$
\begin{aligned}
W &= \int_0^{\frac{\pi}{2}} F(X(t)) \cdot X'(t)dt \\
&= \int_0^{\frac{\pi}{2}} (-2\cos^2 t \sin t)dt \\
&= \left[2\frac{\cos^3 t}{3}\right]_0^{\frac{\pi}{2}} = -\frac{2}{3}
\end{aligned}
$$

□

Example 2.3.3 When a particle is moving along the timelike curve $X(t) = \cos t\mathbf{i} + \sin t\mathbf{j}$, $0 \le t < \frac{\pi}{4}$, find the amount of work done by the force field $F(x, y) = x^2\mathbf{i} + xy\mathbf{j}$ on X.

Solution. Since $x = \cos t, y = \sin t,\ 0 \le t < \frac{\pi}{4}$,

$$
\begin{aligned}
F(X(t)) &= \cos^2 t\mathbf{i} + \cos t \sin t\mathbf{j} \\
X'(t) &= -\sin t\mathbf{i} + \cos t\mathbf{j}.
\end{aligned}
$$

Hence, $X(t)$ is timelike for $0 \le t < \frac{\pi}{4}$ and lightlike at $t = \frac{\pi}{4}$. Thus we extend the line integral to $t = \frac{\pi}{4}$ and calculate

$$
\begin{aligned}
W &= \int_0^{\frac{\pi}{4}} F(X(t)) \cdot X'(t)dt \\
&= \int_0^{\frac{\pi}{4}} (-2\cos^2 t \sin t)dt \\
&= \left[2\frac{\cos^3 t}{3}\right]_0^{\frac{\pi}{4}} = -\frac{\sqrt{2}-2}{6}
\end{aligned}
$$

□

n-Polygonal Curves in $\mathbf{L}^2$

Definition 2.3.4 A regular curve $c : [a, b] \to \mathbf{L}^2$ is called a **timelike n-polygonal curve** if for a partition of $[a, b]$: $a = t_0 < t_1 < t_2 < \cdots < t_n = b$ and for each $i = 1, 2, \ldots\ldots, n$,

(1) $C_i(t) = c(t),\ t \in [t_{i-1}, t_i]$ is an 1-1 timelike regular arc.

(2) $c = C_1 + C_2 + C_3 + \cdots + C_n$, where $C_{i-1} + C_i$ implies that $C_{i-1}(t_{i-1}) = C_i(t_{i-1})$.

(3) c is a simply closed curve with $c(a) = C_1(a) = C_n(b) = c(b)$.

$c(t_i)$ is called a **vertex** of the timelike n-polygonal curve c and C_i is called an **edge** of c. if $n=3$, c is called a **timelike triangle** on $\mathbf{L}^2$. And if $n=4$, c is called a **timelike quadrangle** on $\mathbf{L}^2$.

Similarly, we may define a **spacelike** n**-polygonal curve**, a **spacelike triangle** and a **spacelike quadrangle** on $\mathbf{L}^2$.

Definition 2.3.5 A piecewise regular curve c is **convex** if it lies on one side of each tangent line. If a region $R \subset \mathbf{L}^2$ is enclosed by a timelike(or spacelike) convex curve, R is called the **timelike(or spacelike) convex region**.

Hence, for $n \geq 3$ we decompose R into an $n-2$ timelike(or spacelike) triangles as follows(Figure2.3-15).

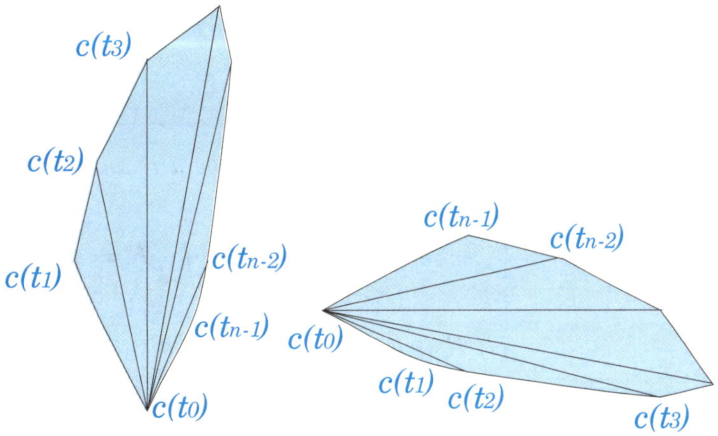

Figure2.3-15 triangulation

Lemma 2.3.6 *Let R be a timelike(or spacelike) n-polygonal convex region with for $n \geq 3$. Then the timelike(or spacelike) convex region R has an $n-2$ timelike(or spacelike) triangle decomposition.*

Green's Theorem in E^2 or L^2

Definition 2.3.7 Let C be a regular curve in $\mathbf{E}^2$ or a nonlightlike curve in $\mathbf{L}^2$ given by $\alpha(t) = (x(t), y(t))$. If f and g are continuous functions in $\mathbf{E}^2$ or $\mathbf{L}^2$, we have the **line integral** of the associated vector function $F(t) = (f(t), g(t))$ on C given by in $\mathbf{E}^2$

$$\begin{aligned}\int_C F &= \int_a^b \left[f(x(t), y(t)) \frac{dx}{dt} + g(x(t), y(t)) \frac{dy}{dt} \right] dt \\ &= \int_C f(x,y)dx + g(x,y)dy\end{aligned}$$

or in $\mathbf{L}^2$

$$\int_C F = \int_a^b \left[f(x(t), y(t))\frac{dx}{dt} - g(x(t), y(t))\frac{dy}{dt} \right] dt$$
$$= \int_C f(x,y)dx - g(x,y)dy.$$

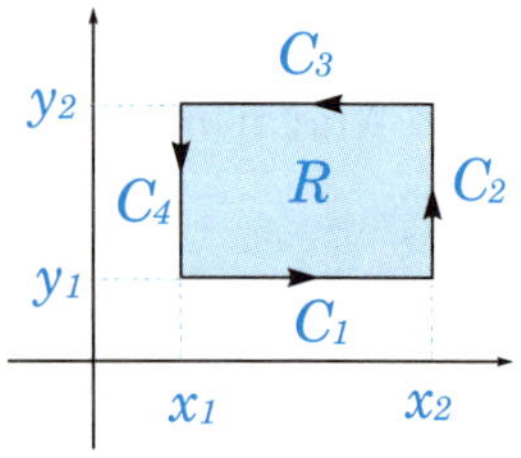

Figure 2.3-16 rectangular region

If R is a rectangle as in Figure2.3-16 and $F = (f,g)$ is a (continuous) vector field on R, then the line integral of F along the boundary ∂R of R is given by in $\mathbf{E}^2$

$$\int_{\partial R} F = \int_{C_1} f(x,y)dx + \int_{C_2} g(x,y)dy + \int_{C_3} f(x,y)dx + \int_{C_4} g(x,y)dy$$

or in $\mathbf{L}^2$

$$\int_{\partial R} F = \int_{C_1} f(x,y)dx - \int_{C_2} g(x,y)dy + \int_{C_3} f(x,y)dx - \int_{C_4} g(x,y)dy.$$

Lemma 2.3.8 *Let R be a rectangle in $\mathbf{E}^2$ and $F = (f,g)$ be a C^1 vector field on R, then*

$$\int_{\partial R} F = \iint_R \left(\frac{\partial g}{\partial x} - \frac{\partial f}{\partial y} \right) dA.$$

If R is a rectangle in $\mathbf{L}^2$, then

$$\int_{\partial R} F = -\iint_R \left(\frac{\partial g}{\partial x} + \frac{\partial f}{\partial y} \right) dA.$$

(In both $\mathbf{E}^2$ and $\mathbf{L}^2$, $dA = dxdy$ is the area element with the respect to the standard coordinates (x,y) for $\mathbf{E}^2$ and $\mathbf{L}^2$.)

Proof. Let $(x_1,y_1),(x_2,y_1),(x_2,y_2),(x_1,y_2)$ be vertices of R. Then

$$\begin{aligned}\int_{\partial R} F &= \int_{C_1} f(x,y)dx + \int_{C_2} g(x,y)dy + \int_{C_3} f(x,y)dx \\ &\quad + \int_{C_4} g(x,y)dy \\ &= \int_{x_1}^{x_2} f(x,y_1)dx + \int_{y_1}^{y_2} g(x_2,y)dy + \int_{x_2}^{x_1} f(x,y_2)dx \\ &\quad + \int_{y_2}^{y_1} g(x_1,y)dy \\ &= \int_{y_1}^{y_2} [g(x_2,y) - g(x_1,y)]dy - \int_{x_1}^{x_2} [f(x,y_2) - f(x,y_1)]dx \\ &= \int_{y_1}^{y_2} \left[\int_{x_1}^{x_2} \frac{\partial g}{\partial x} dx \right] dy - \int_{x_1}^{x_2} \left[\int_{y_1}^{y_2} \frac{\partial f}{\partial y} dy \right] dx\end{aligned}$$

$$= \int_{y_1}^{y_2}\int_{x_1}^{x_2} \frac{\partial g}{\partial x} dxdy - \int_{x_1}^{x_2}\int_{y_1}^{y_2} \frac{\partial f}{\partial y} dydx$$
$$= \int_{x_1}^{x_2}\int_{y_1}^{y_2} \left(\frac{\partial g}{\partial x} - \frac{\partial f}{\partial y}\right) dydx = \iint\limits_R \left(\frac{\partial g}{\partial x} - \frac{\partial f}{\partial y}\right) dA.$$

If R is a rectangle in $\mathbf{L}^2$, by the same manner as in $\mathbf{E}^2$,

$$\begin{aligned}\int_{\partial R} F &= \int_{C_1} f(x,y)dx - \int_{C_2} g(x,y)dy + \int_{C_3} f(x,y)dx \\ &\quad - \int_{C_4} g(x,y)dy \\ &= \int_{x_1}^{x_2} f(x,y_1)dx - \int_{y_1}^{y_2} g(x_2,y)dy + \int_{x_2}^{x_1} f(x,y_2)dx \\ &\quad - \int_{y_2}^{y_1} g(x_1,y)dy \\ &= -\int_{y_1}^{y_2} [g(x_2,y) - g(x_1,y)]dy - \int_{x_1}^{x_2} [f(x,y_2) - f(x,y_1)]dx \\ &= -\int_{y_1}^{y_2} \left[\int_{x_1}^{x_2} \frac{\partial g}{\partial x} dx\right] dy - \int_{x_1}^{x_2} \left[\int_{y_1}^{y_2} \frac{\partial f}{\partial y} dy\right] dx \\ &= -\int_{y_1}^{y_2}\int_{x_1}^{x_2} \frac{\partial g}{\partial x} dxdy - \int_{x_1}^{x_2}\int_{y_1}^{y_2} \frac{\partial f}{\partial y} dydx \\ &= -\int_{x_1}^{x_2}\int_{y_1}^{y_2} \left(\frac{\partial g}{\partial x} + \frac{\partial f}{\partial y}\right) dydx = -\iint\limits_R \left(\frac{\partial g}{\partial x} + \frac{\partial f}{\partial y}\right) dA.\end{aligned}$$
□

Theorem 2.3.9 (Green) *Let R be a region in $\mathbf{E}^2$ with the piecewise C^2 closed counterclockwise oriented boundary $\gamma(t) = \partial R$. Then*

$$\int_\gamma f dx + g dy = \iint\limits_R \left(\frac{\partial g}{\partial x} - \frac{\partial f}{\partial y}\right) dA$$

for differentiable functions f and g defined on R.

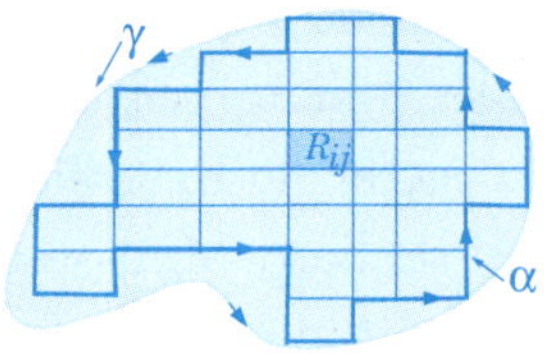

Figure 2.3-17 partition

Proof. To apply the double integral we need a partition of R by rectangular small regions R_{ij} as in Figure2.3-17 and let α be the counterclockwise oriented boundary of the region given by the union of all partitioned rectangular regions. Then we may take a very fine partition so that

$$\int_\gamma F \cdot dX \approx \int_\alpha F \cdot dX.$$

Now, we integrate F along the clockwise boundary of R_{ij} in P and add up all of integrations. Then canceling integrations for adjoint edges against each other, the remaining sum of integrals on P implies that

$$\sum_{R_{ij}\in P} \int_{\partial R_{ij}} F \cdot dX = \int_{\alpha} F \cdot dX.$$

Hence we have

$$\int_{\gamma} F \cdot dX \approx \sum_{R_{ij}\in P} \int_{\partial R_{ij}} F \cdot dX.$$

Using the Lemma 2.3.8,

$$\int_{\gamma} F \cdot dX \approx \sum_{R_{ij}\in P} \iint\limits_{R_{ij}} \left(\frac{\partial g}{\partial x} - \frac{\partial f}{\partial y}\right) dA.$$

As $|P| \to 0$, the right-hand side of the above approximation equality goes to

$$\iint\limits_{R} \left(\frac{\partial g}{\partial x} - \frac{\partial f}{\partial y}\right) dA.$$

□

Using the second integral equality in Lemma 2.3.8 and a partition by Euclidean rectangles, the following version of Green's Theorem in $\mathbf{L}^2$ may be obtained.

Corollary 2.3.10 (Green) *Let R be a convex region in $\mathbf{L}^2$ with the piecewise C^2 closed clockwise oriented boundary $\gamma(t) = \partial R$. Then*

$$\int_{\gamma} f dx - g dy = -\iint\limits_{R} \left(\frac{\partial g}{\partial x} + \frac{\partial f}{\partial y}\right) dA$$

for differentiable functions f and g defined on R.

Green's Theorem for the Hyperbolic Wedge

Motivated by regions of the type useful in General Relativity(e.g., the Rindler wedge), we consider a version of Green's Theorem over a coordinate rectangle more adapted to relativity than the Euclidean rectangle(Figure 2.3-16). In this portion of Section 2.3, let (x, t) denote the standard Minkowski coordinates in place of (x, y) used in Lemma 2.3.8. We consider subsets of $\mathbf{L}^2$ lying in the region $\mathcal{V} = \{(x,t) \in \mathbf{L}^2 |\ t > 0, x < t\}$. Hence the Lorentz distance $r = \sqrt{t^2 - x^2}$ between the point $(0, 0)$ and the point $(t, x) \in \mathcal{V}$ is differentiable(cf. Figure 1.4-39). Also, the **hyperbolic angle** y of (x, t) has been discussed in Sections 1.3 and 1.4. The change of coordinates on $\mathcal{V}$ (analogous to polar coordinates for $\mathbf{E}^2$) is then given by

$$\begin{cases} x = r\sinh(y) \\ t = r\cosh(y). \end{cases}$$

Associated to these coordinates (r, y) on $\mathcal{V}$ are the basis vector fields $\frac{\partial}{\partial r}$ and $\frac{\partial}{\partial y}$ which by the chain rule are related to the standard basis vector fields $\mathbf{i} = \frac{\partial}{\partial x}$ and $\mathbf{j} = \frac{\partial}{\partial t}$ for $\mathbf{L}^2$ by

$$\begin{cases} \frac{\partial}{\partial r} = \sinh(y)\frac{\partial}{\partial x} + \cosh(y)\frac{\partial}{\partial t} \\ \frac{\partial}{\partial y} = r\cosh(y)\frac{\partial}{\partial x} + r\sinh(y)\frac{\partial}{\partial t}. \end{cases}$$

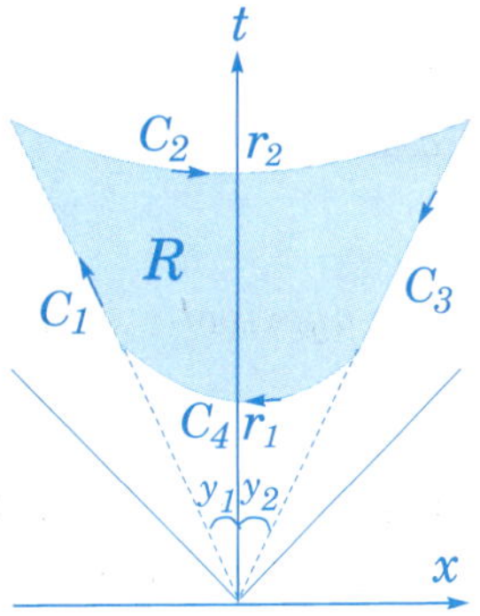

Figure 2.3-18 timelike coordinate rectangle

Because of the identity $\cosh^2(y) - \sinh(y) = 1$, the Lorentz metric g for $\mathbf{L}^2$ and the vector fields $\frac{\partial}{\partial r}$ and $\frac{\partial}{\partial y}$ satisfy $g\left(\frac{\partial}{\partial r}, \frac{\partial}{\partial r}\right) = -1$, $g\left(\frac{\partial}{\partial r}, \frac{\partial}{\partial y}\right) = 0$ and $g\left(\frac{\partial}{\partial y}, \frac{\partial}{\partial y}\right) = r^2$. These formulas are traditionally expressed as

$$g = -dr^2 + r^2 dy^2$$

(in analogy to the formula $< , >= dr^2 + r^2 d\theta^2$ for polar coordinates in $\mathbf{E}^2$). Now let R denote the coordinate rectangle in $\mathcal{V}$(a "hyperbolic wedge") with $0 < r_1 \leq r \leq r_2$ and $y_1 \leq y \leq y_2$(cf. Figure2.3-18) with corresponding edges C_1, C_2, C_3, C_4 as shown.

We need to determine the line integrand in hyperbolic coordinates (r, y) that corresponds to the expression $f(x,t)dx - g(x,t)dt$ for a region in $\mathbf{L}^2$ in the standard coordinates (x, t) as in Lemma 2.3.8. Note that in the calculation before the statement of Lemma 2.3.8, the integrand for the line integral on $\mathbf{L}^2$ is $g(F(t), \alpha'(t))dt$. Hence, we express the vector field F in (r, y) coordinate on $\mathcal{V}$ as

$$F(r, y) = f(r, y)\frac{\partial}{\partial r} + g(r, y)\frac{\partial}{\partial y}$$

and also express the curve $\alpha(t)$ in these coordinates as $\alpha(t) = (r(t), y(t))$. Then

$$\alpha'(t) = r'(t)\frac{\partial}{\partial r}|_{\alpha(t)} + y'(t)\frac{\partial}{\partial y}|_{\alpha(t)}.$$

Recalling $g\left(\frac{\partial}{\partial r}, \frac{\partial}{\partial r}\right) = -1$, $g\left(\frac{\partial}{\partial r}, \frac{\partial}{\partial y}\right) = 0$ and $g\left(\frac{\partial}{\partial y}, \frac{\partial}{\partial y}\right) = r^2$, we obtain $g(F(r(t), y(t)), \alpha'(t)) = -f(r, y)r'(t) + r^2 g(r, y)y'(t)$. Then

$$\begin{aligned} g(F(r(t), y(t)), \alpha'(t))dt &= -f(r, y)r'(t)dt + r^2 g(r, y)y'(t)dt \\ &= -f(r, y)dr + r^2 g(r, y)dy. \end{aligned}$$

Thus line integrals of the form

$$\int_C -f(r,y)dr + r^2 g(r,y)dy$$

are adapted to the hyperbolic wedge R of Figure2.3-18.

Lemma 2.3.11 *Let R be a hyperbolic rectangle in $\mathcal{V} = \{(x,t) \in \mathbf{L}^2 \mid x > 0, t > x\}$ and let $F(r,y) = f(r,y)\frac{\partial}{\partial r} + g(r,y)\frac{\partial}{\partial y}$ be a C^1 vector field on $\mathcal{V}$ expressed in the hyperbolic coordinates (r,y) for $\mathcal{V}$. Then*

$$\begin{aligned}\int_{\partial R} F &= \int_{\partial R} -f(r,y)dr + r^2 g(r,y)dy \\ &= \iint\limits_R \left[\frac{\partial f}{\partial y} + \frac{\partial}{\partial r}\left(r^2 g(r,y)\right)\right] drdy \\ &= \iint\limits_R \left[\frac{1}{r}\frac{\partial f}{\partial y} + 2g(r,y) + r\frac{\partial g}{\partial r}\right] dA\end{aligned}$$

where $dA = rdrdy$ denotes the Minkowski area element in hyperbolic coordinates on $\mathcal{V}$.

Proof. Let $(r_1,y_1),(r_2,y_1),(r_2,y_2),(r_2,y_1)$ be the successive vertices of R with corresponding timelike or spacelike boundary curves C_1, C_2, C_3, C_4. Then

$$\begin{aligned}\int_{\partial R} F &= \int_{\partial R} -f(r,y)dr + r^2 g(r,y)dy \\ &= -\int_{C_1} f(r,y)dr + \int_{C_2} r^2 g(r,y)dy - \int_{C_3} f(r,y)dr \\ &= -\int_{r_1}^{r_2} f(r,y_1)dr + \int_{y_1}^{y_2} r_2^2 g(r_2,y)dy - \int_{r_2}^{r_1} f(r,y_2)dr + \int_{y_2}^{y_1} r_1^2 g(r_1,y)dy \\ &= \int_{r_1}^{r_2} \left(f(r,y_2) - f(r,y_1)\right) dr + \int_{y_1}^{y_2} \left(r_2^2 g(r_2,y) - r_1^2 g(r_1,y)\right) dy \\ &= \int_{r_1}^{r_2} \left(\int_{y_1}^{y_2} \frac{\partial f}{\partial y} dy\right) dr + \int_{y_1}^{y_2} \left(\int_{r_1}^{r_2} \frac{\partial}{\partial r}\left(r^2 g(r,y)\right) dr\right) dy \\ &= \int_{r_1}^{r_2}\int_{y_1}^{y_2} \left(\frac{\partial f}{\partial y} + \frac{\partial}{\partial r}\left(r^2 g(r,y)\right)\right) drdy \\ &= \iint\limits_R \left[\frac{1}{r}\left(\frac{\partial f}{\partial y} + \frac{\partial}{\partial r}\left(r^2 g(r,y)\right)\right)\right] rdrdy \\ &= \iint\limits_R \left[\frac{1}{r}\frac{\partial f}{\partial y} + 2g(r,y) + r\frac{\partial g}{\partial r}\right] dA.\end{aligned}$$

□

If the vector field F is given as

$$F(x,t) = f_1(x,t)\mathbf{i} + f_2(x,t)\mathbf{j}$$

in the standard coordinates (x,t) for $\mathbf{L}^2$, then the $f(r,y)$ and $g(r,y)$ of Lemma 2.3.11 are given as

$$\begin{aligned} f(r,y) =& -\sinh(y) f_1(r\sinh(y), r\cosh(y)) \\ & + \cosh(y) f_2(r\sinh(y), r\cosh(y)) \end{aligned}$$

and

$$\begin{aligned} g(r,y) =& \frac{1}{r}\cosh(y) f_1(r\sinh(y), r\cosh(y)) \\ & - \frac{1}{r}\sinh(y) f_2(r\sinh(y), r\cosh(y)) \end{aligned}$$

using

$$\begin{cases} \dfrac{\partial}{\partial x} = -\sinh(y)\dfrac{\partial}{\partial r} + \dfrac{1}{r}\cosh(y)\dfrac{\partial}{\partial y} \\ \dfrac{\partial}{\partial t} = \cosh(y)\dfrac{\partial}{\partial r} - \dfrac{1}{r}\sinh(y)\dfrac{\partial}{\partial y}. \end{cases}$$

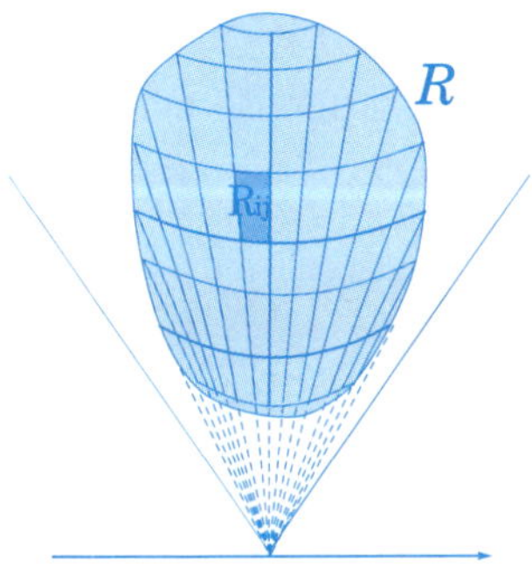

Figure 2.3-19 timelike partition of R

Hence, the Green's theorem on a region(Figure 2.3-19) in $\mathbf{L}^2$ may be proved using the result of Lemma 2.3.11 by the same manner of the proof of Theorem 2.3.9.

Covariant Derivatives

In Section 1.2 we defined the derivative of a real-valued differentiable function f in the direction of a tangent vector v_P at a point P to $\mathbf{E}^3$ (or $\mathbf{L}^3$) by

$$v_P[f] := \frac{d}{dt}(f(P+tv))\Big|_{t=0},$$

equivalently,

$$v_P[f] = \sum_{i=1}^{3} v_i \frac{\partial f}{\partial x_i}(P)$$

where $x_1 = x$, $x_2 = y$, $x_3 = z$. Moreover, the derivative $V[f]$ of a function f with respect to a vector field V on $\mathbf{E}^3$ (or $\mathbf{L}^3$) was defined by $V(P)[f]$ for each point P. In Section 2.1 we also studied the Frenet-Serret Apparatus $\{T(s), N(s), B(s)\}$ along the unit speed (timelike or spacelike) curves $\gamma(s)$, and the derivatives $\{T'(s), N'(s), B'(s)\}$ of the vector fields in Theorem Frenet-Serret along $\gamma(s)$. Now, we extend the derivative of vector fields in the direction of a tangent vector in $\mathbf{E}^3$ (or $\mathbf{L}^3$), which measures the instant rate of change of W at P in the direction v(Figure 2.3-20).

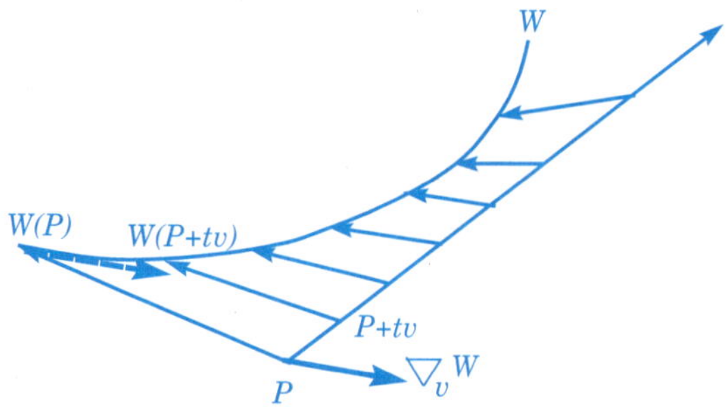

Figure2.3-20

Definition 2.3.12 The **covariant derivative** of a vector field W in the direction of a tangent vector v to $\mathbf{E}^3$ (or $\mathbf{L}^3$) at the point P is a tangent vector at P defined by

$$\nabla_v W = \frac{d}{dt} W(P+tv)\big|_{t=0}.$$

Recall that the standard frame field $\{E_1, E_2, E_3\}$ on $\mathbf{E}^3$ (or $\mathbf{L}^3$) in Section 1.2 is given by

$$E_1(P) := (\mathbf{e}_1)_P = (1,0,0)_P$$
$$E_2(P) := (\mathbf{e}_2)_P = (0,1,0)_P$$
$$E_3(P) := (\mathbf{e}_3)_P = (0,0,1)_P$$

for each $P \in \mathbf{E}^3$ (or $\mathbf{L}^3$) and for $i = 1, 2, 3$ and a differentiable function f on $\mathbf{E}^3$ (or $\mathbf{L}^3$)

$$E_i[f] = \frac{\partial f}{\partial x_i}.$$

Let $W = w_1E_1 + w_2E_2 + w_3E_3$. Then $W(P+tv) = \sum_{i=1}^{3} w_i(P+tv)E_i(P+tv)$, and hence

$$\nabla_v W = W(P+tv)'(0) = \sum_{i=1}^{3} w_i(P+tv)'(0)E_i(P+tv)\big|_{t=0} = \sum_{i=1}^{3} v[w_i]\mathbf{e}_i.$$

In the last equation $w_i(P+tv)'(0) = v[w_i]$ is given by the derivative of the real valued function w_i in the direction v at P. Since for $v = (v_1, v_2, v_3)$ $v[w_i] = \sum_{j=1}^{3} v_j \frac{\partial w_i}{\partial x_j}(P)$ where $x_1 = x,\ x_2 = y,\ x_3 = z$, we also have

$$\nabla_v W = \sum_{i=1}^{3}\sum_{j=1}^{3} v_j \frac{\partial w_i}{\partial x_j}(P)\mathbf{e}_i$$

where $x_1 = x,\ x_2 = y,\ x_3 = z$.

Example 2.3.13 Let $W = x^2yE_1 + yE_2 + yzE_3$ be a vector field and $v = (1,3,2)$ be a tangent vector at $P = (1,1,2)$. Compute $\nabla_v W$.

Solution. Since $\frac{\partial w_1}{\partial x} = 2xy,\ \frac{\partial w_2}{\partial y} = 1,\ \frac{\partial w_3}{\partial z} = y,$

and $v_1\frac{\partial w_1}{\partial x}(P) = 2,\ v_2\frac{\partial w_2}{\partial y}(P) = 3,\ v_3\frac{\partial w_3}{\partial z}(P) = 2,$

we have

$$\nabla_v W = 2\mathbf{e}_1 + 3\mathbf{e}_2 + 2\mathbf{e}_3.$$

□

Now, we prove the properties of the covariant derivatives using the Theorem direction-derivative as follows.

Theorem 2.3.14 *For the tangent vectors v, w to $\mathbb{R}^3$ at p, the vector fields V, W, a differentiable function f on $\mathbb{R}^3$ and $s, k \in \mathbb{R}$,*

(1) $\nabla_{sv+kw}V = s\nabla_v V + k\nabla_w V$

(2) $\nabla_v(sV + kW) = s\nabla_v V + k\nabla_v W$

(3) $\nabla_v(fV) = v[f]V(P) + f(P)\nabla_v V$

(4) $v[< V, W >] =< \nabla_v V, W(P) > + < V(P), \nabla_v W >$

Proof. We leave the proofs of (1), (2) and (3) for exercises, and we just prove (4) here.

Let $V = \sum_{i=1}^{3} v_iE_i$ and $W = \sum_{i=1}^{3} w_iE_i$. Then $< V, W >= \sum_{i=1}^{3} v_iw_i$. Hence by Theorem direction-derivative,

$$\begin{aligned}
v[< V, W >] &= \sum_{i=1}^{3} v[v_iw_i] = \sum_{i=1}^{3} v[v_i]w_i(P) + \sum_{i=1}^{3} v_i(P)v[w_i] \\
&= \left\langle \sum_{i=1}^{3} v[v_i]E_i(P), \sum_{i=1}^{3} w_i(P)E_i(P) \right\rangle \\
&\quad + \left\langle \sum_{i=1}^{3} v_i(P)E_i(P), \sum_{i=1}^{3} v[w_i]E_i(P) \right\rangle \\
&=< \nabla_v V, W(P) > + < V(P), \nabla_v W >
\end{aligned}$$

□

The fist two properties in Theorem covariant-derivative mean that $\nabla_v W$ is the $\mathbb{R}$-linear in the lower slot V and in the upper slot W respectively. the third property is a kind of Leibniz rule for differentiation of a product of a differentiable function and a vector field in the upper slot. The fourth property is the directional derivative of the inner product of two vector fields

We now use the pointwise principal to define an operation of a vector field $\nabla_V W$ of the vector field W respect to the lower slot V whose value is the covariant derivative $\nabla_{V(P)}W$ at each point P.

Hence, we may obtain following corollary corresponding to Theorem 1.2.4.

Corollary 2.3.15 *For vector fields U, V, W on $\mathbb{R}^3$, functions f, g on $\mathbb{R}^3$, and $s, k \in \mathbb{R}$*

(1) $\nabla_{(fU+gV)}W = f\nabla_U W + g\nabla_V W$

(2) $\nabla_U(sV + kW) = s\nabla_U V + k\nabla_U W$

(3) $\nabla_U(fV) = U[f]V + f\nabla_U V$

(4) $U[< V, W >] = \langle \nabla_U V, W\rangle + \langle V, \nabla_U W\rangle$

The proof of this corollary may use the pointwise principal given by $(\nabla_V W)(P) = \nabla_{V(P)}W$. The first property implies the the linearity in the lower slot on the functions on $\mathbb{R}^3$ called f-linear. The second property is the $\mathbb{R}$-linear in the upper slot. The third and fourth properties are also extended from the Theorem 1.2.4 (3) and (4) with respect to the vector field in the lower slot of $\nabla_U V$.

Remark that $\nabla_V W = \sum_{i=1}^{3} V[w_i]E_i$ for $W = w_1E_1 + w_2E_2 + w_3E_3$.

Example 2.3.16 Let $V = (x - y, x, yz), W = (x^2, 0, xy)$. Compute $\nabla_V W$.
Solution. Since W=$x^2E_1 + xyE_3$, we can compute

$$\begin{aligned} V[x^2] &= (x-y)E_1[x^2] + xE_2[x^2] + yzE_3[x^2] \\ &= 2(x^2 - xy) \end{aligned}$$

and

$$\begin{aligned} V[xy] &= (x-y)E_1[xy] + xE_2[xy] + yzE_3[xy] \\ &= (x-y)(y) + x(x) = xy - y^2 + x^2. \end{aligned}$$

Hence

$$\nabla_V W = \sum_{i=1}^{3} V[w_i]E_i = 2(x^2 - xy)E_1 + (x^2 + xy - y^2)E_3.$$ □

For further study, see the following references.

1. O'Neill, B., Elementary Differential Geometry, Academic Press, New York, 1966.

2. Rindler, Wolfgang, Relativity: Special, General and Cosmological, Oxford University Press, Oxford, 2001.

2.4 The Global Theory of Plane Curves

The Orientation

For two ordered bases $\{u_1, u_2, u_3\}$ and $\{v_1, v_2, v_3\}$ in $\mathbf{E}^3$, we have a matrix (a^i_j) given by $v_j = \sum_{i=1}^{3} a^i_j u_i$ for $j = 1, 2, 3$. If $\det(a^i_j) > 0$ (< 0) the bases $\{u_1, u_2, u_3\}$ and $\{v_1, v_2, v_3\}$ give the **same orientation** (**opposite orientation**) to $\mathbf{E}^3$. Moreover, a basis $\{u_1, u_2, u_3\}$ of $\mathbf{E}^3$ is said to be **positively oriented**(or **right-handed**) if $\{u_1, u_2, u_3\}$ and $\{\mathbf{i}, \mathbf{j}, \mathbf{k}\}$ give the same orientation.

Example 2.4.1 Show that $\{(1, -1, 0), (1, 0, -1), (2, 1, 3)\}$ is positively oriented.

Solution.

$$\det(a^i_j) = \begin{vmatrix} 1 & 1 & 2 \\ -1 & 0 & 1 \\ 0 & -1 & 3 \end{vmatrix} = 6 > 0.$$ □

In Section 1.3, we classified the nonspacelike vectors in $\mathbf{L}^3$ into the future-directed and past-directed nonspacelike vectors, which gives the **time orientation** of $\mathbf{L}^3$. Now we may adapt the same time orientation for $\mathbf{L}^2$ as for $\mathbf{L}^3$. Further, we take the oriented hyperbolic angle $\theta \in \mathbb{R}$ from x to y for two future directed timelike vectors $x, y \in \mathbf{L}^2$ if $F_\theta(x) = y$ where $F_\theta : \mathbf{L}^2 \to \mathbf{L}^2$ is the hyperbolic rotation transformation (or boost transformation) with the associated matrix $\begin{pmatrix} \cosh\theta & \sinh\theta \\ \sinh\theta & \cosh\theta \end{pmatrix}$ is given in Section 1.4.

The Plane Curvature in $\mathbf{E}^2$

Suppose that $\beta : (a, b) \to \mathbb{R}^3$ is a plane curve. Redefining coordinates, we may assume that $\beta : (a, b) \to \mathbb{R}^2$ by supposing that β lies in the plane $z = 0$.

Let $\beta : (a, b) \to \mathbf{E}^2$ be a unit speed C^2 curve. The tangent vector field $\mathbf{t}(s)$ to β is given by $\mathbf{t}(s) = \beta'(s)$. The normal vector field $\mathbf{n}(s)$ to β is the unique vector field so that $\{\mathbf{t}(s), \mathbf{n}(s)\}$ is a positively oriented orthonormal basis in $\mathbf{E}^2$. The **plane curvature** $k(s)$ of β is given by $k(s) =< \mathbf{t}'(s), \mathbf{n}(s) >$.

Lemma 2.4.2 *If $\beta(s)$ is a unit speed curve in $\mathbf{E}^2$, then*

(1) $\mathbf{t}'(s) = k(s)\mathbf{n}(s)$

(2) $\mathbf{n}'(s) = -k(s)\mathbf{t}(s)$

(3) If $\beta(s) = (x(s), y(s))$ for two C^2 real functions x and y, then $\mathbf{t}(s) = (x'(s), y'(s))$ and $\mathbf{n}(s) = (-y'(s), x'(s))$.

Proof. (1) We have $\mathbf{t}'(s) =< \mathbf{t}'(s), \mathbf{t}(s) > \mathbf{t}(s) + < \mathbf{t}'(s), \mathbf{n}(s) > \mathbf{n}(s)$. Since $|\mathbf{t}(s)| = 1$, $< \mathbf{t}'(s), \mathbf{t}(s) >= 0$. Thus, $\mathbf{t}'(s) = k(s)\mathbf{n}(s)$.

(2) Since $\mathbf{n}(s)$ is unit, $< \mathbf{n}(s), \mathbf{n}'(s) >= 0$. Hence $\mathbf{n}'(s) = a(s)\mathbf{t}(s)$. Since $< \mathbf{n}(s), \mathbf{t}(s) >= 0$, by differentiating both sides, we have

$$\begin{aligned} 0 &= \frac{d}{ds}(< \mathbf{n}(s), \mathbf{t}(s) >) =< \mathbf{n}'(s), \mathbf{t}(s) > + < \mathbf{n}(s), \mathbf{t}'(s) > \\ &= a(s) < \mathbf{t}(s), \mathbf{t}(s) > + k(s) < \mathbf{n}(s), \mathbf{n}(s) >= a(s) + k(s). \end{aligned}$$

That is, $a(s) = -k(s)$.

(3) Since $\beta(s)$ is a unit speed, $1 = |\beta'(s)| = |(x'(s), y'(s))| = \sqrt{x'(s)^2 + y'(s)^2}$. Hence, $x'(s)^2 + y'(s)^2 = 1$. It suffices to prove that $\{\mathbf{t}(s), (-y'(s), x'(s))\}$ is a positively oriented orthonormal basis in $\mathbf{E}^2$. Clearly, $|(-y'(s), x'(s))| = 1$. Moreover, $< (x'(s), y'(s)), (-y'(s), x'(s)) > = -x'(s)y'(s) + x'(s)y'(s) = 0$. Further,
$\begin{vmatrix} x'(s) & -y'(s) \\ y'(s) & x'(s) \end{vmatrix} = x'(s)^2 + y'(s)^2 = 1 > 0.$ □

Remark 2.4.3 For the unit speed curve $\beta(s)$ in $\mathbf{E}^2$

$$\mathbf{t}(s) = T(s), \;\; \mathbf{n}(s) = \pm N(s)$$

for all s(figure2.4-21).

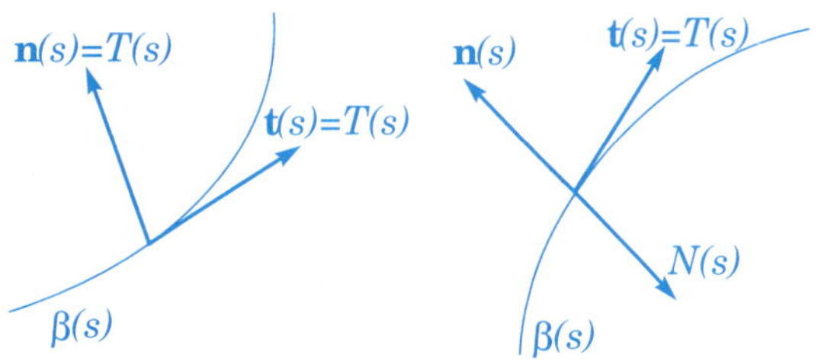

Figure2.4-21

Example 2.4.4 For two curves given by $\alpha(s) = \left(r\cos\frac{s}{r}, r\sin\frac{s}{r}\right)$ and $\beta(s) = \left(r\cos\frac{s}{r}, -r\sin\frac{s}{r}\right)$, find the plane curvatures.

Solution. For $\alpha(s)$, we can compute

$$\mathbf{t}(s) = \left(-\sin\frac{s}{r}, \cos\frac{s}{r}\right),$$

$$\mathbf{n}(s) = \left(-\cos\frac{s}{r}, -\sin\frac{s}{r}\right), \text{ and}$$

$$\mathbf{t}'(s) = \left(-\frac{1}{r}\cos\frac{s}{r}, -\frac{1}{r}\sin\frac{s}{r}\right).$$

Thus, we obtain $k(s) =< \mathbf{t}'(s), \mathbf{n}(s) >= \frac{1}{r}$.

For $\beta(s)$, we may compute similarly

$$\mathbf{t}(s) = \left(-\sin\frac{s}{r}, -\cos\frac{s}{r}\right),$$

$$\mathbf{n}(s) = \left(\cos\frac{s}{r}, -\sin\frac{s}{r}\right), \text{ and}$$

$$\mathbf{t}'(s) = \left(-\frac{1}{r}\cos\frac{s}{r}, \frac{1}{r}\sin\frac{s}{r}\right).$$

Thus, we may obtain $k(s) =< \mathbf{t}'(s), \mathbf{n}(s) >= -\frac{1}{r}$. □

Note that the sign of the plane curvature $k(s)$ is positive(negative) if the plane curve is bending toward(away from) the direction of $\mathbf{n}(s)$.

The Plane Curvature in $\mathbf{L}^2$

The theory of curvature for a unit timelike or a unit spacelike plane curve in $\mathbf{L}^2$ is slightly more complex than for curves in $\mathbf{L}^2$, since no isometry takes a unit timelike to a unit spacelike vectors.

Let $\beta : (a, b) \to \mathbf{L}^2$ be a unit speed timelike or spacelike C^2 curve. The tangent vector field $\mathbf{t}(s)$ to β is given by $\mathbf{t}(s) = \beta'(s)$. If $\beta(s) = (x(s), y(s))$ for two C^2 real functions x and y, then $\mathbf{t}(s) = (x'(s), y'(s))$. The normal vector field $\mathbf{n}(s)$ to β is defined by

$$\mathbf{n}(s) = (y'(s), x'(s)).$$

Thus, the plane curvature $k(s)$ of β in $\mathbf{L}^2$ is defined as follows.

(1) If $\beta(s)$ is timelike, $k(s) =< \mathbf{t}'(s), \mathbf{n}(s) >$.

(2) If $\beta(s)$ is spacelike, $k(s) = - < \mathbf{t}'(s), \mathbf{n}(s) >$.

Corollary 2.4.5 *If $\beta(s)$ is a unit speed timelike (or spacelike) curve in $\mathbf{L}^2$, then*

(1) $\mathbf{t}'(s) = k(s)\mathbf{n}(s)$

(2) $\mathbf{n}'(s) = k(s)\mathbf{t}(s)$

Proof. (1) Let $\beta(s)$ be a unit speed timelike (or spacelike) curve in $\mathbf{L}^2$. Then we have $\mathbf{t}'(s) = - < \mathbf{t}'(s), \mathbf{t}(s) > \mathbf{t}(s) + < \mathbf{t}'(s), \mathbf{n}(s) > \mathbf{n}(s)$ (or $\mathbf{t}'(s) =< \mathbf{t}'(s), \mathbf{t}(s) > \mathbf{t}(s) - < \mathbf{t}'(s), \mathbf{n}(s) > \mathbf{n}(s)$).
Since $|\mathbf{t}(s)| = 1$, $< \mathbf{t}'(s), \mathbf{t}(s) >= 0$. Thus, $\mathbf{t}'(s) = k(s)\mathbf{n}(s)$.

(2) Since $\mathbf{n}(s)$ is unit, $< \mathbf{n}(s), \mathbf{n}'(s) >= 0$. Hence $\mathbf{n}'(s) = a(s)\mathbf{t}(s)$. Since $< \mathbf{n}(s), \mathbf{t}(s) >= 0$, by differentiating both sides, we have

$$\begin{aligned} 0 &= \frac{d}{ds}(< \mathbf{n}(s), \mathbf{t}(s) >) =< \mathbf{n}'(s), \mathbf{t}(s) > + < \mathbf{n}(s), \mathbf{t}'(s) > \\ &= a(s) < \mathbf{t}(s), \mathbf{t}(s) > + k(s) < \mathbf{n}(s), \mathbf{n}(s) >= -a(s) + k(s), \end{aligned}$$

that is, $a(s) = k(s)$. (or for $\beta(s)$ spacelike,

$$\begin{aligned} 0 &= \frac{d}{ds}(< \mathbf{n}(s), \mathbf{t}(s) >) =< \mathbf{n}'(s), \mathbf{t}(s) > + < \mathbf{n}(s), \mathbf{t}'(s) > \\ &= a(s) < \mathbf{t}(s), \mathbf{t}(s) > + k(s) < \mathbf{n}(s), \mathbf{n}(s) >= a(s) - k(s), \end{aligned}$$

that is, $a(s) = k(s)$). □

The Fundamental Theorem of Plane Curves

In the Section 2.2 we observed that a regular Euclidean or non-lightlike curve has its curvature and torsion defined along the curve. Thus, the question arises: given any two continuous functions $\kappa, \tau : (a, b) \to \mathbb{R}^3$, can we find a unit speed curve so that its curvature is $\kappa(s)$ and its torsion is $\tau(s)$ for all $s \in (a, b)$? The answer is known to be affirmative. In this section, we prove an analogous result for the simpler case of plane curves.

Theorem 2.4.6 *Given any continuous function $k : (a, b) \to \mathbb{R}$, there exists a curve $\beta : (a, b) \to \mathbf{E}^2$ parametrized by arc length such that $k(s)$ is the plane curvature of $\beta(s)$ for all $s \in (a, b)$. If $\bar{\beta} : (a, b) \to \mathbf{E}^2$ has the same plane curvature $k(s)$, then it differs from β by a rotation and a translation.*

Proof. In view of equation (1) of Lemma 2.4.2 for $\mathbf{t}(s)$, consider the differential equation for the function $F(s) = (f(s), g(s))$ given by

$$(f'(s), g'(s)) = k(s)(-g(s), f(s)),$$

i.e., $F'(s) = \begin{pmatrix} 0 & -k(s) \\ k(s) & 0 \end{pmatrix} F(s)$ with the initial condition $F(s_0) = u$ and $|u| = 1$. Then this solution $F(s)$ is unit, since

$$\begin{aligned}\frac{d}{ds}|F(s)|^2 &= \frac{d}{ds}(f(s)^2 + g(s)^2) \\ &= 2f(s)f'(s) + 2g(s)g'(s) \\ &= 2 < (f(s), g(s)), (f'(s), g'(s)) > \\ &= 2 < (f(s), g(s)), (-k(s)g(s), k(s)f(s)) > \\ &= 0\end{aligned}$$

and hence $|F(s)| = 1$ from the initial condition. Thus we have a unique solution $F(s) = (f(s), g(s))$ by the uniqueness and existence of solutions to systems of linear differential equations. Now, from the equation $T'(s) = k(s)N(s)$ we have

$$T'(s) = \begin{pmatrix} 0 & -k(s) \\ k(s) & 0 \end{pmatrix} T(s). \qquad (*)$$

Let $T(s)$ be a solution to the above equations (*) and let

$$\beta(s) = \int_{s_0}^{s} T(\sigma)d\sigma.$$

Equivalently, let $x(s) = \int_{u=s_0} f(u)du$, $y(s) = \int_{u=s_0} g(u)du$, and $\beta(s) := (x(s), y(s))$. Then $\beta(s)$ is a plane curve in $\mathbf{E}^2$ with $\beta(s_0) = (0, 0)$ and its plane curvature is the given function $k(s)$.

Suppose that $\bar{\beta}(s)$ is a second unit speed plane curve in $\mathbf{E}^2$ with the given curvature $k(s)$ and $\bar{\beta}(s_0) = w$. Since $\beta'(s_0)$ and $\bar{\beta}'(s_0)$ are both unit vectors, regarding them as being based at $(0, 0)$, there exists an angle θ so that

$$\bar{\beta}'(s_0) = \begin{pmatrix} \cos\theta & -\sin\theta \\ \sin\theta & \cos\theta \end{pmatrix} \beta'(s_0).$$

Equivalently, if $\bar{T}(s) = \bar{\beta}'(s)$, then $\bar{T}(s_0) = \begin{pmatrix} \cos\theta & -\sin\theta \\ \sin\theta & \cos\theta \end{pmatrix} T(s_0)$. Hence $\bar{T}(s) = \begin{pmatrix} \cos\theta & -\sin\theta \\ \sin\theta & \cos\theta \end{pmatrix} T(s)$ is a solution of (*) with initial condition $\bar{T}(s_0) = \bar{\beta}'(s_0)$. Thus by the uniqueness of solution to (*), we obtain on integrating $\bar{\beta}'(s) = \bar{T}(s)$ that

$$\bar{\beta}(s) = \begin{pmatrix} \cos\theta & -\sin\theta \\ \sin\theta & \cos\theta \end{pmatrix} \beta(s) + w.$$

□

Crucial to the second part of Theorem 2.4.6 is the fact that given any two vectors v, w of the same length based at $(0,0)$ in $\mathbf{E}^2$, there exists a rotation matrix

$$R = \begin{pmatrix} \cos\theta & -\sin\theta \\ \sin\theta & \cos\theta \end{pmatrix}$$

such that $w = Rv$. The theory for $\mathbf{L}^2$ is more complicated, hence Corollary 2.4.8 is more complex than Theorem 2.4.6. In Section 2.3, a future timelike region and hyperbolic wedges were used in Lemma 2.3.11 to formulate a version of Green's Theorem for $\mathbf{L}^2$. Now we define two spacelike wedges in $\mathbf{L}^2$ corresponding to spacelike vectors based at origin: the left hand wedge

$$\mathcal{L} = \{(x_1, x_2) \in \mathbf{L}^2 |\ x_1 < 0, x_1^2 - x_2^2 > 0\}$$

and the right hand wedge

$$\mathcal{R} = \{(x_1, x_2) \in \mathbf{L}^2 |\ x_1 > 0, x_1^2 - x_2^2 > 0\}.$$

In Section 1.4 we considered *boost transformations* of the form

$$F_\theta(x) = \begin{pmatrix} \cos\theta & -\sin\theta \\ \sin\theta & \cos\theta \end{pmatrix} \begin{pmatrix} x_1 \\ x_2 \end{pmatrix}.$$

For any two future(resp., past) unit timelike vectors $v, w \in \mathcal{L}$ or for $v, w \in \mathcal{R}$, there exists a unique θ with $w = F_\theta(v)$. But since $\cosh(\theta) > 0$, it is easily checked for $v = (0,1)$ and $w = (0,-1)$ or for $v = (1,0)$ and $w = (-1,0)$, that there is no θ such That $w = F_\theta(v)$. Since the composition of boost transformations is a boost transformation(cf., Proposition 5.1.5-(4)), these examples imply that there is no boost transformation carrying any future unit timelike vector to any past unit timelike vector, or carrying any unit spacelike vector in $\mathcal{L}$ to any unit spacelike vector in $\mathcal{R}$. Since $< F_\theta(x), F_\theta(y) >=< x, y >$, no boost transformation carries any unit timelike vector to any unit spacelike vector, and conversely.

These considerations suggest defining a linear transformation $A : \mathbf{L}^2 \to \mathbf{L}^2$ given by $A((x_1, x_2)) = (x_2, x_1)$ which takes future timelike vectors(resp., spacelike vectors in $\mathcal{R}$) based at $(0,0)$ to spacelike vectors in $\mathcal{R}$(resp., future timelike vectors). this function also has the property that $< Av, Aw >= - < v, w >$, hence might be termed an **anti-isometry**. Further, A preserves the curvature function in the following sense.

Proposition 2.4.7 *Let $A : \mathbf{L}^2 \to \mathbf{L}^2$ be the anti-isometry $A((x_1, x_2)) = (x_2, x_1)$. Let $\beta_1 : [a, b) \to \mathbf{L}^2$ be a unit speed timelike(resp., spacelike) curve with curvature function $k_1(s)$. Then $\beta_2 : [a, b) \to \mathbf{L}^2$ with $\beta_2(s) = A(\beta_1(s))$ has curvature function $k_2(s) = k_1(s)$ for all $s \in [a, b)$.*

Proof. We give the proof for β_1 timelike, since the spacelike case is similar. If $\beta_1(s) = (x(s), y(s))$, then $\mathbf{t}_1(s) = \beta_1'(s) = (x'(s), y'(s))$, $\mathbf{n}_1(s) = (y'(s), x'(s))$ and thus

$$\begin{aligned} k_1(s) &=< \mathbf{t}_1'(s), \mathbf{n}_1(s) >=< (x''(s), y''(s)), (y'(s), x'(s)) > \\ &= x''(s)y'(s) - x'(s)y''(s). \end{aligned}$$

But then $\beta_2(s) = (y(s), x(s))$ and hence $\mathbf{t}_1'(s) = (y''(s), x''(s))$ and $\mathbf{n}_2(s) = (x'(s), y'(s))$. Thus

$$\begin{aligned} k_2(s) &= - < \mathbf{t}_2'(s), \mathbf{n}_2(s) >= -(y''(s)x'(s) - y'(s)x''(s)) \\ &= k_1(s). \end{aligned}$$

□

Now we formulate the Minkowski space analogue of Theorem 2.4.6 for the cases corresponding to future unit timelike vectors and the right spacelike wedge $\mathcal{R}$. To consider $\mathcal{L}$ or past timelike vectors, it would also be necessary to employ the reflection $T : \mathbf{L}^2 \to \mathbf{L}^2$ given by $T((x, y)) = (x, -y)$.

Corollary 2.4.8 *Let $k : (a, b) \to \mathbb{R}$ be any continuous function. Then*
(1) there exists a future unit timelike curve $\beta : (a, b) \to \mathbf{L}^2$ such that the given function $k(s)$ is the curvature of $\beta(s)$ for all $s \in (a, b)$. Further, if $\bar{\beta} : (a, b) \to \mathbf{L}^2$ is a second future unit timelike curve with curvature function $k(s)$, then $\bar{\beta}$ differs from β by a boost transformation and a translation. If $\bar{\beta} : (a, b) \to \mathbf{L}^2$ is a unit spacelike curve with tangents $\bar{\beta}'(s)$ in the right spacelike wedge $\mathcal{R}$ and with curvature function $k(s)$, then $\bar{\beta}$ differs from β by composing with the anti-isometry $A : \mathbf{L}^2 \to \mathbf{L}^2$ given in Proposition 2.4.7 after performing a boost transformation and a translation.

(2) there exists a unit spacelike curve $\beta : (a, b) \to \mathcal{R}$ with tangent vectors $\beta'(s)$ in $\mathcal{R}$ such that the given function $k(s)$ is the curvature of $\beta(s)$ for all $s \in (a, b)$. Further, if $\bar{\beta} : (a, b) \to \mathcal{R}$ is a second unit spacelike curve with curvature function $k(s)$, then $\bar{\beta}$ differs from β by a boost transformation and a translation. If $\bar{\beta} : (a, b) \to \mathbf{L}^2$ is a unit future timelike curve with tangents $\bar{\beta}'(s)$ in the right spacelike wedge $\mathcal{R}$ and with curvature function $k(s)$, then $\bar{\beta}$ differs from β by composing with the anti-isometry $A : \mathbf{L}^2 \to \mathbf{L}^2$ after performing a boost transformation and a translation.

Proof. Since the proof of (1) and (2) are similar, we only present the proof for (1). Consider a differential equation for the function $F(s) = (f(s), g(s))$ given by

$$(f'(s), g'(s)) = k(s)(g(s), f(s)),$$

for all $s \in (a, b)$, i.e., $F'(s) = \begin{pmatrix} 0 & k(s) \\ k(s) & 0 \end{pmatrix} F(s)$ with the initial condition $F(s_0) = u$ and $|u| = 1$. Then this solution $F(s)$ is unit, since

$$\begin{aligned} \frac{d}{ds}|F(s)|^2 &= \frac{d}{ds}(f(s)^2 - g(s)^2) \\ &= 2f(s)f'(s) - 2g(s)g'(s) \\ &= 2 < (f(s), g(s)), (f'(s), g'(s)) > \\ &= 2 < (f(s), g(s)), (k(s)g(s), k(s)f(s)) > \\ &= 0 \end{aligned}$$

and hence $|F(s)| = 1$ from the initial condition. Thus we have a unique solution $F(s) = (f(s), g(s))$ by the uniqueness and existence of solutions to differential equations. Now, from the equation $T'(s) = k(s)N(s)$ we have

$$T'(s) = \begin{pmatrix} 0 & k(s) \\ k(s) & 0 \end{pmatrix} T(s). \tag{*}$$

Let $T(s)$ be a solution to the above equations (*) and let

$$\beta(s) = \int_{s_0}^{s} T(\sigma) d\sigma,$$

or equivalently, put $x(s) = \displaystyle\int_{\sigma=s_0}^{s} f(\sigma)d\sigma$, $y(s) = \displaystyle\int_{\sigma=s_0}^{s} g(\sigma)d\sigma$, and $\beta(s) = (x(s), y(s))$. Then $\beta(s_0) = (0, 0)$ and $\beta'(s) = T(s)$ so tat $|\beta'(s)| = 1$ and $T'(s) = (f'(s), g'(s)) = k(s)N(s)$, whence $\beta(s)$ has curvature function $k(s)$. If $\bar{\beta} : (a, b) \to \mathbf{L}^2$ is a second future timelike curve with curvature function $k(s)$ and the hyperbolic angle θ is calculated so that

$$F_\theta\left(\bar{\beta}'(s_0)\right) = u,$$

then

$$\bar{T}(s) = \begin{pmatrix} \cosh\theta & \sinh\theta \\ \sinh\theta & \cosh\theta \end{pmatrix} T(s)$$

is another solution of (*) with $T(s_0) = \bar{\beta}'(s_0)$. Hence, $\bar{T}(s) = \bar{\beta}'(s)$ for all $s \in (a, b)$ by uniqueness of solution to (*). In analogy to the Euclidean case,

$$\bar{\beta}(s) = F_\theta(\beta(s)) + \bar{\beta}(s_0).$$

Finally, suppose $\bar{\beta} : (a,b) \to \mathcal{R}$ is a second unit spacelike curve with tangents in $\mathcal{R}$ and curvature function $k(s)$. Then $A \circ \bar{\beta}$ is a future unit timelike curve with curvature function $k(s)$ by Proposition 2.4.7. Hence,

$$A \circ \bar{\beta}(s) = F_\theta(\beta(s)) + A \circ \bar{\beta}(s_0)$$

so that as $A^2 = I$, the given spacelike curve

$$\begin{aligned}\bar{\beta}(s) &= A(F_\theta(\beta(s)) + A \circ \bar{\beta}(s_0)) \\ &= A(F_\theta(\beta(s))) + \bar{\beta}(s_0).\end{aligned}$$

□

Closed Curves in $\mathbf{E}^2$

A regular curve $\beta(t)$ is said to be **closed** if there is a constant $a > 0$ such that $\beta(t+a) = \beta(t)$ for all t. The **period** of β is the least number a such that $\beta(t+a) = \beta(t)$. Note that $\mathbf{t}(0) = \mathbf{t}(a)$.

Proposition 2.4.9 *Let $\alpha(t)$ be a closed curve with period l and $\beta(s)$ be a reparametrization by arc length of $\alpha(t)$ starting at $t = 0$. Then $\beta(s)$ is a closed curve with period $L = \int_0^l |\alpha'(t)|dt$.*

Proof.

$$\begin{aligned}s(t+l) &= \int_0^{t+l} |\alpha'(t)|dt \\ &= \int_0^{l} |\alpha'(t)|dt + \int_l^{t+l} |\alpha'(t)|dt \\ &= L + \int_0^{t} |\alpha'(t)|dt \\ &= L + s(t).\end{aligned}$$

Thus, we have

$$\begin{aligned}\beta(s+L) &= \beta(s(t)+L) = \beta(s(t+l)) \\ &= \alpha(t+l) = \alpha(t) = \beta(s(t)) = \beta(s).\end{aligned}$$

Since l is the least positive number such that $\alpha(t+l) = \alpha(t)$, L is the least positive number such that $\beta(s+L) = \beta(s)$ □

Let $\alpha(t)$ be a regular curve. Then $\alpha(t)$ is said to be **simple** if α is a 1-1 function or $\alpha(t)$ is a closed curve of period l such that $\alpha(t_1) = \alpha(t_2)$ if and only if $t_1 - t_2 = nl$ for some integer n.

Example 2.4.10 The unit circle $\alpha(t) = (\cos t, \sin t)$ is a simple closed curve of period 2π.

The Rotation Index

For use in the proof of the Gauss-Bonnet Theorem in Chapter 5, a concept important in both the topology and geometry of plane curves, the **rotation index**, is now introduced. It has been shown in topology that closed curves c_1 and c_2 in $\mathbb{R}^2 - \{(0,0)\}$ which may be continuously deformed one onto the other("homotopic as closed curves") have the same winding number. A celebrated result of the 1930's, the Whitney-Graustein Theorem, asserts the converse that two regular closed curves with the same rotation index are regularly homotopic. Also in the 1930's, H. Hopf proved the so-called "Theorem on turning tangents" giving a geometric proof using the curvature and the polar angle of a plane curve that the rotation index of a simple, closed plane curve is ± 1.

A closed plane curve $\beta(s)$ of period L satisfies $\mathbf{t}(0) = \mathbf{t}(L)$, and the unit tangent vector field $\mathbf{t}(s)$ rotates along $\beta(s)$ in the direction of increasing s. To measure this rotation we assume that the image of $\beta(s)$ lies in the upper half plane with $\beta(0) = (0,0)$ so that $\mathbf{t}(0)$ is on the x axis. Define the angle $\theta(s)$ to be the angle between the positive x axis and $\mathbf{t}(s)$ measured counterclockwise. Thus $\cos\theta(s) =< \mathbf{t}'(s), \mathbf{i} >$. Then $\theta : [0, L] \to \mathbb{R}$ is a continuous function with $\theta(0) = 0$. Since $\mathbf{t}(s)$ is differentiable, $\theta(s)$ is also differentiable. Moreover, $\theta(L)$ need not be 0 but is an integral multiple of 2π.

Definition 2.4.11 The **rotation index** i_β of a closed unit speed plane curve $\beta(s)$ is the integer $i_\beta = \dfrac{\theta(L) - \theta(0)}{2\pi} = \dfrac{\theta(L)}{2\pi}$.

Since $\mathbf{t}(s) =< \mathbf{t}(s), \mathbf{i} >\mathbf{i}+ < \mathbf{t}(s), \mathbf{j} >\mathbf{j} = (\cos\theta(s), \sin\theta(s))$, $\mathbf{t}'(s) = k(s)\mathbf{n}(s)$ and since $(-\sin\theta(s), \cos\theta(s))\theta'(s) = \theta'(s)\mathbf{n}(s)$, we obtain $\theta'(s) = k(s)$ and thus $|\theta'(s)| = \kappa(s)$ if $\beta(s)$ is not necessarily oriented counterclockwise.

Theorem 2.4.12 (**Rotation Index Theorem**) *The rotation index of a simple closed plane curve* $\beta : [0, L] \to \mathbf{E}^2$ *is* ± 1.

Proof. Assume that $\beta : [0, L] \to \mathbf{E}^2$ is parametrized by arc length of parameter s. Let L be a period of the simple closed plane curve $\beta(s)$ and let

$$A(u, v) = \frac{\beta(v) - \beta(u)}{|\beta(v) - \beta(u)|}, \quad 0 \le u < v \le L,$$

$$A(u, u) = \mathbf{t}(u),$$

$$A(0, L) = -\mathbf{t}(0) = -\mathbf{t}(L).$$

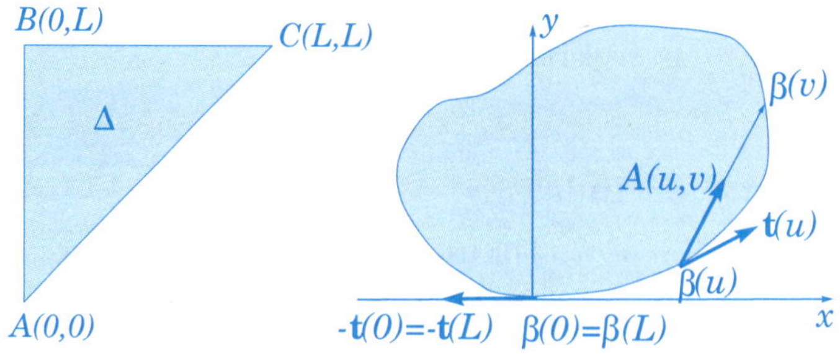

Figure2.4-22

Then A is a C^2 function in the region $\triangle$ of Figure 2.4-22. If $\gamma(u,v)$ is the angle between x axis and $A(u,v)$, then $\gamma(u,v)$ is also a C^2 function on $\triangle$ with $\gamma(u,u)=\theta(u)$.

Since $i_\beta = \dfrac{\theta(L)-\theta(0)}{2\pi}$ and since $\gamma(u,u)=\theta(u)$ and $\theta(0)=0$,

$$2\pi i_\beta = \int_0^L \left(\frac{d\theta}{ds}\right) ds = \int_\beta d\theta = \int_{\overline{AC}} d\gamma.$$

Now, since $d\gamma = \dfrac{\partial\gamma}{\partial u}du + \dfrac{\partial\gamma}{\partial v}dv$ and since $\dfrac{\partial^2\gamma}{\partial u\partial v} = \dfrac{\partial^2\gamma}{\partial v\partial u}$,

by Green's theorem in Section 2.3, $\displaystyle\int_{\overline{AC}+\overline{CB}+\overline{BA}} d\gamma = 0.$

Hence,

$$\int_{\overline{AC}} d\gamma = \int_{\overline{AB}} d\gamma + \int_{\overline{BC}} d\gamma.$$

Since $\displaystyle\int_{\overline{AB}} d\gamma = \pm\pi$ and $\displaystyle\int_{\overline{BC}} d\gamma = \pm\pi$, we obtain

$$2i_\beta = \pm 2\pi. \text{ i.e., } i_\beta = \pm 1.$$

□

Let $\alpha : [a,b] \to E^2$ be an n-polygonal curve with vertices $\alpha(s_0)$, $\alpha(s_1)$,..., $\alpha(s_n)=\alpha(s_0)$ and let $\mathbf{t}^-(s_i) = \lim_{s\to s_i^-} \mathbf{t}(s)$ and $\mathbf{t}^+(s_i) = \lim_{s\to s_i^+} \mathbf{t}(s)$.

Also let θ_i be the angle from $\mathbf{t}^-(s_i)$ to $\mathbf{t}^+(s_i)$) and φ_i be the angle through which $\mathbf{t}(s)$ rotates on the segment $\alpha([s_i, s_{i+1}])$.

Since $\varphi_i = \displaystyle\int_{s_i}^{s_{i+1}} k(s)ds$, we may also define the rotation index of the n-polygonal curve α by $i_\alpha = \dfrac{1}{2\pi}\left(\displaystyle\int_{s_i}^{s_{i+1}} k(s)ds + \sum_{i=0}^{n-1}\theta_i\right)$. Using this definition, H. Hopf showed the same result as Theorem 2.4.12 in 1935.

For further study, see the following references.

1. Millman, R. S. and G. D. Parker,Elements of Differential Geometry, Prentice-Hall, Englewood Cliffs, NJ, 1977.

2. McCleary, J., Geometry from a Differentiable Viewpoint, Cambridge University Press, 1994.

3. O'Neill, B., Elementary Differential Geometry, Academic Press, New York, 1966.

Chapter 3
Calculus of Surfaces

3.1 Definition of Surfaces

When we studied the curve theory, we supposed the curve was regular. Now for the surface theory, we also formulate the concept of regular surface for similar reasons. The surface theory is more complicated than the curve theory since there are no natural parametrizations on a regular surface which correspond to the reparametrization of a regular curve by arc length.

The theory of surfaces originated in the calculus of variations problem of finding the shortest arc between between two points on a surface, so-called "geodesics". The Bernoulli brothers, Euler, Jacob Hermann (1678-1733), and Clairaut all found specific geodesics on various particular surfaces, with Euler in 1728 deriving the differential equations satisfied by such curves. However, in 1760 Euler introduced a more general theory of surfaces, especially introducing the concept that the coordinates of any point on a surface should be represented by 2 parameters, hence the idea of a "surface patch" $X(u, v)$ as discussed in this section. Monge and his students studied more specific types of surfaces, such as developable surfaces. Hence, in this section we also encounter the commonly used terminology of a "Monge patch". Finally, Gauss in his celebrated 1827 paper "Disquisitiones Generales circa Superficies Curvas" (General Investigations of Curved Surfaces) not only made a definitive treatment of surfaces lying in 3-dimensional Euclidean space, but also introduced the concept of a surface as an abstract 2-dimensional space in its own right. This work of Gauss was motivated by his extensive studies of geodesy and map-making, including actual surveying of mountain ranges in Germany.

The Coordinate Patch

Let D be an open subset of $\mathbb{R}^2$ and let $X : D \to \mathbb{R}^3$ be a differentiable function given by $X(u, v) = (f_1(u, v), f_2(u, v), f_3(u, v))$. Then the Jacobian matrix of X is given by

$$J(X) = \begin{pmatrix} \dfrac{\partial f_1}{\partial u} & \dfrac{\partial f_1}{\partial v} \\ \dfrac{\partial f_2}{\partial u} & \dfrac{\partial f_2}{\partial v} \\ \dfrac{\partial f_3}{\partial u} & \dfrac{\partial f_3}{\partial v} \end{pmatrix}$$

If $rank(J(X))$ = dimension of $D = 2$ on D, then $X : D \to \mathbf{E}^3$ is said to be **regular**. Moreover, if X is a 1-1 regular map, then X is said to be a **coordinate patch**.

Example 3.1.1 Investigate the regularity for $X(u,v) = (u^2 - v^2, 2uv)$ for $(u,v) \in \mathbb{R}^2$.

Solution. Since $J(X(u,v)) = \begin{pmatrix} \dfrac{\partial f_1}{\partial u} & \dfrac{\partial f_1}{\partial v} \\ \dfrac{\partial f_2}{\partial u} & \dfrac{\partial f_2}{\partial v} \end{pmatrix}\Bigg|_{(u,v)} = \begin{pmatrix} 2u & -2v \\ 2v & 2u \end{pmatrix} = 2\begin{pmatrix} u & -v \\ v & u \end{pmatrix}$, the $rank(J(X(u,v))) = 2$ for all $(u,v) \neq (0,0)$ in $\mathbb{R}^2$ but $rank(J(X(u,v))) = 0$ for $(u,v) = (0,0)$. □

The Tangent Plane and Unit Normal Vector

For a coordinate patch $X(u,v) = (f_1(u,v), f_2(u,v), f_3(u,v))$ for a surface in $\mathbf{E}^3$, we have

$$X_u(u,v) = \frac{\partial X}{\partial u} = \left(\frac{\partial f_1}{\partial u}, \frac{\partial f_2}{\partial u}, \frac{\partial f_3}{\partial u}\right)$$

and

$$X_v(u,v) = \frac{\partial X}{\partial v} = \left(\frac{\partial f_1}{\partial v}, \frac{\partial f_2}{\partial v}, \frac{\partial f_3}{\partial v}\right).$$

Since

$$\begin{aligned} X_u \times X_v &= \begin{vmatrix} \mathbf{i} & \mathbf{j} & \mathbf{k} \\ \dfrac{\partial f_1}{\partial u} & \dfrac{\partial f_2}{\partial u} & \dfrac{\partial f_3}{\partial u} \\ \dfrac{\partial f_1}{\partial v} & \dfrac{\partial f_2}{\partial v} & \dfrac{\partial f_3}{\partial v} \end{vmatrix} \\ &= \begin{vmatrix} \dfrac{\partial f_2}{\partial u} & \dfrac{\partial f_3}{\partial u} \\ \dfrac{\partial f_2}{\partial v} & \dfrac{\partial f_3}{\partial v} \end{vmatrix} \mathbf{i} - \begin{vmatrix} \dfrac{\partial f_1}{\partial u} & \dfrac{\partial f_3}{\partial u} \\ \dfrac{\partial f_1}{\partial v} & \dfrac{\partial f_3}{\partial v} \end{vmatrix} \mathbf{j} + \begin{vmatrix} \dfrac{\partial f_1}{\partial u} & \dfrac{\partial f_2}{\partial u} \\ \dfrac{\partial f_1}{\partial v} & \dfrac{\partial f_2}{\partial v} \end{vmatrix} \mathbf{k}, \end{aligned}$$

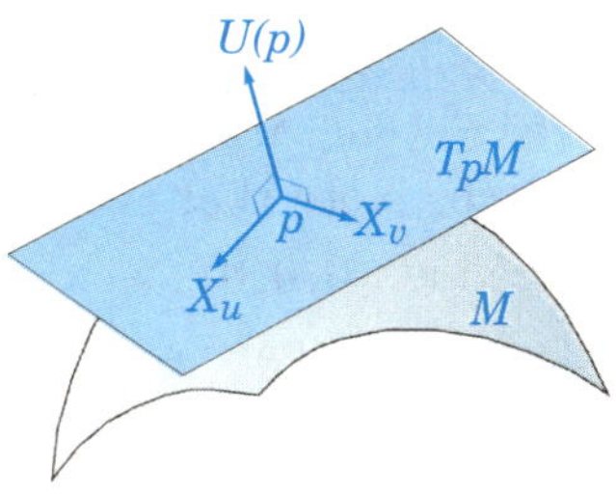

Figure3.1-1

for any $x = (u,v) \in D$ we may check that $rank(J(X(u,v))) = 2$ if and only if $X_u \times X_v \neq 0$ if and only if X_u and X_v are linearly independent. Hence the **tangent plane** at the point $p = X(a,b)$ to a coordinate patch $X : D \to \mathbf{E}^3$ is defined by the vector subspace of $\mathbb{R}^3$ of vectors normal to $X_u(a,b) \times X_v(a,b)$ and the **unit normal vector** to the coordinate patch at P is defined by $U(p) = \dfrac{X_u \times X_v}{|X_u \times X_v|}\Bigg|_{(a,b)}$ (Figure3.1-1). Thus $\{X_u(a,b), X_v(a,b)\}$ is a basis for the tangent plane at p to the coordinate patch.

In $\mathbf{L}^3$

$$X_u \times_g X_v = \begin{vmatrix} \mathbf{i} & \mathbf{j} & -k \\ \frac{\partial f_1}{\partial u} & \frac{\partial f_2}{\partial u} & \frac{\partial f_3}{\partial u} \\ \frac{\partial f_1}{\partial v} & \frac{\partial f_2}{\partial v} & \frac{\partial f_3}{\partial v} \end{vmatrix}$$
$$= \begin{vmatrix} \frac{\partial f_2}{\partial u} & \frac{\partial f_3}{\partial u} \\ \frac{\partial f_2}{\partial v} & \frac{\partial f_3}{\partial v} \end{vmatrix} \mathbf{i} - \begin{vmatrix} \frac{\partial f_1}{\partial u} & \frac{\partial f_3}{\partial u} \\ \frac{\partial f_1}{\partial v} & \frac{\partial f_3}{\partial v} \end{vmatrix} \mathbf{j} - \begin{vmatrix} \frac{\partial f_1}{\partial u} & \frac{\partial f_2}{\partial u} \\ \frac{\partial f_1}{\partial v} & \frac{\partial f_2}{\partial v} \end{vmatrix} \mathbf{k}.$$

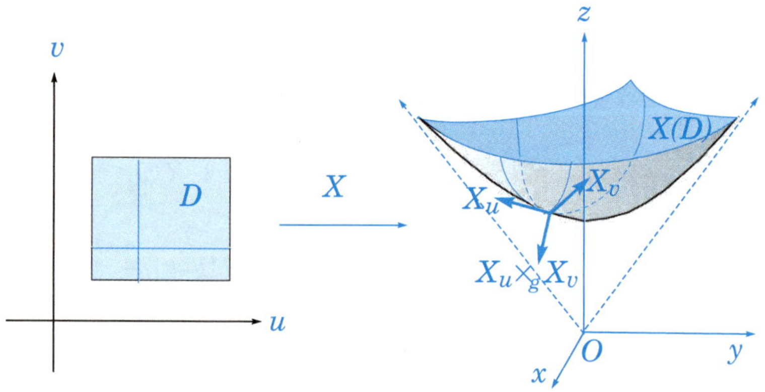

Figure3.1-2

For any $x = (u, v) \in D$ we may also check that $rank(J(X(u, v))) = 2$ if and only if $X_u \times_g X_v \neq 0$ if and only if X_u and X_v are linearly independent. Hence the tangent plane at $p = X(a, b)$ to a coordinate patch $X : D \to \mathbf{L}^3$ is the vector subspace of $\mathbb{R}^3$ of vectors normal to $X_u(a, b) \times_g X_v(a, b)$. Note that if X_u and X_v are spacelike and $X_u(a, b) \times_g X_v(a, b)$ is timelike, then the tangent plane is left handed oriented(Figure3.1-2). Moreover, the unit normal vector to the coordinate patch at p is $U(p) = \left.\frac{X_u \times_g X_v}{|X_u \times_g X_v|}\right|_{(a,b)}$. However, $X_u(a, b) \times X_v(a, b)$ need not be timelike, cf. Example 3.1.4.

Example 3.1.2 Show that $X : D \to \mathbf{E}^3$ given by $X(u, v) = (u, v, \sqrt{1 - u^2 - v^2})$ where $D = \{(u, v) \in \mathbb{R}^2 | u^2 + v^2 < 1\}$ is a coordinate patch. Note that $X(D)$ is the northern hemisphere of the unit sphere in $\mathbf{E}^3$ centered at $(0, 0, 0)$.

Solution. $X_u = \left(1, 0, \frac{-u}{\sqrt{1 - u^2 - v^2}}\right)$, $X_v = \left(0, 1, \frac{-v}{\sqrt{1 - u^2 - v^2}}\right)$, $< X_u, X_u >= \frac{1 - v^2}{1 - u^2 - v^2} > 0$ and $< X_v, X_v >= \frac{1 - u^2}{1 - u^2 - v^2} > 0$ on D. Since $X_u \times X_v = \left(\frac{u}{\sqrt{1 - u^2 - v^2}}, \frac{v}{\sqrt{1 - u^2 - v^2}}, 1\right) \neq 0$, X is regular. Clearly, X is an 1-1 map.

□

Definition 3.1.3 In $\mathbf{L}^3$ we define the 3 types of coordinate patches as follows: X is **timelike** if X_u is timelike on D and X_v is spacelike on D(or conversely), X is **spacelike** if X_u and X_v are both spacelike on D, and X is **lightlike** if X_u(or X_v) is lightlike on D.

Example 3.1.4 Show that $X : D \to L^3$ given by $X(u,v) = (u-v, u^2+v^2, u+v)$ where $D = \mathbb{R}^2 - \{(0,0)\}$ is a spacelike coordinate patch.

Solution. $X_u = (1, 2u, 1)$, $X_v = (-1, 2v, 1)$, $< X_u, X_u >= 4u^2 > 0$ and $< X_v, X_v >= 4v^2 > 0$ on D. Thus, X_u, X_v are spacelike vectors. Since $X_u \times_g X_v = 2(u-v, -1, -u-v) \neq 0$, X is regular. Now suppose $X(u,v) = X(u_1, v_1)$. Then $u-v = u_1-v_1,\ u+v = u_1+v_1$. Hence $u = u_1,\ v = v_1$ and thus $(u,v) = (u_1, v_1)$. Therefore X is an 1-1 map. □

Note that the normal vector field $X_u \times_g X_v = 2(u-v, -1, -u-v) \neq 0$ on the coordinate patch in Example 3.1.4 may be timelike, spacelike or lightlike.

Example 3.1.5 Show that $X : D \to L^3$ given by $X(u,v) = (u,v,v)$ where $D = \mathbb{R}^2$ is a lightlike coordinate patch.

Solution. $X_u = (1,0,0)$, $X_v = (0,1,1)$ so that X_u is spacelike and X_v is lightlike. Since $X_u \times_g X_v = (0,-1,-1) \neq 0$, X is regular. As X is clearly 1-1, X is a coordinate patch. □

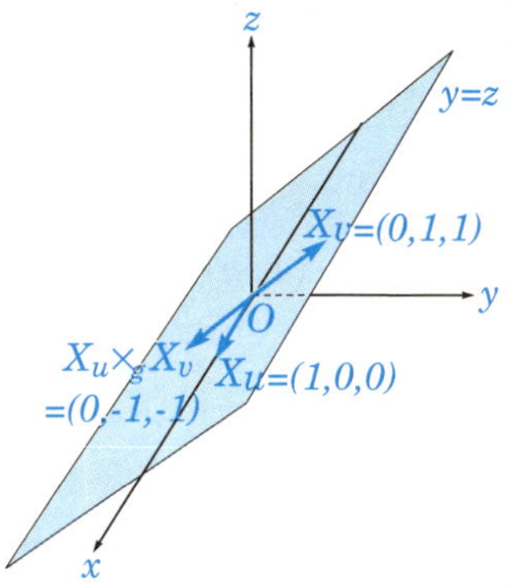

Figure3.1-3

In this case, the normal vector field $X_u \times_g X_v = -X_v$ is lightlike and actually lies in the tangent plane at each point of the surface $M = X(D)$. Note that M is the plane $y = z$ (or $y - z = 0$) in $\mathbf{L}^3$. In Section 2.2 just before the proof of Theorem 2.2.2, we claimed that this plane should have normal $N_1 = \mathbf{j} - (-\mathbf{k}) = \mathbf{j} + \mathbf{k}$; however, $N_1 = -X_u \times_g X_v$(Figure3.1-3) so that these two different techniques are consistent.

This same lightlike plane $M = X(D)$ may also be parametrized by the global spacelike coordinate patch $Y : D \to \mathbf{L}^3$ given by $Y(u,u) = (u+v, v, v)$ where $D = \mathbb{R}^2$ as before. (Given $(x,y,y) \in M$, set $v = y$ and $u = x - y$.) Here $Y_u = (1,0,0)$ and $Y_v = (1,1,1)$ are both spacelike. Thus a surface in $\mathbf{L}^3$ is not characterized by the type of coordinate patch which describes a neighborhood of p. However, reflecting Proposition 3.1.11 to be proved in the next subsection, note that $Y_u \times_g Y_v = (0,-1,-1)$ which is parallel to $X_u \times_g X_v$. Hence Definition 3.1.3 may be regarded as a preliminary step in the development of Lorentz

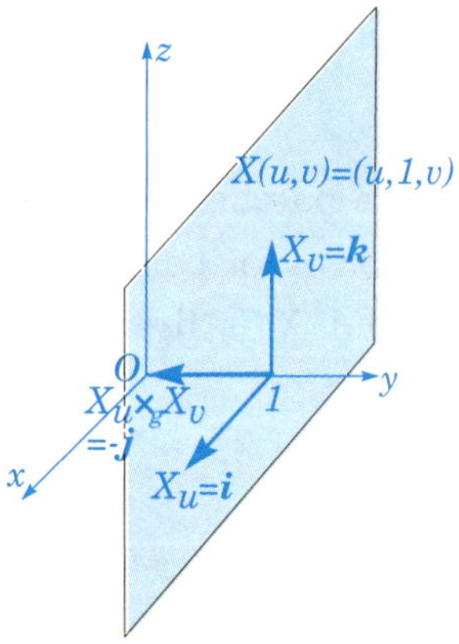

Figure3.1-4

surface theory as we work toward the more satisfactory Proposition 3.1.11 (b), Definition 3.1.12 and Theorem 3.1.17.

A second simple example is furnished by the plane $\mathcal{P}$ given by $y = 1$ in $\mathbf{L}^3$. First, $\mathcal{P}$ admits a timelike patch $X : \mathbb{R}^2 \to \mathbf{L}^3$ given by $X(u, v) = (u, 1, v)$ with $X_u = (1, 0, 0)$ spacelike and $X_v = (0, 0, 1)$ timelike(Figure 3.1-4). But $\mathcal{P}$ also admits the lightlike coordinate patch $Y : \mathbb{R}^2 \to \mathbf{L}^3$ given by $Y(u, v) = (u + v, 1, v)$ with $Y_u = (1, 0, 0)$ spacelike and $Y_v = (1, 0, 1)$ lightlike. Further, $\mathcal{P}$ also admits the coordinate patch $Z : \mathbb{R}^2 \to \mathbf{L}^3$ given by $Z(u, v) = (u, 1, u + v)$ with $Z_u = (1, 0, 1)$ lightlike and $Z_v = (0, 0, 1)$ timelike. In all three cases, $X_u \times_g X_v = Y_u \times_g Y_v = Z_u \times_g Z_v = -\mathbf{j}$. But also $\mathcal{P}$ admits the coordinate patch $\bar{X} : \mathbb{R}^2 \to \mathbf{L}^3$ given by $\bar{X}(u, v) = (u + v, 1, u - v)$ for which $\bar{X}_u = (1, 0, 1)$ and $\bar{X}_v = (1, 0, -1)$ are both lightlike. In this case, $\bar{X}_u \times_g \bar{X}_v = 2\mathbf{j}$ does not equal to the other three Lorentz cross products, but is a multiple of them. This plane $\mathcal{P}$ is an example of a timelike surface in the sense of Definition 3.1.12 below. The Lorentz metric g for $\mathbf{L}^3$ restricted to any tangent space of $\mathcal{P}$ induces an inner product on that tangent space of signature $(-, +)$.

The Proper Patch, Monge Patch and Surfaces

Definition 3.1.6 If a coordinate patch $X : D \to \mathbf{E}^3$ (or $\mathbf{L}^3$) has a continuous inverse function $X^{-1} : X(D) \to D$, then X is said to be a **proper patch**. Further, we call M a **surface** in $\mathbf{E}^3$ (resp., surface in $\mathbf{L}^3$) if for each point $p \in M$ there is a proper patch $X : D \to \mathbf{E}^3$(resp., a timelike, lightlike, or spacelike proper patch in $\mathbf{L}^3$) containing an open neighborhood of p. Especially, for a given differentiable function $f : D \to \mathbb{R}$ the coordinate patch $X : D \to \mathbf{E}^3$ (or $\mathbf{L}^3$) defined by $X(u, v) = (u, v, f(u, v))$ is said to be a **Monge patch** and a surface covered by a Monge patch is called a **simple surface**.

For example, if $D = \mathbb{R}^2$ and $f : D \to \mathbb{R}$ is given by $f(x, y) = y$, then $X(u, v) = (u, v, f(u, v)) = (u, v, v)$ is a Monge patch for the simple surface which is the lightlike plane of Example 3.1.5.

We recall without proof two basic theorems from Advanced Calculus which are useful in surface theory. The Inverse Function Theorem shows that a differentiable function at p has an differentiable inverse function on some open set containing p under the condition of the nonvanishing of the Jacobian determinant at p.

Theorem 3.1.7 (**Inverse Function Theorem**) *Let $f : D \to \mathbb{R}^n$ be differentiable for an open set D in $\mathbb{R}^n$. If $\det J(f(p)) \neq 0$ at $p \in D$, then there exist two open sets V containing p and W containing $f(p)$ such that $f^{-1}|_W : W \to V$ is differentiable.*

The Implicit Function Theorem shows that if a differentiable implicit function $f(x_1, x_2, \cdots, x_n) = 0$ has nonvanishing gradient vector at p then there is a differentiable explicit function g so that near p the level set $f(x_1, x_2, \cdots, x_n) = 0$ is given as the graph of g.

Theorem 3.1.8 (**Implicit Function Theorem**) *Let $f : \mathbb{R}^{n+1} \to \mathbb{R}$ be a differentiable function with grad$f(p) \neq 0$ for $p \in \mathbb{R}^{n+1}$(we may assume $\dfrac{\partial f}{\partial x_i} \neq 0$ for the ith coordinate in $(x_1, x_2, \cdots, x_{n+1})$.). Then there is a differentiable function $g : V \to \mathbb{R}$ for some open set V containing $\hat{p} = (p_1, p_2, \cdots, p_i, p_{i+1}, \cdots, p_{n+1})$ in $\{(x_1, x_2, \cdots, x_{i-1}, x_{i+1}, \cdots, x_{n+1}) \mid x_j \in \mathbb{R}, j = 1, 2, \cdots, n+1, j \neq i\}$ such that $f(x_1, x_2, \cdots, x_{n+1}) = 0$ if and only if $x_i = g((x_1, x_2, \cdots, x_{i-1}, x_{i+1}, \cdots, x_{n+1}))$ for $(x_1, x_2, \cdots, x_{i-1}, x_{i+1}, \cdots, x_n)$ in V.*

Change of Coordinates

If we take a point on a surface M in $\mathbf{E}^3$ or $\mathbf{L}^3$, we may find a second proper patch $Y : E \to M$ containing p different from the proper patch $X : D \to M$ containing p. The the following Theorem 3.1.9 shows that the two proper patches containing p overlap smoothly.

Theorem 3.1.9 *Let M be a surface and $X : D \to M$ and $Y : E \to M$ be two coordinate patches with $W = X(D) \cap Y(E)$ on M nonempty. Then the change of coordinate $Y^{-1} \circ X : X^{-1}(W) \to Y^{-1}(W)$ and its inverse $X^{-1} \circ Y : Y^{-1}(W) \to X^{-1}(W)$ are differentiable(Figure 3.1-5).*

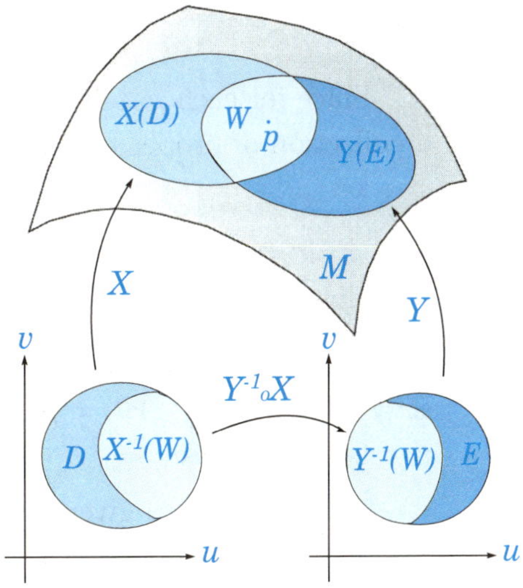

Figure 3.1-5

Proof. Let $(u_0, v_0) \in X^{-1}(W)$ with $p = X(u_0, v_0)$. At $Y^{-1}(p) = (\bar{u}_0, \bar{v}_0)$ we write the proper patch $Y(\bar{u},\bar{v}) = (f(\bar{u},\bar{v}), g(\bar{u},\bar{v}), h(\bar{u},\bar{v}))$. Then

$$J(Y) = \begin{pmatrix} \frac{\partial f}{\partial \bar{u}} & \frac{\partial f}{\partial \bar{v}} \\ \frac{\partial g}{\partial \bar{u}} & \frac{\partial g}{\partial \bar{v}} \\ \frac{\partial h}{\partial \bar{u}} & \frac{\partial h}{\partial \bar{v}} \end{pmatrix}$$

has the rank 2. We may assume that $\frac{\partial f}{\partial \bar{u}}\frac{\partial g}{\partial \bar{v}} - \frac{\partial f}{\partial \bar{v}}\frac{\partial g}{\partial \bar{u}} \neq 0$ at $(\bar{u}_0, \bar{v}_0)$. Now extend Y to the function $F : E \times \mathbb{R} \to \mathbb{R}^3$ defined by $F(\bar{u},\bar{v},t) = Y(\bar{u},\bar{v}) + (0,0,t)$. Clearly, F is differentiable and $F(\bar{u},\bar{v},0) = Y(\bar{u},\bar{v})$. Moreover,

$$J(Y) = \begin{pmatrix} \frac{\partial f}{\partial \bar{u}} & \frac{\partial f}{\partial \bar{v}} & 0 \\ \frac{\partial g}{\partial \bar{u}} & \frac{\partial g}{\partial \bar{v}} & 0 \\ \frac{\partial h}{\partial \bar{u}} & \frac{\partial h}{\partial \bar{v}} & 1 \end{pmatrix}$$

has nonzero determinant at $(\bar{u}_0, \bar{v}_0, 0)$. By the Inverse function Theorem, there is an open set $V \subset \mathbb{R}^3$ containing $F(\bar{u}_0, \bar{v}_0, 0)$ on which F^{-1} is differentiable. Since X is continuous on D, there is an open set $U \subset D$ containing (u_0, v_0) such that $X(U) \subset V \cap M$. Further, since $F^{-1} \circ X\big|_U = Y^{-1} \circ X\big|_U$ and since $F^{-1} \circ X$ is differentiable at (u_0, v_0) as a composition of differentiable functions, $Y^{-1} \circ X$ is differentiable at (u_0, v_0). Since $(u_0, v_0) \in X^{-1}(W)$ is arbitrary, $Y^{-1} \circ X$ is differentiable at all points of $X^{-1}(W)$. Similarly, we may also show that $X^{-1} \circ Y : Y^{-1}(W) \to X^{-1}(W)$ is differentiable. □

Since for each $p \in M$ there is a proper patch containing p, M is covered by a collection of proper patches.

Example 3.1.10 Let $S^2(r) = \{(x,y,z) \in \mathbb{R}^3 | x^2 + y^2 + z^2 = r^2\}$ be the sphere of radius r in $\mathbf{E}^3$. Then $S^2(r)$ is a surface covered by 6 proper patches

Solution. Consider the open sets $\mathcal{U} = \{(y,z) \in \mathbb{R}^2 | y^2 + z^2 < r^2\}$, $\mathcal{V} = \{(x,z) \in \mathbb{R}^2 | x^2 + z^2 < r^2\}$. $\mathcal{W} = \{(x,y) \in \mathbb{R}^2 | x^2 + y^2 < r^2\}$, Then we have 6 proper patches as follows:

$X^+ : \mathcal{U} \to S^2(r)$ given by $X^+(y,z) = (\sqrt{r^2 - y^2 - z^2}, y, z)$,
$X^- : \mathcal{U} \to S^2(r)$ given by $X^-(y,z) = (-\sqrt{r^2 - y^2 - z^2}, y, z)$,
$Y^+ : \mathcal{V} \to S^2(r)$ given by $Y^+(x,z) = (x, \sqrt{r^2 - x^2 - z^2}, z)$,
$Y^- : \mathcal{V} \to S^2(r)$ given by $Y^-(x,z) = (x, -\sqrt{r^2 - x^2 - z^2}, z)$.
$Z^+ : \mathcal{W} \to S^2(r)$ given by $Z^+(x,y) = (x, y, \sqrt{r^2 - x^2 - y^2})$,

$Z^- : \mathcal{W} \to S^2(r)$ given by $Z^-(x, y) = (x, y, -\sqrt{r^2 - x^2 - y^2})$,

Now, any point of the sphere is contained in at least one of these patches. Second, we show that the change of coordinate on the overlapping of two proper patches is differentiable. It suffices to prove that $(Y^+)^{-1} \circ X^+ : (X^+)^{-1}(X^+(\mathcal{U}) \cap Y^+(\mathcal{V})) \to (Y^+)^{-1}(X^+(\mathcal{U}) \cap Y^+(\mathcal{V}))$ is differentiable. It is easy to check that $((Y^+)^{-1} \circ X^+)(y, z) = (Y^+)^{-1}(\sqrt{r^2 - y^2 - z^2}, y, z) = (\sqrt{r^2 - y^2 - z^2}, z)$, and thus $(Y^+)^{-1} \circ X^+$ is differentiable on $\mathcal{U}$ since $y^2 + z^2 < r^2$. Also, $((X^+)^{-1} \circ Y^+)(x, z) = (X^+)^{-1}(x, \sqrt{r^2 - x^2 - z^2}, z) = (\sqrt{r^2 - x^2 - z^2}, z)$, and thus $(X^+)^{-1} \circ Y^+$ is differentiable on $\mathcal{V}$. □

Now, we need to show that the unit normal vector to M in $\mathbf{E}^3$ at p is independent of choice of proper patch containing p and establish an analogous result for surfaces in $\mathbf{L}^3$.

Proposition 3.1.11 *(a) Let p be a point of a surface M in $\mathbf{E}^3$. If $X : D \to M$ and $Y : E \to M$ are two coordinate patches containing p, then*

$$\left.\frac{X_u \times X_v}{|X_u \times X_v|}\right|_p = \pm \left.\frac{Y_{\bar{u}} \times_g Y_{\bar{v}}}{|Y_{\bar{u}} \times_g Y_{\bar{v}}|}\right|_p.$$

(b) Let p be a point of a surface M in $\mathbf{L}^3$. If $X : D \to M$ and $Y : E \to M$ are two coordinate patches containing p, then $X_u \times_g X_v$ is lightlike if and only if $Y_{\bar{u}} \times_g Y_{\bar{v}}$ is lightlike, and in this case, $X_u \times_g X_v$ and $Y_{\bar{u}} \times_g Y_{\bar{v}}$ are parallel. Further, $X_u \times_g X_v$ is timelike(resp., spacelike) if and only if $Y_{\bar{u}} \times_g Y_{\bar{v}}$ is timelike(resp., spacelike), and in these cases,

$$\left.\frac{X_u \times X_v}{|X_u \times X_v|}\right|_p = \pm \left.\frac{Y_{\bar{u}} \times_g Y_{\bar{v}}}{|Y_{\bar{u}} \times_g Y_{\bar{v}}|}\right|_p.$$

Proof. Let $W = X(D) \cap Y(E)$ on M as in Figure3.1-5. Then $X^{-1}(W) \subset D$, $Y^{-1}(W) \subset E$ and the map $Y^{-1} \circ X : X^{-1}(W) \to Y^{-1}(W)$ given by $Y^{-1} \circ X(u, v) = (\bar{u}(u, v), \bar{v}(u, v))$ is an diffeomorphism.* By the Chain Rule, $X_u = Y_{\bar{u}} \dfrac{\partial \bar{u}}{\partial u} + Y_{\bar{v}} \dfrac{\partial \bar{v}}{\partial u}$ and $X_v = Y_{\bar{u}} \dfrac{\partial \bar{u}}{\partial v} + Y_{\bar{v}} \dfrac{\partial \bar{v}}{\partial v}$. Hence taking either the Euclidean or Lorentz cross product,

$$\begin{aligned} X_u \times X_v &= \left(Y_{\bar{u}} \frac{\partial \bar{u}}{\partial u} + Y_{\bar{v}} \frac{\partial \bar{v}}{\partial u} \right) \times \left(Y_{\bar{u}} \frac{\partial \bar{u}}{\partial v} + Y_{\bar{v}} \frac{\partial \bar{v}}{\partial v} \right) \\ &= \left(\frac{\partial \bar{u}}{\partial u} \frac{\partial \bar{v}}{\partial v} - \frac{\partial \bar{v}}{\partial u} \frac{\partial \bar{u}}{\partial v} \right) Y_{\bar{u}} \times Y_{\bar{v}} \\ &= \det(J(Y^{-1} \circ X)) Y_{\bar{u}} \times Y_{\bar{v}} \end{aligned}$$

*For two open sets D, E in $\mathbb{R}^2$ the map $f : D \to E$ is called a diffeomorphism if f and f^{-1} are differentiable.

in both $\mathbf{E}^3$ and $\mathbf{L}^3$. Since $Y^{-1} \circ X$ is a diffeomorphism, $det(J(Y^{-1} \circ X)) \neq 0$. Therefore in $\mathbf{L}^3$, this last equation implies that $X_u \times_g X_v$ is lightlike (resp., timelike, spacelike) if and only if $Y_{\bar{u}} \times_g Y_{\bar{v}}$ is lightlike(resp., timelike, spacelike). Further, in $\mathbf{E}^3$ or $\mathbf{L}^3$ with $X_u \times_g X_v$ nonlightlike,

$$\frac{X_u \times X_v}{|X_u \times X_v|} = \frac{\det(J(Y^{-1} \circ X))}{|\det(J(Y^{-1} \circ X))|} \frac{Y_{\bar{u}} \times Y_{\bar{v}}}{|Y_{\bar{u}} \times Y_{\bar{v}}|} = \pm \frac{Y_{\bar{u}} \times Y_{\bar{v}}}{|Y_{\bar{u}} \times Y_{\bar{v}}|}.$$

□

Recalling that we previously defined the tangent plane to a proper patch at p as the plane in $\mathbf{E}^3$ or $\mathbf{L}^3$ with normal vector $X_u \times_g X_v$, we obtain the following consequence from Proposition 3.1.11:

(a) The tangent plane to the proper patch $X : D \to M$ at p is equal to the tangent plane to the proper patch $Y : E \to M$ at p.

(b) The normal vector to the proper patch $X : D \to M$ at p is parallel to the normal vector to the proper patch $Y : E \to M$ at p.

For any point p of a surface M the tangent plane to M at p is given by the tangent plane at p to a proper patch $X : D \to M$ containing a neighborhood of p and denoted by T_pM.

With Proposition 3.1.11 established, we may also make the following definition, remedying the ambiguities present in Definition 3.1.3.

Definition 3.1.12 A surface M in $\mathbf{L}^3$ is said to be **timelike**(resp., **spacelike, lightlike**) if given any $p \in M$, there exists a coordinate patch $X : D \to \mathbf{L}^3$ for M with $p \in X(D)$ such that $X_u \times_g X_v$ is spacelike(resp., timelike, lightlike) at p.

The surface described in Example 3.1.4, which is the image of a spacelike coordinate patch, is neither timelike, spacelike, nor lightlike since $\bar{X} = X_u \times_g X_v$ is timelike at some points, spacelike at some points, and lightlike at other points. The plane $y = z$ in $\mathbf{L}^3$ of Example 3.1.5 is a lightlike surface in the sense of Definition 3.1.12. This example also illustrates a general feature of lightlike surfaces in $\mathbf{L}^3$ that not only is the lightlike vector $X_u \times_g X_v$ normal to T_pM but also $X_u \times_g X_v$ is contained in T_pM. Anticipating Proposition 3.1.13 below, we may also show in this example, for which M is the plane $y = z$ in $\mathbf{L}^3$ (as is generally true for lightlike surfaces) that a nonzero vector in T_pM is either parallel to the lightlike normal vector $N = X_u \times_g X_v = (0, -1, -1)$ or else spacelike.For given $v \in T_pM$, let $\alpha : (-\epsilon, \epsilon) \to M$ be a smooth curve with $\alpha'(0) = v$ by Proposition 3.1.13. Representing $\alpha(t) = (x(t), y(t), z(t))$, since $\alpha((-\epsilon, \epsilon)) \subset M$ we have $y(t) = z(t)$ for all t and hence $\alpha(t) = (x(t), y(t), y(t))$.

Then $v = \alpha'(0) = (x'(0), y'(0), y'(0))$ so that $< v, v >= (x'(0))^2 + (y'(0))^2 + (y'(0))^2 = (\alpha'(0))^2$. Thus if $x'(0) \neq 0$, the tangent vector v is spacelike and if $x'(0) = 0$, then $v = (0, y'(0), y'(0)) = -y'(0)N$ is parallel to N.

A Realization of a Tangent Vector to a Curve

The following Proposition 3.1.13 gives an alternative method of representing tangent vectors in TpM.

Proposition 3.1.13 *$v \in T_pM$ in $\mathbf{E}^3$ (or $\mathbf{L}^3$) if and only if for some $\epsilon > 0$ there is a curve $\alpha : (-\epsilon, \epsilon) \to X(D) \subset M$ with $\alpha(0) = p$ and $\alpha'(0) = v$ where $X : D \to M \subset \mathbf{E}^3$ (or $\mathbf{L}^3$) is a proper patch containing p.*

Proof. Take $\epsilon > 0$ sufficiently small that $\alpha : (-\epsilon, \epsilon) \to M$ is a curve with $\alpha(0) = p$ lying on a proper patch $X : D \to M$ containing p. Then $X^{-1} \circ \alpha : (-\epsilon, \epsilon) \to D$ is also a curve on $D \subset \mathbb{R}^2$. If we express the coordinates of $X^{-1} \circ \alpha(t)$ in D by $X^{-1} \circ \alpha(t) = (u(t), v(t))$, then $\alpha(t) = X(u(t), v(t))$ with $p = X(u(0), v(0))$(Figure3.1-6). By the Chain Rule, $\alpha'(t) = X_u u'(t) + X_v v'(t)$.

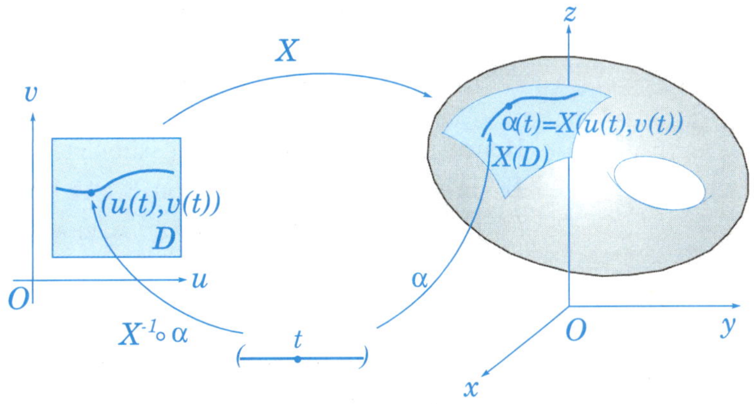

Figure3.1-6

Thus, $v = \alpha'(0) = u'(0)X_u(u(0), v(0)) + v'(0)X_v(u(0), v(0))$, which is a linear combination of $X_u(u(0), v(0))$ and $X_v(u(0), v(0))$. Therefore, $v = \alpha'(0)$ lies in T_pM and v is orthogonal to $X_u(u_0, v_0) \times_g X_v(u_0, v_0)$.

Conversely, suppose that $v \in T_pM$. Then $v = aX_u(u_0, v_0) + bX_v(u_0, v_0)$ where $p = X(u_0, v_0)$. Taking $\epsilon > 0$ small enough so that $(u_0 + at, v_0 + bt) \in D$ we may consider the curve $\beta(t) = X(u_0 + at, v_0 + bt)$ for $t \in (-\epsilon, \epsilon)$. Then $\beta'(0) = v$. Thus, we may realize any vector in T_pM as a tangent vector to a curve passing through p on M. □

Simple Surfaces

Recall that from Definition 3.1.6 that a surface covered by a single Monge patch $X(u,v) = (u, v, f(u,v))$ is said to be a **simple surface**. We now give the proof that this class of objects in $\mathbf{E}^3$ or $\mathbf{L}^3$ obtained by taking the graph of a differentiable function is indeed a proper patch and hence these are indeed surfaces in $\mathbf{E}^3$ or $\mathbf{L}^3$.

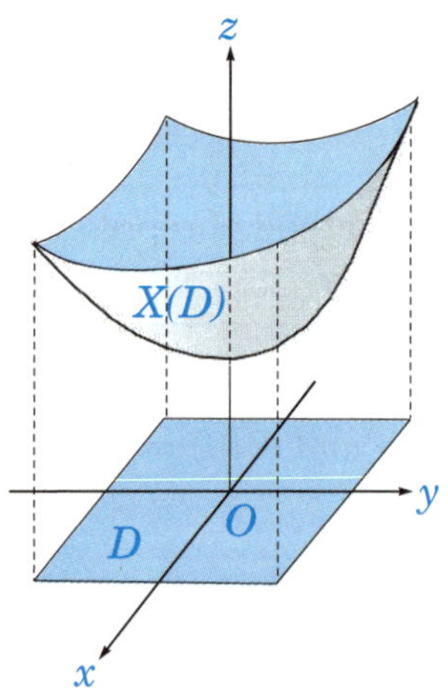

Figure3.1-7

Theorem 3.1.14 *For a differentiable function $f : D \to \mathbb{R}$ the set $M = \{(x, y, f(x,y)) \in \mathbf{E}^3 | (x,y) \in D\}$ is a surface in $\mathbf{E}^3$ (resp., the set $M = \{(x, y, f(x,y)) \in \mathbf{L}^3 | (x,y) \in D\}$ is a surface in $\mathbf{L}^3$).*

Proof. Let's first show that $X : D \to \mathbf{E}^3$ defined by $X(u,v) = (u, v, f(u,v))$ is a proper patch. Since $X_u = (1, 0, f_u)$, $X_v = (0, 1, f_v)$ and $X_u \times X_v = (-f_u, -f_v, 1) \neq 0$, X is regular. It is obvious that X is 1-1. Finally, the inverse function X^{-1} of X is defined by $X^{-1}(x,y,z) = (x,y)$ and thus X^{-1} is continuous. Hence X is a proper patch. For $X : D \to \mathbf{L}^3$ still given by $X(u,v) = (u, v, f(u,v))$, we still have $X_u = (1, 0, f_u)$, $X_v = (0, 1, f_v)$ but now $X_u \times_g X_v = (-f_u, -f_v, -1) \neq 0$ so the same arguments apply as for $\mathbf{E}^3$. □

In the case of $\mathbf{L}^3$, note that

$$g(X_u \times_g X_v, X_u \times_g X_v) = (f_u)^2 + (f_v)^2 - 1.$$

Hence the Monge patch is a timelike (resp., spacelike or lightlike) surface in $\mathbf{L}^3$ in the sense of Definition 3.1.12 provided that $(f_u)^2 + (f_v)^2 > 1$ (resp., $(f_u)^2 + (f_v)^2 < 1$, $(f_u)^2 + (f_v)^2 = 1$) at all points of D.

Level Surfaces

In addition to the Monge patch construction, a second important method of constructing surfaces is as level sets of differentiable functions with nonvanishing gradients. Recall from Section 2.1 that if $f : \mathbf{E}^3 \to \mathbb{R}$ is a differentiable function, then the **gradient** $grad(f)$ of f is the vector field defined by $grad(f) = \frac{\partial f}{\partial x}\mathbf{i} + \frac{\partial f}{\partial y}\mathbf{j} + \frac{\partial f}{\partial z}\mathbf{k} = \left(\frac{\partial f}{\partial x}, \frac{\partial f}{\partial y}, \frac{\partial f}{\partial z}\right)$. This vector field has the defining property that $< v, grad(f) >= v(f)$ for any tangent vector $v \in \mathbf{E}^3 = (\mathbb{R}^3, < , >)$.

Theorem 3.1.15 *Let $f : \mathbf{E}^3 \to \mathbb{R}$ be a differentiable function and $M = \{(x,y,z) \in \mathbf{E}^3 | f(x,y,z) = c\}$ be a level set with $grad(f)(p) \neq O$ at each point $p \in M$. Then M is a (smooth) surface in $\mathbf{E}^3$. Moreover, for each point $p \in M$ the normal vector to T_pM is a scalar multiple of $grad(f)(p)$.*

Proof. For each $p \in M$ with $grad(f)(p) \neq O$, we may suppose that $\frac{\partial f}{\partial z}(p) \neq 0$. Let $p = (x_0, y_0, z_0)$. By the Implicit Function Theorem, there exists a neighborhood D of (x_0, y_0) in $\mathbb{R}^2$ and a differentiable function $z = h(x, y), h : D \to \mathbb{R}$ such that $f(x, y, h(x, y)) = c$ for all $(x, y) \in D$. Then the Monge patch $X : D \to \mathbf{E}^3$ given by $X(u, v) = (u, v, h(u, v))$ is a proper patch containing a neighborhood of p. Since M is covered by the proper patches, M is a surface.

Moreover, given that M is a surface, by Proposition 3.1.13 for any vector v tangent at p to M there is a regular curve $\alpha : (-\epsilon, \epsilon) \to M$ such that $\alpha(0) = p$ and $\alpha'(0) = v$. Let $\alpha(t) = (x(t), y(t), z(t))$ as M is contained in $\mathbf{E}^3$. Since $f(\alpha(t)) = f(x(t), y(t), z(t)) = c$, differentiating both sides with respect to t we have $\dfrac{\partial f}{\partial x}\dfrac{dx}{dt} + \dfrac{\partial f}{\partial y}\dfrac{dy}{dt} + \dfrac{\partial f}{\partial z}\dfrac{dz}{dt} = 0$. Setting $t = 0$, we obtain $< grad(f)(p), \alpha'(0) >=< grad(f)(p), v >= 0$. Thus, $grad(f)(p)$ is orthogonal to T_pM. □

Example 3.1.16 (a) Let M be a subset of $\mathbf{E}^3$ given as the graph $z = x^2 + y^2$, a paraboloid. Then M is a simple surface in $\mathbf{E}^3$ since M has the global Monge patch $X : \mathbb{R}^2 \to \mathbf{E}^3$ given by $X(u, v) = (u, v, u^2 + v^2)$. But $z = x^2 + y^2$ if and only if $z^3 = (x^2 + y^2)^3$. Hence, if $f : \mathbf{E}^3 \to \mathbb{R}$ is the differentiable function $f(x, y, z) = (x^2 + y^2)^3 - z^3$, then M is also given as the level surface

$$M = \{(x, y, z) \in \mathbf{E}^3 \mid f(x, y, z) = 0\}.$$

Note that $grad(f) = (6x(x^2+y^2)^2, 6y(x^2+y^2)^2, -3z^2)$ so that $grad(f)(0, 0, 0) = O$ even though M is a (smooth) surface at $(0, 0, 0)$.

(b) The double cone C in $\mathbf{E}^3$ given by $z^2 = x^2 + y^2$ is a (smooth) surface except at the point $(0, 0, 0)$ in C at which the dimension collapses. Define $f(x, y, z) = x^2 + y^2 - z^2$. Then C is given as a level set by

$$C = \{(x, y, z) \in \mathbf{E}^3 \mid f(x, y, z) = 0\}.$$

Since $grad(f) = (2x, 2y, -2z)$, Theorem 3.1.15 implies that C is a surface except possibly at $(0, 0, 0)$. For $(0, 0, 0)$, we have $grad(f)(0, 0, 0) = O$ and also C fails to be a surface.

To give a version of Theorem 3.1.15 for surfaces in $\mathbf{L}^3$, we need to first recall the Lorentz gradient $grad_L(f)$ for a function $f : \mathbf{L}^3 \to \mathbb{R}$. If g denotes the Lorentz metric for $\mathbf{L}^3 = (\mathbb{R}^3, g)$, then $grad_L(f)$ is characterized by the property that $g(grad_L(f), v) = v(f)$ for any tangent vector v in $\mathbf{L}^3$. Taking $v = \mathbf{i}, \mathbf{j}$ and $\mathbf{k}$, it follows that

$$\begin{aligned} grad_L(f) &= \frac{\partial f}{\partial x}\mathbf{i} + \frac{\partial f}{\partial y}\mathbf{j} - \frac{\partial f}{\partial z}\mathbf{k} \\ &= \left(\frac{\partial f}{\partial x}, \frac{\partial f}{\partial y}, -\frac{\partial f}{\partial z}\right). \end{aligned}$$

Especially, note that $grad_L(f)(p) \neq O$ if and only if $grad(f)(p) \neq O$.

Theorem 3.1.17 *Let $f : \mathbf{L}^3 \to \mathbb{R}$ be a differentiable function and $M = \{(x, y, z) \in \mathbf{L}^3 \mid f(x, y, z) = c\}$ be a level set with $grad_L(f)(p) \neq O$ at each point $p \in M$. Then M is a (smooth) surface in $\mathbf{L}^3$. Moreover, for any $p \in M$ the normal vector to T_pM is a scalar multiple of $grad_L(f)(p)$.*

Proof. Given $grad_L(f)(p) \neq 0$ at all points of M in $\mathbf{L}^3$, by our above remarks on $grad_L(f)$ and $grad(f)$, $grad(f)(p) \neq O$ at all points of M in $\mathbf{E}^3$. Hence M is a surface from Theorem 3.1.15.

Moreover, given that M is a surface and v is tangent to M at p, take $\alpha(t) = (x(t), y(t), z(t))$ as in the proof of Theorem 3.1.15 with $\alpha'(0) = v$. Again

$$\frac{\partial f}{\partial x}\frac{dx}{dt} + \frac{\partial f}{\partial y}\frac{dy}{dt} + \frac{\partial f}{\partial z}\frac{dz}{dt} = 0.$$

But because the minus sign in the third component of $grad_L(f)(p)$ corresponds to that in the Lorentz inner product g, setting $t = 0$ we still obtain

$$g(grad_L(f)(p), \alpha'(0)) = g(grad_L(f)(p), v) = 0.$$

Hence $grad_L(f)(p)$ is orthogonal to T_pM. □

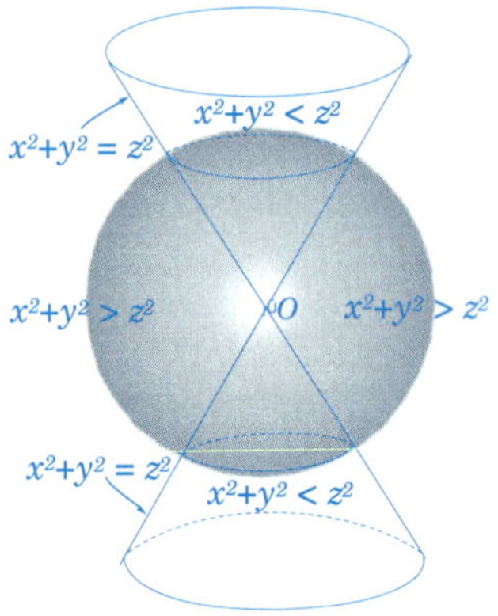

Figure3.1-8

The round Euclidean sphere S^2 is the level surface of the differentiable function $f : \mathbf{L}^3 \to \mathbb{R}$ given by $f(x, y, z) = x^2 + y^2 + z^2$ and $M = \{(x, y, z) \in \mathbf{L}^3 \mid f(x, y, z) = 1\}$. Now $grad_L(f) = (2x, 2y, -2z)$ so that

$$g(grad_L(f), grad_L(f)) = 4(x^2 + y^2 - z^2).$$

Hence $grad_L(f)$ is timelike (resp., spacelike, lightlike) if $x^2 + y^2 < z^2$ (resp., $x^2 + y^2 > z^2$, $x^2 + y^2 = z^2$)(Figure 3.1-8). Hence, the gradient vector field $grad_L(f)$ to a level surface in $\mathbf{L}^3$ need not have a consistent causal character. If so, however, we may obtain the following level surface version of Definition 3.1.12 since $grad_L(f)(p)$ is normal to T_pM.

Corollary 3.1.18 *Let $f : \mathbf{L}^3 \to \mathbb{R}$ be a differentiable function and $M = \{(x, y, z) \in \mathbf{L}^3 \mid f(x, y, z) = c\}$ be a level set of f with $grad_L(f)(p) \neq O$ at all points p of M. Then*

(a) M is a timelike surface if and only if $grad_L(f)(p)$ is spacelike at all points p of M.

(b) M is a spacelike surface if and only if $grad_L(f)(p)$ is timelike at all points p of M.

(c) M is a lightlike surface if and only if $grad_L(f)(p)$ is lightlike at all points p of M.

The lightlike surface M of Example 3.1.5 may be used as an illustration of Corollary 3.1.18 (c) with $f : \mathbf{L}^3 \to \mathbb{R}$ given by $f(x, y, z) = y - z$. Then $M = \{(x, y, z) \in \mathbf{L}^3 \mid f(x, y, z) = 0\}$ and $grad_L(f)(x, y, z) = (0, 1, 1)$ is everywhere null. Cases (a) and (b) are illustrated by the following example.

Example 3.1.19 Show that the hyperbolic sphere $M = \{p = (x, y, z) \in L^3 \mid |p| = 1\}$ is a surface(Figure3.1-9).

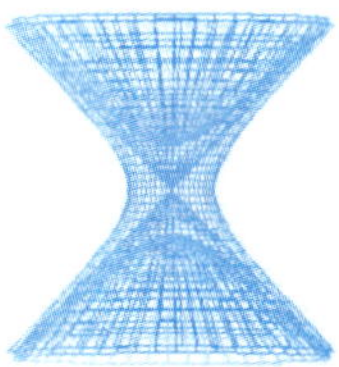

Figure3.1-9

Solution. (1) $U(p)$ is timelike.

Now $|p| = 1$ and $U(p)$ is timelike implies $-x^2 - y^2 + z^2 = 1$. So if we define $f : \mathbf{L}^3 \to \mathbb{R}$ by $f(x, y, z) = -x^2 - y^2 + z^2$, then this portion of the surface is given as $M_1 = \{(x, y, z) \in \mathbf{L}^3 \mid f(x, y, z) = 1\}$. Now $grad_L(f) = 2(-x, -y, -z)$ so that $g(grad_L(f), grad_L(f)) = 4(x^2 + y^2 - z^2) = -4$ on M_1. Since $grad_L(f)$ is everywhere nonzero, M_1 is a surface and since $grad_L(f)$ is everywhere timelike on M_1, this surface is spacelike.

(2) $U(p)$ is spacelike.

In this case, $M_2 = \{(x, y, z) \in \mathbf{L}^3 \mid f(x, y, z) = -1\}$ and $g(grad_L(f), grad_L(f)) = 4(x^2 + y^2 - z^2) = 4$ is everywhere spacelike on M_2, so M_2 is a timelike surface. □

For further study, see the following references.

1. Millman, R. S. and G. D. Parker,Elements of Differential Geometry, Prentice-Hall, Englewood Cliffs, NJ, 1977.

2. McCleary, J., Geometry from a Differentiable Viewpoint, Cambridge University Press, 1994.

3. O'Neill, B., Elementary Differential Geometry, Academic Press, New York, 1966.

3.2 Examples of Surfaces

Surface Plots with Maple in $\mathbf{E}^3$

In this section we first use the command *plot3d* in Maple to graph various surfaces. The *plot3d* command computes the plot of a surface in the three-dimensional space. The first type of surface plot in Cartesian coordinates is given by $plot3d(f(x,y),\ x=a..b,\ y=c..d)$ where $x=a..b$ specifies $a \le x \le b$ and $y=c..d$ implies $c \le y \le d$. Sometimes, the range $c..d$ is expressions in x. The second type of surface plot in parametric coordinates, which we use in this section frequently, is given by $plot3d([f(s,t),\ g(s,t),\ h(s,t)],\ s=a..b,\ t=c..d)$. Any additional arguments are interpreted as options, which are specified as equations of the form option = value. For example, the option $grid=[m,n]$ where m and n are positive integers specifies that the plot is to be constructed on an m by n grid at equally spaced points in the ranges $a..b$ and $c..d$ respectively. By default, a 25 by 25 grid is used and 625 points are generated. Other options include specification of alternate coordinate systems and rendering styles. See the help page plot3d[options] such as color, coords, grid, linestyle, numpoints, shading, style, symbol, symbolsize and thickness. Sometimes,surfaces are given by an implicit function which may be hard to express by parametric coordinates. In this case we may also draw the surfaces by the command $implicitplot3d(f(x,y,z)=0,\ x=a..b,\ y=c..d,\ z=p..q,\ options)$. Since we have 3 variables x,y,z, the option is given by $grid=[m,n,r]$. By default a 10 by 10 by 10 grid is used. The remaining options are the same as those used in plot3d.

The following 12 surfaces are well-known and useful to study the theory of surfaces. We plot those surface by using the Maple programming.

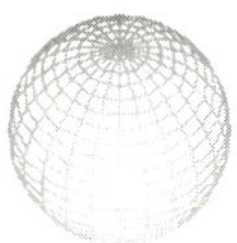

Figure3.2-10

Example 3.2.1 (**Sphere**)

Nonparametric equation: $x^2+y^2+z^2=r^2$.

Coordinate patch: $X(u,v)=(r\cos u\sin v, r\sin u\sin v, r\cos v)$.

Maple plot(Figure3.2-10): with $r=1$

*plot3d([cos(u) * sin(v), sin(u) * sin(v), cos(v)], u = 0..2 * Pi, v = 0..Pi, scaling = constrained);*

Figure3.2-11

Example 3.2.2 (**Cylinder**)

Coordinate patch: $X(u,v)=(r\cos u, r\sin u, v)$.

Maple plot(Figure 3.2-11): with $r = 1$

plot3d([cos(*u*), sin(*u*), *v*], *u* = 0..2 ∗ *Pi*, *v* = 0..*Pi*, *scaling* = *constrained*, *grid* = [35, 35]);

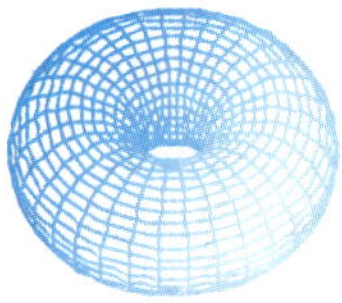

Figure3.2-12

Example 3.2.3 (**Torus**)

Coordinate patch: $X(u, v) = ((R + r\cos u)\cos v, (R + r\cos u)\sin v, r\sin u)$ $R > r > 0$.

Maple plot(Figure 3.2-12): with $R = 3, r = 2$

plot3d([(3 + 2 ∗ cos(*u*)) ∗ cos(*v*), (3 + 2 ∗ cos(*u*)) ∗ sin(*v*), 2 ∗ sin(*u*)], *u* = 0..2 ∗ *Pi*, *v* = 0..2 ∗ *Pi*, *scaling* = *constrained*, *grid* = [35, 35]);

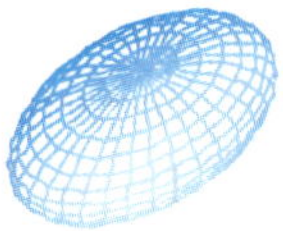

Figure3.2-13

Example 3.2.4 (**Ellipsoid**)

Nonparametric equation: $\dfrac{x^2}{a^2} + \dfrac{y^2}{b^2} + \dfrac{z^2}{c^2} = 1.$

Coordinate patch: $X(u, v) = (a\cos u\cos v, b\sin u\cos v, c\sin v)$.

Maple plot(Figure 3.2-13):

plot3d([5 ∗ cos(*u*) ∗ cos(*v*), 3 ∗ sin(*u*) ∗ cos(*v*), 2 ∗ sin(*v*)], *u* = 0..2 ∗ *Pi*, *v* = 0..2 ∗ *Pi*, *scaling* = *constrained*);

Example 3.2.5 (**Paraboloid**)

Nonparametric equation: $z = a(x^2 + y^2)$.

Coordinate patch: $X(u, v) = (u, v, a(u^2 + v^2))$.

In this case, the parameters u, v may be all real numbers. But this paraboloid changes more sharply as $|u|, |v|$ are getting bigger, so we restrict to smaller u, v-intervals in the 3 plots below.

Maple plot(Figures 3.2-14, 3.2-15, 3.2-16):

plot3d([*u*, *v*, 2 ∗ ($u^2 + v^2$)], *u* = −0.5..0.5, *v* = −0.5..0.5, *scaling* = *constrained*);

plot3d([*u*, *v*, 2 ∗ ($u^2 + v^2$)], *u* = −1..1, *v* = −1..1, *scaling* = *constrained*);

plot3d([*u*, *v*, 2 ∗ ($u^2 + v^2$)], *u* = −1.5..1.5, *v* = −1.5..1.5, *scaling* = *constrained*);

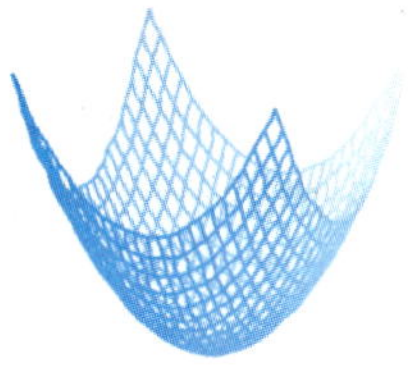

Figure3.2-14

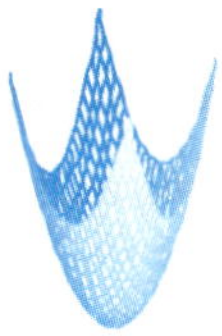

Figure3.2-15

Figure3.2-16

Example 3.2.6 (**Hyperboloid of one sheet**)

Nonparametric equation: $\dfrac{x^2}{a^2}+\dfrac{y^2}{b^2}-\dfrac{z^2}{c^2}=1.$

Coordinate patch: $X(u,v)=(a\cos u\cosh v, b\sin u\cosh v, c\sinh v)$.

We may start with the domain of X at first as

$$0<u<2\pi,\ -1<v<1.$$

However, if the interval of the variable v is extended as follows, then the corresponding surface changes closer to the shape of a cone.

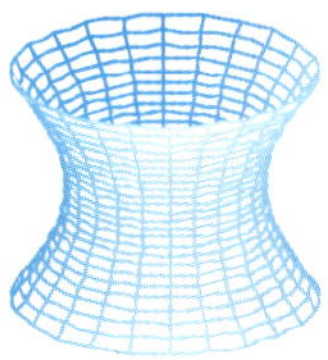

Figure3.2-17

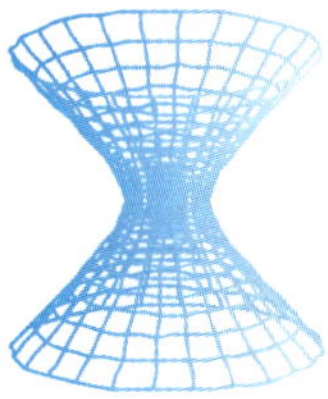

Figure3.2-18

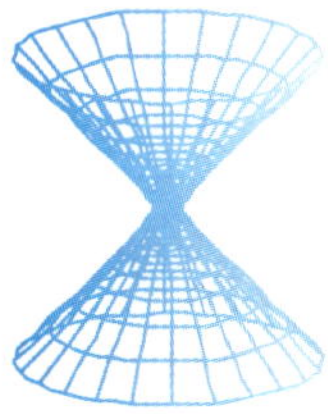

Figure3.2-19

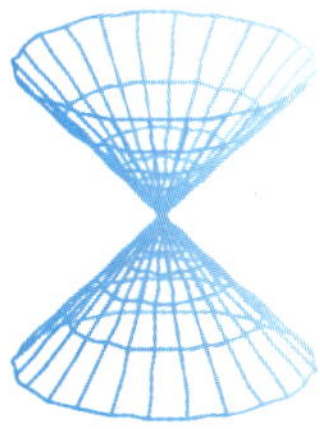

Figure3.2-20

Maple plot(Figures3.2-17, 3.2-18, 3.2-19, 3.2-20):

plot3d([cos(u)∗cosh(v), sin(u)∗cosh(v), sinh(v)], u = 0..2∗*Pi*, v = −1..1, *scaling* = *constrained*);

plot3d([cos(u)∗cosh(v), sin(u)∗cosh(v), sinh(v)], u = 0..2∗*Pi*, v = −2..2, *scaling* = *constrained*);

plot3d([cos(u)∗cosh(v), sin(u)∗cosh(v), sinh(v)], u = 0..2∗*Pi*, v = −3..3, *scaling* = *constrained*);

plot3d([cos(u)∗cosh(v), sin(u)∗cosh(v), sinh(v)], u = 0..2∗*Pi*, v = −5..5, *scaling* = *constrained*);

Example 3.2.7 (**Hyperboloid of two sheets**)

Nonparametric equation: $\dfrac{x^2}{a^2} - \dfrac{y^2}{b^2} - \dfrac{z^2}{c^2} = 1.$

Coordinate patch: $X(u,v) = (a\cosh u \cosh v, b\sinh u \cosh v, c\sinh v)$.

We may extend the domain of the variables u, v as follows.

Figure3.2-21

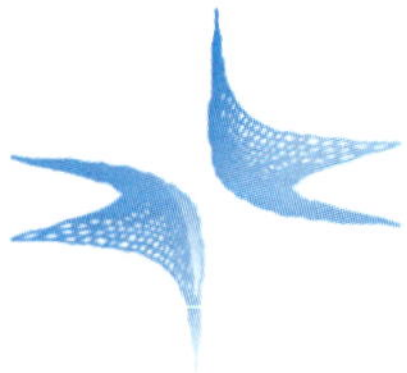

Figure3.2-22

Figure3.2-23

Maple plot(Figures3.2-21, 3.2-22, 3.2-23):

plot3d({[cosh(*u*) ∗ cosh(*v*), sinh(*u*) ∗ cosh(*v*), sinh(*v*)], [− cosh(*u*) ∗ cosh(*v*), sinh(*u*) ∗ cosh(*v*), sinh(*v*)]}, *u* = −1..1, *v* = −1..1, *scaling* = *constrained*);

plot3d({[cosh(*u*) ∗ cosh(*v*), sinh(*u*) ∗ cosh(*v*), sinh(*v*)], [− cosh(*u*) ∗ cosh(*v*), sinh(*u*) ∗ cosh(*v*), sinh(*v*)]}, *u* = −1.5..1.5, *v* = −1.5..1.5, *scaling* = *constrained*);

plot3d({[cosh(*u*) ∗ cosh(*v*), sinh(*u*) ∗ cosh(*v*), sinh(*v*)], [− cosh(*u*) ∗ cosh(*v*), sinh(*u*) ∗ cosh(*v*), sinh(*v*)]}, *u* = −2..2, *v* = −2..2, *scaling* = *constrained*);

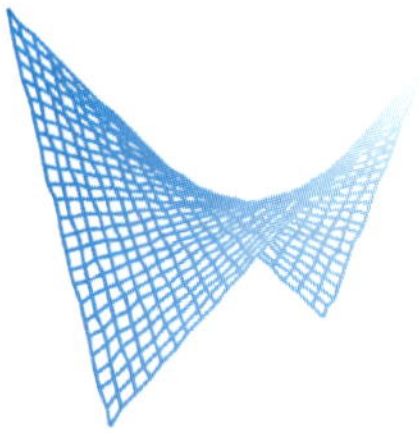

Figure3.2-24

Example 3.2.8 (Hyperbolic paraboloid)

Nonparametric equation: $z = xy$.

Coordinate patch: $X(u,v) = (u, v, uv)$.

Maple plot(Figure3.2-24):

plot3d([*u*, *v*, *u* ∗ *v*], *u* = −5..5, *v* = −5..5);

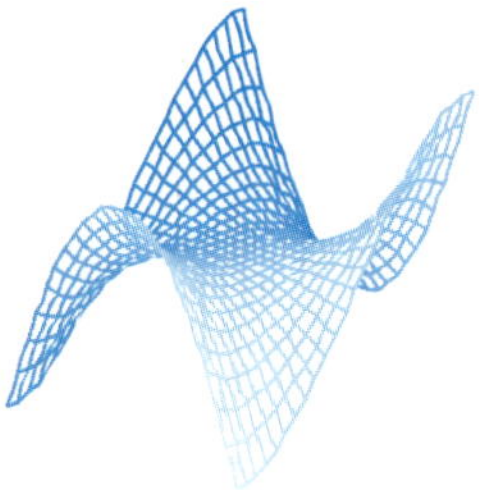

Figure3.2-25

Example 3.2.9 (Monkey saddle)

Nonparametric equation: $z = x^3 - 3xy^2$.

Coordinate patch: $X(u,v) = (u, v, u^3 - 3uv^2)$.

Maple plot(Figure3.2-25):

plot3d([*u*, *v*, *u*ˆ3 − 3 ∗ *u* ∗ *v*ˆ2], *u* = −1.5..1.5, *v* = −1.5..1.5);

Example 3.2.10 (Helicoid)

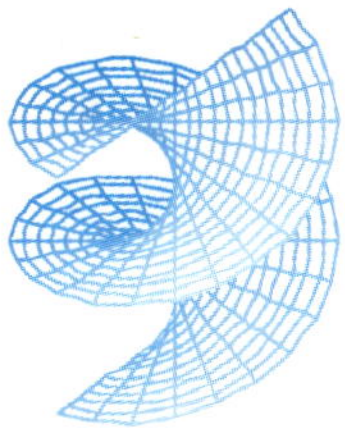

Figure3.2-26

Coordinate patch: $X(u,v) = (v\cos u, v\sin u, bu)$.

Maple plot(Figure3.4-44):

plot3d([$v*cos(u)$, $v*sin(u)$, u], $u = 0..2*Pi$, $v = -1.5..1.5$);

This surface satisfies the cartesian relation $x\sin\left(\frac{z}{b}\right) = y\cos\left(\frac{z}{b}\right)$.

Example 3.2.11 (Catenoid)

Coordinate patch: $X(u,v) = (a\cos u\cosh\frac{v}{a}, a\sin u\cosh\frac{v}{a}, v)$.

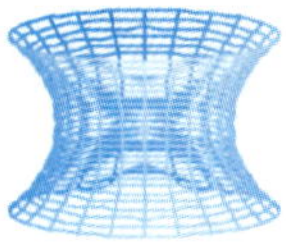

Figure3.2-27

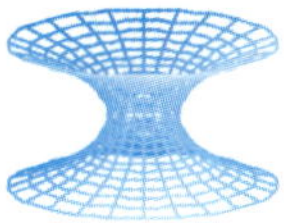

Figure3.2-28

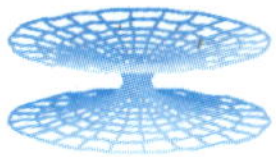

Figure3.2-29

Maple plot(Figures3.2-27, 3.2-28, 3.2-29):

plot3d([$0.5*\cos(v)*\cosh(v/0.5), 0.5*\sin(u)*\cosh(v/0.5)$, v], $u = 0..2*Pi$, $v = -0.5..0.5$, *scaling* = *constrained*);

plot3d([$0.5*\cos(u)*\cosh(v/0.5), 0.5*\sin(u)*\cosh(v/0.5)$, v], $u = 0..2*Pi$, $v = -1..1$, *scaling* = *constrained*);

plot3d([$0.5*\cos(u)*\cosh(v/0.5), 0.5*\sin(u)*\cosh(v/0.5)$, v], $u = 0..2*Pi$, $v = -1.5..1.5$, *scaling* = *constrained*);

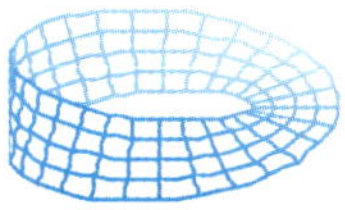

Figure3.2-30

Example 3.2.12 (Möbius strip)

Coordinate patch: $X(u,v) = a\left(\cos u + v\cos\left(\frac{u}{2}\right)\cos u,\right.$
$\left.\sin u + v\cos\left(\frac{u}{2}\right)\sin u,\ v\sin\left(\frac{u}{2}\right)\right)$.

Maple plot(Figure3.2-30):

plot3d([$\cos(u) + v*\cos(u/2)*\cos(u)$,
$\sin(u) + v*\cos(u/2)*\sin(u)$, $v*\sin(u/2)$], $u = 0..2*Pi$,
$v = -0.3..0.3$, *scaling* = *constrained*, *grid* = [30, 5]);

If we extend the interval of v, the width of the Möbius strip is greater.

2-sphere in $\mathbb{L}^3$

Now, we plot the hyperbolic sphere and null cones by Maple as follows(Figure 3.2-31).

$a :=$ plot3d([cos(u) $*$ cosh(v), sin(u) $*$ cosh(v), sinh(v)], $u = Pi..2 * Pi, v = -2..2$, scaling=constrained) :

$b :=$ plot3d({[cos(u) $*$ sinh(v), sin(u) $*$ sinh(v), cosh(v)], [$-$ cos(u) $*$ sinh(v), $-$ sin(u) $*$ sinh(v), $-$ cosh(v)]}, $u = Pi..2 * Pi, v = -2..2$, scaling=constrained) :

$c :=$ plot3d([cos(u)$*$sinh(v), sin(u)$*$sinh(v), $-$ sinh(v)], $u = 11*Pi/6..19*Pi/6, v = -2..0$, scaling=constrained) :

$d :=$ plot3d([$-$ cos(u)$*$sinh(v), $-$ sin(u)$*$sinh(v), sinh(v)], $u = 5*Pi/6..13*Pi/6, v = -2..0$, scaling=constrained) :

with(plots):

display($\{a, b, c, d\}$,style=wireframe);

Figure3.2-31

Ruled Surfaces

Definition 3.2.13 A ruled surface M in $\mathbb{R}^3$ is a surface with a coordinate patch of the form $X(u, v) = \alpha(u) + v\beta(u)$ generated by a straight line $\alpha(u) + v\beta(u)$ called the **ruling** of the surface moving along a **base curve** $\alpha(u)$ with $\alpha' \neq 0$ where $\beta(u)$ is called the **director curve**.

Example 3.2.14 Consider the circle $x^2 + y^2 = 1, z = 1$ as base curve with parametrization $\alpha(u) = (\cos u, \sin u, 0)$. For a constant director curve $\beta(u)$ given by $\beta(u) = (0, 0, 1)$, the corresponding ruled surface is the cylinder(Figure 3.2-11) with the coordinate patch

$$X(u, v) = (\cos u, \sin u, v).$$

Since $X_u \times X_v = (\cos u, \sin u, 0)$ is a unit vector, this parametrization has $X_u \times X_v$ everywhere nonzero. Hence, this coordinate patch is a regular surface as defined in Section 3.1.

Motivated by Example 3.2.14, a **generalized cylinder** is defined to be a ruled surface with a coordinate patch of the form $X(u,v) = \alpha(u) + vP$ where P is a constant vector. (Sometimes the director curve $\alpha(u)$ is assumed to be a closed curve in $\mathbb{R}^3$.) Here $X_u \times X_v = \alpha'(u) \times P$ so the generalized cylinder is a regular surface provided that $\alpha'(u)$ is never parallel to P.

Example 3.2.15 Using the ruled surfaces, we may generate the null cone in $\mathbf{L}^3$(Figure 3.2-32). Consider the parametric equations for the circle given by $\alpha(u) = (\cos u, \sin u, 0)$. Take a curve $\beta(u) = (\cos u, \sin u, 1)$ Define a ruled surface by the coordinate patch $X(u,v) = (1+v)\beta(u) + P$ with $P = (0,0,-1)$ which has Cartesian equation $x^2 + y^2 = (z+1)^2$ and thus is a Euclidean cone with vertex at $(0,0,-1)$. Since

$$X_u \times_g X_v = ((1+v)\cos u, (1+v)\sin u, 1+v),$$

this surface is regular for $v \neq -1$. Also, since $X_u \times_g X_v$ is a lightlike vector for $v \neq -1$, the ruled surface (apart from the point $(0,0,-1)$ is a lightlike surface in the sense of Definition 3.1.12.

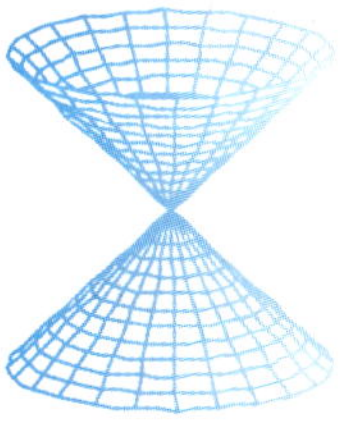

Figure3.2-32

In Example 3.2.15, by substituting $\bar{u} = u$ and $\bar{v} = v+1$, the coordinate patch $X(u,v) = ((1+v)\cos u, (1+v)\sin u, v)$ may be represented by $X(\bar{u},\bar{v}) = (0,0,-1) + \bar{v}(\cos\bar{u}, \sin\bar{u}, 1) = (0,0,-1) + \bar{v}\beta(\bar{u})$.

The example of this particular null cone in $\mathbf{L}^3$ suggests defining a **generalized cone** in $\mathbf{E}^3$ or $\mathbf{L}^3$ as

$$X(u,v) = P + v\beta(u)$$

where P is a constant vector. Because

$$X_u \times_g X_v = v\beta'(u) \times_g \beta(u),$$

this surface is regular in $\mathbf{L}^3$ provided $v \neq 0$ and $\beta(u) \times_g \beta'(u) \neq 0$. The surface is lightlike in $\mathbf{L}^3$ provided that $\beta(u) \times_g \beta'(u)$ is everywhere lightlike.

The hyperboloid of one sheet, already graphed in Figure3.2-17, may also be shown to be a ruled surface.

Example 3.2.16 (Hyperboloid of one sheet reconsidered) Consider the parametrization in $\mathbb{R}^3$ for the ellipse $\dfrac{x^2}{a^2} + \dfrac{y^2}{b^2} = 1$ in the xy-plane given by $\alpha(u) = (a\cos u, b\sin u, 0)$. Take a director curve $\beta(u)$ given by $\beta(u) = \alpha'(u) +$

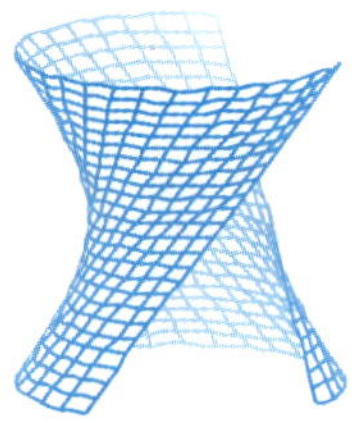

Figure3.2-33

$(0,0,c)$. Then the coordinate patch for $\frac{x^2}{a^2}+\frac{y^2}{b^2}-\frac{z^2}{c^2}=1$ described as a ruled surface(Figure3.2-33) is

$$X(u,v)=(a(\cos u - v\sin u), b(\sin u + v\cos u), cv).$$

Another class of ruled surfaces studied in differential geometry have been formulated using the Frenet apparatus of a unit speed curve in $\mathbf{E}^3$. An issue, which originated in map making, was to find surfaces which could be flattened out onto a plane without distortion. Work on these so-called "developable surfaces" goes back to Euler and to Garspard Monge. Especially, Euler showed that the family of tangents to any space curve forms a developable surface. In the notation of Section 2.1, a surface called the **tangent developable** for a unit speed curve $\alpha(u)$ is given by

$$X(u,v)=\alpha(u)+vT(u)$$

where $T(u)$ is the unit tangent vector of $\alpha(u)$. Similarly, we may define the **normal developable** and **binormal developable** for a unit speed curve $\alpha(u)$ by the coordinate patch

$$Y(u,v)=\alpha(u)+vN(u) \text{ and } Z(u,v)=\alpha(u)+vB(u)$$

where $N(u)$ and $B(u)$ are the unit normal vector and the unit binormal vector of $\alpha(u)$ respectively.

Example 3.2.17 Given the circular helix $\gamma(u)=(a\cos u, a\sin u, bu)$, construct the tangent developable, the normal developable and the binormal developable for $\gamma(u)$.

Solution. Since $\gamma(u)$ is not a unit speed curve, reparametrize to $\alpha(u)=(a\cos(cu), a\sin(cu), bcu)$ where $c=\frac{1}{\sqrt{a^2+b^2}}$. Then

$$\begin{aligned} T(u) &= c(-a\sin cu, a\cos cu, b),\\ N(u) &= (-\cos cu, -\sin cu, 0),\\ B(u) &= c(b\sin cu, -b\cos cu, a). \end{aligned}$$

Hence the parametrizations for the tangent,normal, and binormal developable surfaces are given by

$$\begin{aligned} X(u,v) &= (a\cos cu, a\sin cu, bu) + vc(-a\sin cu, a\cos cu, b),\\ Y(u,v) &= (a\cos cu, a\sin cu, bu) + v(-\cos cu, -\sin cu, 0), \text{ and}\\ Z(u,v) &= (a\cos cu, a\sin cu, bcu) + vc(b\sin cu, b\cos cu, a). \end{aligned}$$ □

Surfaces of Revolution

Let C be a regular curve given by $c(u) = (a(u), b(u), 0)$ with $c'(u) \neq 0$ and $b(u) > 0$ in the xy-plane in $\mathbf{E}^3$. By revolving the curve C called the **profile curve** about x axis, a parametrization of the surface M, said to be the **surface of revolution**, is obtained as

$$X(u,v) = (a(u), b(u)\cos v, b(u)\sin v),$$

when the curve C rotates from xy-plane to $X(u,v)$ by the angle v(Figure3.2-34). Then we have

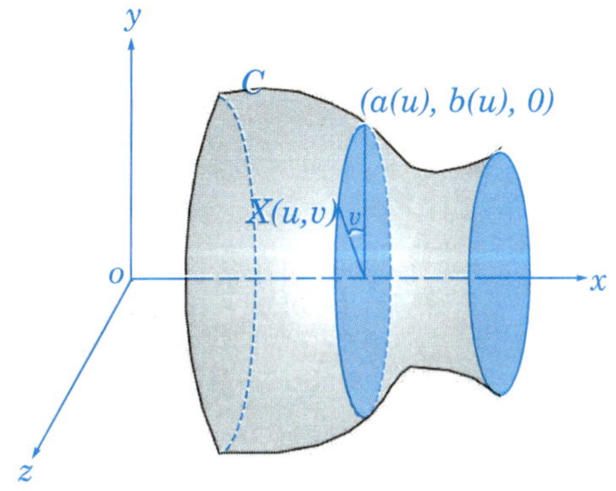

Figure3.2-34

$$\begin{aligned}
X_u &= X_u(u,v) = (a'(u), b'(u)\cos v, b'(u)\sin v),\\
X_v &= X_v(u,v) = (0, -b(u)\sin v, b(u)\cos v),\\
X_u \times X_v &= (b(u)b'(u), -b(u)a'(u)\cos v, -b(u)a'(u)\sin v),\\
< X_u \times X_v, X_u \times X_v > &= (b(u))^2\left((b'(u))^2 + (a'(u))^2\right) = (b(u))^2|c'(u)|^2.
\end{aligned}$$

Hence the assumption $b(u) > 0$ makes this surface of revolution a regular surface parametrization.

Now, we construct the **timelike (lightlike or spacelike, resp.) surfaces of revolution** by revolving a given timelike (lightlike or spacelike, resp.) curve C in xz-plane in $\mathbf{L}^3$ about the time axis z. Corresponding to the Euclidean case, we take the curve $c(u)$ of the form $c(u) = (b(u), 0, a(u))$ with $c'(u) \neq 0$ and $b(u) > 0$. Then the parametrization of the surface M(Figure3.2-35) is obtained by

$$X(u,v) = (b(u)\cos v, b(u)\sin v, a(u)).$$

If c is a timelike curve, $(b'(u))^2 - (a'(u))^2 < 0$.

If c is a lightlike curve, $(b'(u))^2 - (a'(u))^2 = 0$.

If c is a spacelike curve, $(b'(u))^2 - (a'(u))^2 > 0$.

Thus, we have

$$X_u = (b'(u)\cos v, b'(u)\sin v, a'(u)),$$

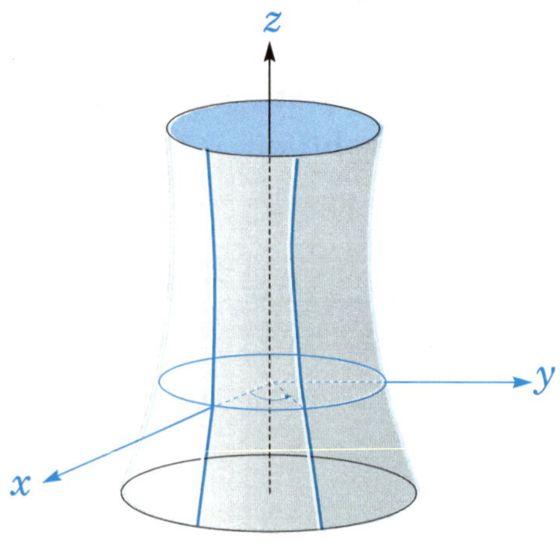

Figure3.2-35

$$X_v = (-b(u)\sin v, b(u)\cos v, 0),$$
$$X_u \times_g X_v = (-b(u)a'(u)\cos v, -b(u)a'(u)\sin v, -b(u)b'(u)),$$

$$\begin{aligned} g(X_u \times_g X_v, X_u \times_g X_v) &= -(b(u))^2 \left((b'(u))^2 - (a'(u))^2\right) \\ &= -(b(u))^2 g(c'(u), c'(u)). \end{aligned}$$

Therefore, by Definition 3.1.12, we obtain a timelike (lightlike, spacelike resp.) surface of revolution according to whether the profile curve C is timelike (lightlike, spacelike resp.).

Figure3.2-36

Example 3.2.18 Let M be a surface of revolution in x, z-plane revolving the profile curve $c(u) = (u, 0, \cos u)$ for $0 < u < \dfrac{\pi}{2}$ about z-axis. Then show that M is a spacelike surface of revolution.

Solution. Since $c'(u) = (1, 0, -\sin u)$ and since $g(c'(u), c'(u)) = 1 - \sin^2 u = \cos^2 u > 0$ for $0 < u < \dfrac{\pi}{2}$, c is a spacelike profile curve. Hence the parametrization of M given by $X(u, v) = (u\cos v, u\sin v, \cos u)$ is a spacelike surface of revolution. □

Note that $X_v = 0$ at $u = 0$, so the point $(0, 0, 1)$ must be excluded from this surface parametrization.

For further study, see the following references.

1. McCleary, J., Geometry from a Differentiable Viewpoint, Cambridge University Press, 1994.

2. O'Neill, B., Elementary Differential Geometry, Academic Press, New York, 1966.

3.3 The First Fundamental Form

Definition 3.3.1 Let M be a regular surface in $(\mathbb{R}^3, <\ ,\ >)$. At a point $p \in M$ the **first fundamental form**

$$I_p : T_pM \times T_pM \to \mathbb{R}$$

is defined by $I_p(v, w) = < v, w >$ for $v, w \in T_pM$, where $<\ ,\ >$ denotes the Euclidean metric for $\mathbb{R}^3$.

Suppose that M is a regular surface in $\mathbf{L}^3 = (\mathbb{R}^3, g)$. Since the the first fundamental form is defined by the inner product in $\mathbf{E}^3$, it is natural to analogously define the first fundamental form $I_p : T_pM \times T_pM \to \mathbb{R}$ at a point $p \in M$ by $I_p(v, w) = g(v, w)$ for $v, w \in T_pM$, where g denotes the Minkowski metric for $\mathbb{R}^3$.

To study local properties of a surface, we may calculate with a coordinate patch about a point of the surface. Let $X : D \subset \mathbb{R}^2 \to M \subset \mathbf{E}^3$ be a coordinate patch about p. Since M is a regular surface, X_u, X_v are linearly independent in T_pM. Hence, for $v \in T_pM$ we may choose a regular curve $c : (-a, a) \to M$ such that $v = c'(0)$ in T_pM. Since p is contained in the coordinate patch $X(D)$,we have $c(t) = X(u(t), v(t))$ with $p = X(u(0), v(0))$ and $v = c'(0) = (X_u)_p u'(0) + (X_v)_p v'(0)$(Figure3.3-37). Then we may compute $I_p(v, v)$ by

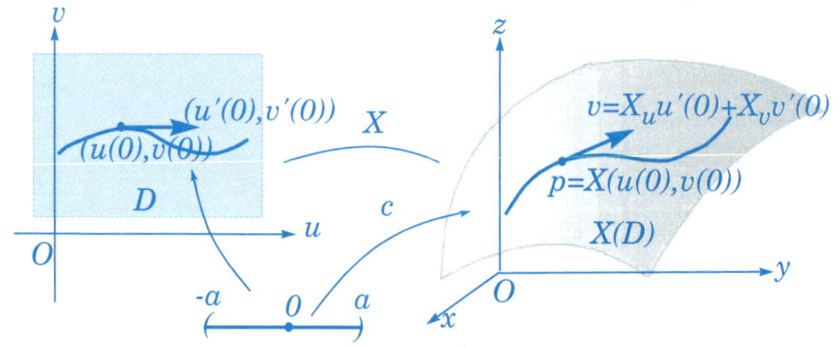

Figure3.3-37

$$\begin{aligned} I_p(v, v) &= < c'(0), c'(0) > \\ &= (u'(0), v'(0)) \begin{pmatrix} E & F \\ F & G \end{pmatrix} \begin{pmatrix} u'(0) \\ v'(0) \end{pmatrix} \end{aligned}$$

where $E = < X_u, X_u >_p, F = < X_u, X_v >_p, G = < X_v, X_v >_p$ which are called the **coefficients of the first fundamental form*** at p on the coordinate patch

*The notation E, F and G for the coefficient functions of the first fundamental form was introduced by Gauss himself in his celebrated paper of 1827 "Disquisitiones generales circa superficies curvas"="General investigations of curved surfaces."

$X(D)$, and the matrix

$$\begin{pmatrix} E & F \\ F & G \end{pmatrix}$$

is the representation of the first fundamental form I_p. Because of the properties of the inner product, the matrix is symmetric, nonsingular and positive definite with respect to the basis $\{(X_u)_p, (X_v)_p\}$ for T_pM.

In $\mathbf{L}^3$, the coefficients of the first fundamental form are given by $E = g(X_u, X_u)_p, F = g(X_u, X_v)_p, G = g(X_v, X_v)_p$, where g is the Minkowski metric for $\mathbb{R}^3$.

Let $c(t)$ be any regular curve in $X(D) \subset M$ in $\mathbf{E}^3$ such that $c(t) = X(u(t), v(t))$. For the arc-length

$$s(t) = \int_a^t |c'(\sigma)| d\sigma$$

we have $\dfrac{ds}{dt} = |c'(t)| = \sqrt{E(u'(t))^2 + 2Fu'(t)v'(t) + G(v'(t))^2}$. That is,

$$\left(\frac{ds}{dt}\right)^2 = \left\langle \frac{dX}{dt}, \frac{dX}{dt} \right\rangle = E\left(\frac{du}{dt}\right)^2 + 2F\frac{du}{dt}\frac{dv}{dt} + G\left(\frac{dv}{dt}\right)^2.$$

Using the notation $\dfrac{ds}{dt}dt = ds$, $\dfrac{du}{dt}dt = du$ and $\dfrac{dv}{dt}dt = dv$, we may write this equation symbolically as

$$ds^2 = E(du)^2 + 2Fdudv + G(dv)^2$$

which is independent of the curve $c(t)$. Hence, the first fundamental form $I = ds^2$ on $X(D)$ is given by $I(p) = I_p$ for all $p \in X(D)$.

Similarly, if $c(t)$ is any regular curve in $X(D) \subset M$ in $\mathbf{L}^3$ and we represent $c(t) = X(u(t), v(t))$, then

$$g(c'(t), c'(t)) = E(u'(t))^2 + 2Fu'(t)v'(t) + G(v'(t))^2$$

and one can symbolically write

$$I = ds^2 = E(du)^2 + 2Fdudv + G(dv)^2.$$

But now

$$|c'(t)| = \sqrt{|E(u'(t))^2 + 2Fu'(t)v'(t) + G(v'(t))^2|}.$$

Example 3.3.2 Compute the first fundamental form I of the Monge patch $X(u, v) = (u, v, f(u, v))$ given in Example 3.1.14 in $\mathbf{E}^3$.

Solution. Since $X_u = (1, 0, f_u)$, $X_v = (0, 1, f_v)$, we have $E = 1 + (f_u)^2$, $F = f_u f_v$ and $G = 1 + (f_v)^2$. Thus,

$$I = ds^2 = \left(1 + (f_u)^2\right)(du)^2 + 2f_u f_v du dv + \left(1 + (f_v)^2\right)(dv)^2. \qquad \square$$

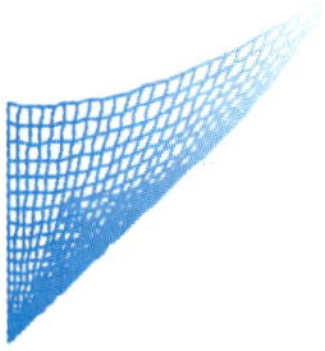

Figure3.3-38

Example 3.3.3 Compute the first fundamental form of the coordinate patch $X(u,v) = (u^2, uv, v^2)$ in $\mathbf{E}^3$(Figure3.3-38).

Solution. Since $X_u = (2u, v, 0)$, $X_v = (0, u, 2v)$, we have $E = 4u^2 + v^2$, $F = uv$ and $G = u^2 + 4v^2$. Thus,

$$I = ds^2 = \left(4u^2 + v^2\right)(du)^2 + 2uv du dv + \left(u^2 + 4v^2\right)(dv)^2. \qquad \square$$

Example 3.3.4 Show that the first fundamental forms of the following coordinate patches in $\mathbf{L}^3$ are equivalent.

(1) **The timelike plane**: $X(u,v) = P + uA + vB$ where A is a unit spacelike vector and B is a unit timelike vector such that $g(A,B) = 0$, and P is a constant vector(Figure3.3-39).

(2) **The generalized timelike cylinder**: $X(u,v) = c(u) + vB$ where $c(u)$ is a unit speed spacelike curve and B is a unit timelike vector such that $g(c'(u), B) = 0$(Figure3.3-40).

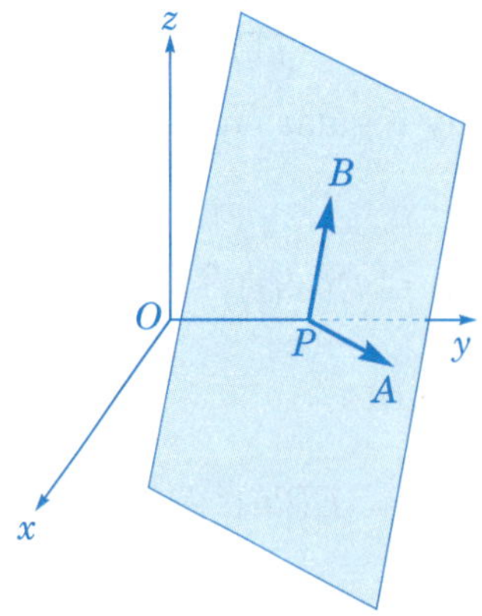

Figure3.3-39

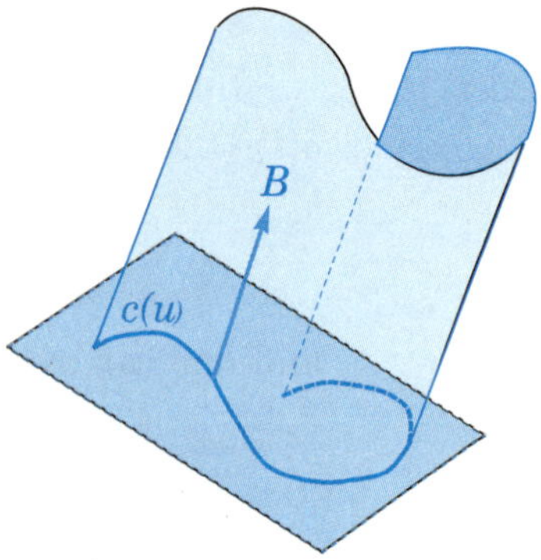

Figure3.3-40

Solution. We may compute both first fundamental forms as $I = (du)^2 - (dv)^2$. $\square$

Lemma 3.3.5 *$EG - F^2 = |X_u \times X_v|^2$ on D for a regular surface in $\mathbf{E}^3$.*

Proof. If θ denotes the angle between the nonzero vectors X_u and X_v, then

$$\begin{aligned} EG - F^2 &= |X_u|^2|X_v|^2 - (< X_u, X_v >)^2 \\ &= |X_u|^2|X_v|^2 \left(1 - \frac{(< X_u, X_v >)^2}{|X_u|^2|X_v|^2}\right) \\ &= |X_u|^2|X_v|^2(1 - \cos^2\theta) \\ &= |X_u|^2|X_v|^2 \sin^2\theta \\ &= |X_u \times X_v|^2. \end{aligned}$$

□

This identity in Lemme 3.3.5 may also be derived from Theorem 1.3.1-(5).

Remark 3.3.6 In $\mathbf{L}^3$, if X_u, X_v are future directed timelike as in Section 1.3, Theorem 1.3.12, and $\theta \geq 0$ is the (unoriented) hyperbolic angle between X_u and X_v, then

$$\begin{aligned} EG - F^2 &= |X_u|^2|X_v|^2 - (< X_u, X_v >)^2 \\ &= |X_u|^2|X_v|^2 \left(1 - \frac{(< X_u, X_v >)^2}{|X_u|^2|X_v|^2}\right) \\ &= |X_u|^2|X_v|^2(1 - \cosh^2\theta) \\ &= |X_u|^2|X_v|^2(-\sinh^2\theta) \\ &= -|X_u \times_g X_v|^2. \end{aligned}$$

Note also that $|X_u \times_g X_v| = |X_u||X_v| \sinh\theta$.

But for Theorem 3.3.11, a version of the identity is needed for general X_u, X_v not necessarily timelike. This may be obtained in all cases from Theorem 1.3.12-(1).

Lemma 3.3.7 *$EG - F^2 = -g(X_u \times_g X_v, X_u \times_g X_v)$ on D for a regular surface in L^3.*

Proof. Applying Theorem 1.3.12-(1) we have

$$\begin{aligned} g(X_u \times X_v, X_u \times X_v) &= \left[g(X_u, X_u)g(X_v, X_v) - (g(X_u, X_v))^2\right] \\ &= -(EG - F^2). \end{aligned}$$

□

In the context of Definition 3.3.10 and Theorem 3.3.11 characterizing surface patches in $\mathbf{L}^3$ in terms of $EG-F^2$, a strengthening of the wrong way Schwarz inequality(Theorem 1.3.6-(5)) from timelike to causal vectors is appropriate. Here a nonzero vector x in $\mathbf{L}^3$ is said to be **causal** if x is either timelike or lightlike. We will give proofs in the next several results which while stated for $\mathbf{L}^3$ are equally valid in more general Lorentz spaces. Especially Theorem 1.3.6-(4) as well as Lemma 3.3.8-(a), both of which fail to be valid in the more general pseudo-Riemannian, non-Lorentz case, play key roles.

Lemma 3.3.8 *In* $\mathbf{L}^3$,

(a) orthogonal lightlike vectors are collinear,

(b) orthogonal nonspacelike vectors are lightlike, hence collinear,

(c) there exists no 2 dimensional subspace V of $\mathbf{L}^3$ on which the inner product g is identically zero.

Proof. (a) Let n be a given lightlike vector and e_3 a unit timelike vector in $\mathbf{L}^3$. By projecting n onto the orthogonal complement $(e_3)^{\perp}$ which is a spacelike subspace of $\mathbf{L}^3$, we may find a Lorentz orthonormal basis $\{e_1, e_2, e_3\}$ for $\mathbf{L}^3$ with $n = a_2e_2 + a_3e_3$. Then as $g(n,n) = 0$, we have $a_2^2 = a_3^2$. Hence by rescaling, we may take $n = e_2 + e_3$ or $n = e_2 - e_3$. Since the arguments are similar, it suffices to show for $n = e_2 + e_3$ that if m is a second lightlike vector in $\mathbf{L}^3$ with $g(n,m) = 0$, then n and m are collinear. Expand $m = b_1e_1 + b_2e_2 + b_3e_3$. Then $0 = g(n,m) = g(e_2+e_3, b_1e_1+b_2e_2+b_3e_3) = b_2-b_3$ so that $b_2 = b_3$. Hence m is of the form $m = b_1e_1+b_3e_2+b_3e_3$. now calculating $0 = g(m,m) = b_1^2+b_3^2-b_3^2 = b_1^2$, we have $b_1 = 0$ so that $m = b_3e_2 + b_3e_3 = b_3n$.

(b) From Theorem 1.3.6, if v is timelike and w is timelike or lightlike, then $g(v,w) \neq 0$. Hence if causal vectors are orthogonal, they must both be lightlike and hence proportional from part (a).

(c) Suppose V were a 2-dimensional subspace of $\mathbf{L}^3$ on which $g = 0$. Then there exist non-collinear lightlike vectors $v, w \in V$. If V were a subspace, then $v + w \in V$ whence $g(v+w, v+w) = 0$. On the other hand from part (a),

$$g(v+w, v+w) = g(v,v) + 2g(v,w) + g(w,w) = 2g(v,w) \neq 0,$$

in contradiction. □

Combining Theorem 1.3.6 and Lemma 3.3.8, we may obtain the complete version of the wrong way Schwarz inequality for $\mathbf{L}^3$.

Corollary 3.3.9 (Wrong Way Schwarz Inequality) *Let v, w be causal vectors in $\mathbf{L}^3$. Then*

$$|g(v,w)| \geq |v||w|$$

and equality holds if and only if v, w are collinear lightlike vectors or v, w are timelike vectors with $w = kv$ for some $k > 0$.

Proof. If v, w are timelike, this is the content of Theorem 1.3.6-(5). If one of v or w is lightlike, then $|g(v,w)| \geq 0 = |v||w|$, so the inequality is valid. Also from Theorem 1.3.6, if one of v, w is timelike and the other is lightlike, then $g(v,w) > 0$ so equality fails to hold. But then if both v, w are lightlike and $g(v,w) = 0$, then Lemma 3.3.8-(a) gives the desired conclusion. □

In considering the type of behavior that the first fundamental form can exhibit at a point p of a coordinate patch $X(D)$, two plausible possibilities are that $I_p(v,v) > 0$ for any nonzero $v \in T_pM$, or that I_p take on both positive and negative values on T_pM. By Lemma 3.3.8-(c), the possibility that $I_p(v,v) = 0$ for all $v \in T_pM$ is precluded, since T_pM is a 2-dimensional subspace of $\mathbf{L}^3$. Example 3.1.5 in Section 3.1 of the lightlike coordinate patch $X(u,v) = (u,v,v)$ and $D = \mathbb{R}^2$ provides an example of a third type of behavior for I_p. Here the tangent plane(which was the given lightlike plane itself) contained a single lightlike direction and all other vectors in the tangent plane were spacelike. in order to characterize this type of surface behavior in terms of the first fundamental form I_p, the concept of the **radical** or **null space** of T_pM with respect to the bilinear form I_p has been formulated as

$$Rad_p(M) = \{v \in T_pM; I_p(v,w) = 0 \text{ for all } w \in T_pM\}.$$

(By Lemma 3.3.8-(c), we have $\dim(Rad_p(M)) \leq 1$.)
In the case of the lightlike plane $y = z$ with surface patch $X(u,v) = (u,v,v)$, we have $X_u = (1,0,0)$ and $X_v = (0,1,1)$. Hence $E = 1, F = G = 0$ so $EG - F^2 = 0$ which is consistent with our remark in Example 3.1.5 that $X_u \times X_v$ is lightlike. For tangent vectors $v = a_1 X_u|_p + b_1 X_v|_p$ and $w = a_2 X_u|_p + b_2 X_v|_p$, we have

$$I_p(v,v) = Ea_1a_2 + F(a_1b_2 + a_2b_1) + Gb_1b_2 = a_1a_2.$$

Hence for $n = X_v = (0,1,1)$, we have $I_p(n,n) = 0$. Also, $I_p(w,w) = a_2^2 \geq 0$ and $I_p(n,w) = 0 \cdot a_2 = 0$ for any $w \in T_pM$. Thus $Rad_p(M) = \text{Span}\{X_v|_p\}$ is 1-dimensional. With this motivation in hand, we are ready for the basic definition.

Definition 3.3.10 At $p \in X(D) \subset M$, I_p is **positive definite** if $I_p(v,v) > 0$ for any nonzero vector $v \in T_pM$; I_p is **indefinite** if I_p has both positive and negative values on T_pM; and I_p is **isotropic** if I_p has rank 1 on T_pM, i.e., $\dim(Rad_p(M)) = 1$.

Generally a surface in $\mathbf{L}^3$ will not have all points of a given type in the sense of Definition 3.3.10. The usual round 2-sphere $S^2 : x^2 + y^2 + z^2 = 1$ in $\mathbf{L}^3$ is positive definite if $|z| > \dfrac{1}{\sqrt{2}}$ and indefinite if $|z| < \dfrac{1}{\sqrt{2}}$ for instance, yet classes of surface patches of the type considered in Definition 3.1.12 of Section 3.1 do manifest one of the three types of behavior of Definition 3.3.10 at each of their points, and this may even be characterized in terms of the behavior of the quadratic form $EG - F^2$. Also, recall from Definition 3.1.12 that a surface patch $X(u,v)$ is timelike(resp., spacelike, lightlike) if $X_u \times_g X_v$ is spacelike(resp., timelike, lightlike) at all points of D.

Theorem 3.3.11 *Let $X : D \to \mathbf{L}^3$ be a coordinate patch. Then at all points $p \in X(D)$:*

(a) I_p is positive definite if and only if $EG - F^2 > 0$ if and only if X is spacelike.

(b) I_p is indefinite if and only if $EG - F^2 < 0$ if and only if X is timelike.

(c) I_p is isotropic if and only if $EG - F^2 = 0$ if and only if X is lightlike.

Proof. From previous considerations, given $v \in T_pM$ for any regular curve $c(t) = X(u(t), v(t))$ with $c'(0) = v$ we have

$$I_p(v,v) = E(p)(u'(0))^2 + 2F(p)u'(0)v'(0) + G(p)(v'(0))^2$$

(although we will usually write $E = E(p), F = F(p), G = G(p)$ omitting the variable p). If $E(p) \neq 0$, then

$$I_p(v,v) = E\left(u'(0) + \frac{F}{E}v'(0)\right)^2 + \frac{EG - F^2}{E}(v'(0))^2$$

which explains the appearance of the discriminant term $EG - F^2$ in the statement of Theorem 3.3.11. Also, because $EG - F^2 = -g(X_u \times_g X_v, X_u \times_g X_v)$, the equivalence of the second and third conditions in parts (a), (b), (c) is is immediate from Definition 3.1.12. Thus we will prove the equivalence of the first and second conditions.

(a) First, given I_p is positive definite, $E = I_p(X_u, X_u) > 0$ and we can form the factorization

$$I_p(v,v) = E\left(u'(0) + \frac{F}{E}v'(0)\right)^2 + \frac{EG - F^2}{E}(v'(0))^2.$$

Take $v = -FX_u|_p + EX_v|_p \in T_pM$, i.e., $u'(0) = -F$ and $v'(0) = E$. Then $I_p(v,v) = E(EG - F^2) > 0$ which implies $EG - F^2 > 0$ since $E > 0$.

Conversely, given $EG - F^2 > 0$, then $X_u \times_g X_v$ is timelike at all points p in $X(D)$. Since X_u is nonzero and Lorentz orthogonal to the timelike vector $X_u \times_g X_v$, we must have X_u spacelike from Theorem 1.3.6-(4). Hence $E = g(X_u, X_v) > 0$ and from the factorization

$$I_p(v,v) = E\left(u'(0) + \frac{F}{E}v'(0)\right)^2 + \frac{EG - F^2}{E}(v'(0))^2,$$

it is easily shown that $I_p(v,v) > 0$ if v is nonzero.

(b) If I_p takes on both positive and negative values on T_pM, then from pat (a) we have $EG - F^2 \leq 0$. Suppose first that $EG - F^2 = 0$. If $E = 0$, then $F = 0$ also and hence $I_p(v,v) = G(v'(0))^2$ can not take on both positive and negative values. Hence $E \neq 0$ and we have the reduced factorization

$$I_p(v,v) = E\left(u'(0) + \frac{F}{E}v'(0)\right)^2$$

so again I_p can not take on both positive and negative values, in contradiction. Hence, $EG - F^2 < 0$.

Conversely, if $EG - F^2 < 0$, then $EG < 0$. Hence both E and G are nonzero and have opposite signs. Thus $I_p(X_u, X_u) = E$ and $I_p(X_v, X_v) = G$ have opposite signs as required and I_p is indefinite.

(c) First suppose I_p is isotropic and take any lightlike vector $n \in T_pM$ with $Rad_p(M) = \text{Span}\{n\}$. Since n is lightlike, $I_p(n,n) = 0$. Now for any $w \neq cn$ in T_pM, we have $g(n,w) = 0$ as $n \in Rad_p(M)$. But then as n is lightlike, and w is not collinear to n, we must have w spacelike or zero, by Lemma 3.3.8-(b). Thus $I_p(w,w) \geq 0$ and $I_p(n,n) = 0$. Since $I_p(n,n) = 0$, I_p is not positive definite on T_pM so $EG - F^2 > 0$ is impossible from part (a). Since I_p does not take on both positive and negative values, from part (b) also $EG - F^2 < 0$ is impossible. Hence $EG - F^2 = 0$.

Finally, suppose $EG - F^2 = 0$ and hence $X_u \times X_v$ is lightlike. Since X_u and X_v are both nonzero tangent vectors, both X_u and X_v are either lightlike or spacelike in view of Lemma 3.3.8. Hence, $E, G \geq 0$. First suppose $E = 0$. Then

$$I_p(v,v) = G(v'(0))^2$$

and X_u is null. By polarization, if $v \in T_pM$ is represented by the curve $c_1(t) = X(u_1(t), v_1(t))$ and $w \in T_pM$ is represented by the curve $c_2(t) = X(u_2(t), v_2(t))$, then

$$I_p(v,w) = Gv_1'(0)v_2'(0).$$

Hence if $n = X_u|_p$ and $u_1(t) = t$, $v_1(t) = 0$, we have

$$I_p(n,w) = G \cdot 0 \cdot v_2'(0) = 0$$

for any $w \in T_pM$ and $n = X_u|_p$ spans $Rad_p(M)$. Thus if $E(p) = 0$ and $EG - F^2 = 0$, I_p is isotropic at p.

Now suppose $EG - F^2 = 0$ and $E > 0$. Then

$$\begin{aligned} I_p(v, v) &= E\left(u'(0) + \frac{F}{E}v'(0)\right)^2 \\ &= E(u'(0))^2 + 2Fu'(0)v'(0) + \frac{F^2}{E}(v'(0))^2. \end{aligned}$$

Taking $n = -FX_u|_p + EX_v|_p$, i.e., $u'(0) = -F$ and $v'(0) = E$, we obtain $I_p(n, n) = 0$ and n is a lightlike vector. We need to show that n spans $Rad_p(M)$. By polarizing the above identity, we obtain for $v, w \in T_pM$ that

$$I_p(v, w) = Eu_1'(0)u_2'(0) + F\left(u_1'(0)v_2'(0) + u_2'(0)v_1'(0)\right) + \frac{F^2}{E}v_1'(0)v_2'(0).$$

Hence, taking $u_1'(0) = -F$ and $v_1'(0) = E$, we have

$$\begin{aligned} I_p(n, w) &= -EFu_2'(0) + F(-Fv_2'(0) + Eu_2'(0)) + \frac{F^2}{E} \cdot Ev_2'(0) \\ &= -EFu_2'(0) - F^2v_2'(0) + FEu_2'(0) + F^2v_2'(0) \\ &= 0. \end{aligned}$$

Thus $I_p(n, w) = 0$ for any $w \in T_pM$ and M is isotropic at p. □

Definition 3.3.12 In the vector calculus, the **surface area** of a bounded region R contained in a Euclidean coordinate patch $X(D)$ is given by

$$\text{Area}(R) = \iint_R dA = \iint_{X^{-1}(R)} |X_u \times X_v|\, dudv.$$

Hence, by Lemma 3.3.5,

$$\text{Area}(R) = \iint_{X^{-1}(R)} \sqrt{EG - F^2}\, dudv.$$

If R is a bounded region in a timelike coordinate patch $X(D)$, then, in view of Proposition 1.4.3 and Remark 3.3.6(1), we also define the surface area of R by

$$\text{Area}(R) = \iint_R dA = \iint_{X^{-1}(R)} |X_u \times_g X_v|\, dudv.$$

If R is a bounded region in a spacelike coordinate patch $X(D)$, then, we similarly define the area of R by

$$\text{Area}(R) = \iint_R dA = \iint_{X^{-1}(R)} |X_u \times_g X_v|\, dudv.$$

Proposition 3.3.13 *The surface area of a bounded region is independent of choice of coordinate patch. That is, for another coordinate path $\bar{X}(\bar{D})$ such that $R \subset X(D) \cap \bar{X}(\bar{D})$*

$$\iint\limits_{\bar{X}^{-1}(R)} |\bar{X}_{\bar{u}} \times \bar{X}_{\bar{v}}|\, d\bar{u}d\bar{v} = \iint\limits_{X^{-1}(R)} |X_u \times X_v|\, dudv.$$

Proof. From Theorem 3.1.9, $h = X^{-1} \circ \bar{X} : \bar{X}^{-1}(R) \to X^{-1}(R)$ is a diffeomorphism and from the vector calculus, the area element $dudv$ of D is transformed to the area element $|\det(J(h)| d\bar{u}d\bar{v}$ of $\bar{D}$. In the proof of Proposition 3.1.11, we have

$$\bar{X}_{\bar{u}} \times \bar{X}_{\bar{v}} = \det(J(h)) X_u \times X_v.$$

Hence,

$$\begin{aligned} \iint\limits_{\bar{X}^{-1}(R)} |\bar{X}_{\bar{u}} \times \bar{X}_{\bar{v}}|\, d\bar{u}d\bar{v} &= \iint\limits_{\bar{X}^{-1}(R)} |X_u \times X_v| |\det(J(h)| d\bar{u}d\bar{v} \\ &= \iint\limits_{X^{-1}(R)} |X_u \times X_v|\, dudv. \end{aligned}$$

□

Example 3.3.14 Compute the first fundamental form I and the area of the torus T in Example 3.2.3.

Solution. Since $X_u = (-r\sin u \cos v, -r \sin u \sin v, r\cos u)$ and $X_v = (-(R + r\cos u)\sin v, (R+r\cos u)\cos v, 0)$, we have $E = r^2$, $F = 0$ and $G = (R+r\cos u)^2$. Hence

$$I = E(du)^2 + 2Fdudv + G(dv)^2 = r^2(du)^2 + (R + r\cos u)^2(dv)^2.$$

Further, we have $EG - F^2 = r^2(R + r\cos u)^2$. Therefore,

$$\begin{aligned} \text{Area}(T) &= \iint\limits_{X^{-1}(T)} \sqrt{EG - F^2}\, dudv \\ &= \iint\limits_{X^{-1}(T)} r(R + r\cos u)\, dudv \\ &= \int_0^{2\pi} \left(\int_0^{2\pi} r(R + r\cos u) du \right) dv \\ &= 2\pi r R \int_0^{2\pi} dv \\ &= 4\pi^2 r R. \end{aligned}$$

□

For further study, see the following references.

1. Kühnel, Wolfgang, Differential Geometry, Curves-Surfaces-Manifolds, AMS, Student Mathematical Library Volume 16, 2002.

2. McCleary, J., Geometry from a Differentiable Viewpoint, Cambridge University Press, 1994.

3. Pressley, Andrew, Elementary Differential Geometry, Springer Undergraduate Mathematics Series, 2002.

3.4 The Gauss Map and the Second Fundamental Form

The curvature of a regular curve is measured by changes in the normal vector. To measure the curvature of a patch for a surface, Carl Friedrich Gauss(1777-1855) measured the changes of a normal vector field to the patch. A plane in $\mathbf{E}^3$ or $\mathbf{L}^3$ is determined by a point and a normal vector to the given plane. Since a differentiable surface is well approximated locally by its tangent plane, this normal vector field (now often called the Gauss map) is fundamental in studying the geometry of surfaces.

Gauss map

Let M be an orientable surface and let S^2 be the unit sphere in $\mathbf{E}^3$ centered at the origin. Then the **Gauss map** is a map $U : M \to S^2$ given by associating to each point p in M the unit normal vector $U(p)$ at p (translated to the origin in $\mathbb{R}^3$). From now on, we write $U(p) = U_p$(Figure3.4-41).

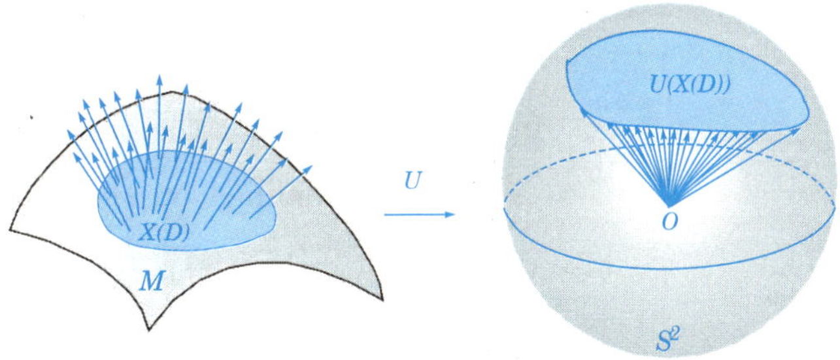

Figure3.4-41

For a coordinate patch $X(u, v)$ containing p in the surface M, we have the unit normal vector field

$$U = \frac{X_u \times X_v}{|X_u \times X_v|}.$$

Since M is an orientable surface, the Gauss map does not depend on the choice of coordinate patches by Proposition 3.1.11.

Moreover, the Gauss map U is differentiable by the regularity of M. Hence, it has a differential map $dU_p : T_pM \to T_{U_p}S^2$. Since T_pM is a tangent plane at p normal to U_p and since $T_{U_p}S^2$ is the tangent plane at U_p normal to U_p, T_pM and $T_{U_p}S^2$ are parallel, so again by translation, we may take $T_pM = T_{U_p}S^2$(Figure 3.4-42). Thus, dU_p is a linear map $dU_p : T_pM \to T_pM$ defined as follows. For $w \in T_pM$ there is a curve $\alpha : (-\epsilon, \epsilon) \to M$ with $\alpha(0) = p$ and $\alpha'(0) = w$. Then we put $dU_p(w) := (U \circ \alpha)'(0)$. If $U = (u_1, u_2, u_3)$, $U \circ \alpha(t) = (u_1 \circ \alpha(t), u_2 \circ \alpha(t), u_3 \circ \alpha(t))$.

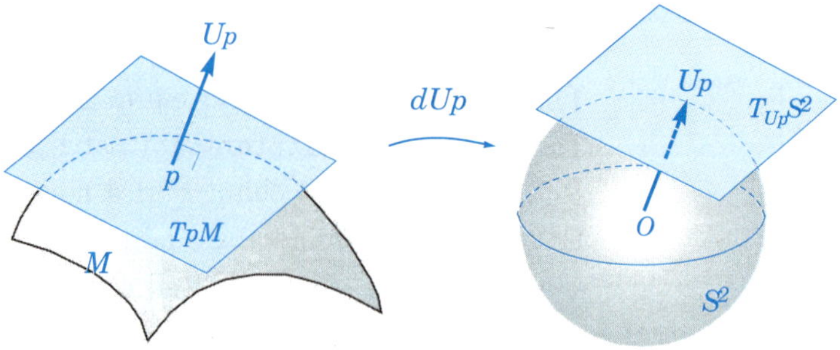

Figure3.4-42

Hence,

$$\begin{aligned}dU_p(w) &= (U \circ \alpha)'(0)\\ &= \big((u_1 \circ \alpha)'(0), (u_2 \circ \alpha)'(0), (u_3 \circ \alpha)'(0)\big)\\ &= \big(\alpha'(0)[u_1], \alpha'(0)[u_2], (\alpha'(0)[u_3]\big)\\ &= \sum_{i=1}^{3} w[u_i]\mathbf{e}_i(p).\end{aligned}$$

In $\mathbf{L}^3 = (\mathbf{R}^3, g)$, for the timelike(resp. spacelike) unit normal vector field $U = \dfrac{X_u \times_g X_v}{|X_u \times_g X_v|}$ defined on the spacelike(resp. timelike) surface M, we may also define the timelike(resp. spacelike) Gauss map $U : M \to H^2$ where $H^2 = \{x \in \mathbf{L}^3 \mid |x|_g = 1\}$. Further, we have the differential map $dU_p : T_pM \to T_pM$ as in $\mathbf{E}^3$.

Surface patches as in Example 3.4.11 may be considered for which $X_u \times_g X_v$ may be timelike, spacelike, or lightlike at different points of M. In the lightlike case, as $|X_u \times_g X_v| = 0$, one can not define $\dfrac{X_u \times_g X_v}{|X_u \times_g X_v|}$ as in the timelike or spacelike cases.

Hence in the rest of this section, we will restrict attention to open subsets of a chart domain $X : D \to M$ in $\mathbf{L}^3$ for which $X_u \times_g X_v$ is not lightlike.

Lemma 3.4.1 *$dU_p : T_pM \to T_pM$ is linear.*

Proof.

$$\begin{aligned}dU_p(v + w) &= \sum_{i=1}^{3}(v + w)[u_i]\mathbf{e}_i(p)\\ &= \sum_{i=1}^{3} v[u_i]e_i(p) + \sum_{i=1}^{3} w[u_i]\mathbf{e}_i(p)\\ &= dU_p(v) + dU_p(w)\end{aligned}$$

$$\begin{aligned} dU_p(kv) &= \sum_{i=1}^{3} (kv)[u_i]\mathbf{e}_i(p) \\ &= k\sum_{i=1}^{3} v[u_i]\mathbf{e}_i(p) \\ &= kdU_p(v) \end{aligned}$$

□

Example 3.4.2 Let $M = \{(x, y, z) \in \mathbf{L}^3 \mid ax + by + cz + d = 0,\ a^2 + b^2 < c^2\}$ be a spacelike plane in $\mathbf{L}^3$. Then find the timelike unit normal vector field and compute the differential of the timelike Gauss map on M.

Solution. Let $h(x, y, z) = ax + by + cz + d$. Then $\nabla h = (a, b, -c)$ recalling the discussion of the Lorentz gradient prior to Theorem 3.1.17 and $g(\nabla h, \nabla h) = a^2 + b^2 - c^2 < 0$. Since $|\nabla h|_g = \sqrt{-a^2 - b^2 + c^2}$, we have $U = \dfrac{(a, b, c)}{\sqrt{-a^2 - b^2 + c^2}} =$ constant. Therefore, $dU = 0$. □

Example 3.4.3 Let $M = S^2 = \{(x, y, z) \in \mathbf{E}^3 \mid x^2 + y^2 + z^2 = 1\}$ be the unit sphere in $\mathbf{E}^3$. Then compute dU on M.

Solution. For a differentiable curve $c(t) = (x(t), y(t), z(t))$ satisfying $(x(t))^2 + (y(t))^2 + (z(t))^2 = 1$, $2x(t)x'(t) + 2y(t)y'(t) + 2z(t)z'(t) = 0$, or $< c(t), c'(t) >= 0$. Thus, $c(t)$ is normal to $c'(t)$. Further considering all tangent vectors to S^2 at $c(t)$ for t fixed, one may deduce that the unit normal vector $U(t) = U(c(t)) = c(t)$ defined along $c(t)$. It follows that

$$dU_{c(t)}(c'(t)) = c'(t).$$

Therefore, for any $p \in M$ and $w \in T_pM$, $dU_p(w) = w$. That is, $dU = Id$. □

Example 3.4.4 Let $M = \{(x, y, z) \in \mathbf{L}^3 \mid x^2 + y^2 = 1\}$ be a timelike cylinder in $\mathbf{L}^3$. Then compute dU on M.

Solution. For each point $p \in M$, we may find a proper patch $X(u, v) = (\cos u, \sin u, v)$ about p. Hence,

$X_u = (-\sin u, \cos u, 0) = (-y, x, 0)$, $X_v = (0, 0, 1)$ and the coefficients of the first fundamental form for X are

$E = g(X_u, X_u) = x^2 + y^2 = 1 > 0$

$F = (X_u, X_v) = 0,$

$G = (X_v, X_v) = -1 < 0.$

Hence $EG - F^2 = -1 = -g(X_u \times_g X_v, X_u \times_g X_v)$. Since $U = X_u \times_g X_v = (x, y, 0)$,

$$dU(X_u) = (\frac{\partial}{\partial u}[\cos u], \frac{\partial}{\partial u}[\sin u], \frac{\partial}{\partial u}[0]) = (-\sin u, \cos u, 0) = X_u$$

and

$$dU(X_v) = (\frac{\partial}{\partial v}[\cos u], \frac{\partial}{\partial v}[\sin u], \frac{\partial}{\partial v}[0]) = (0, 0, 0). \qquad \square$$

Self-adjoint maps

For the rest of this section, we are assuming for a surface patch $X(u, v)$ in $\mathbf{L}^3$ that the normal $X_u \times_g X_v$ is timelike or spacelike.

Definition 3.4.5 Let V be a vector space with inner product $< , >$. A linear map $f : V \to V$ is said to be **self-adjoint** if $< f(v), w > = < v, f(w) >$ for any $v, w \in V$.

Theorem 3.4.6 *$dU_p : T_pM \to T_pM$ is a self-adjoint map for a surface in $\mathbf{E}^3$ or $\mathbf{L}^3$.*

Proof. For a proper patch $X : D \to M$ about p, we know that $\{X_u, X_v\}$ is a basis for T_pM. Let $< , >$ denote either the Euclidean or Lorentz inner product. Since $< U, X_u > = 0$, $< U, X_v > = 0$, we have on calculating $\frac{\partial}{\partial u}$ and $\frac{\partial}{\partial v}$ of the first and second quantities respectively that

$$< U_v, X_u > + < U, X_{uv} > = 0,$$

$$< U_u, X_v > + < U, X_{vu} > = 0.$$

Hence, $< U_v, X_u > = < U_u, X_v >$ because $X_{uv} = X_{vu}$ as the surface patch is differentiable. Here, $U_u = \frac{\partial}{\partial u} U(X(u, v)) = dU(X_u)$, $U_v = \frac{\partial}{\partial v} U(X(u, v)) = dU(X_v)$.

Thus,

$$< dU(X_v), X_u > = < dU(X_u), X_v > = < X_v, dU(X_u) > .$$

Now, since $v = aX_u + bX_v$, $w = cX_u + dX_v$ for any $v, w \in T_pM$,

$$
\begin{aligned}
< dU(v), w >= &< dU(aX_u + bX_v), cX_u + dX_v > \\
=& < adU(X_u) + bdU(X_v), cX_u + dX_v > \\
=& ac < dU(X_u), X_u > + ad < dU(X_u), X_v > \\
&+ bc < dU(X_v), X_u > + bd < dU(X_v), X_v > \\
=& ac < X_u, dU(X_u) > + ad < X_u, dU(X_v) > \\
&+ bc < X_v, dU(X_u) > + bd < X_v, dU(X_v) > \\
=& < aX_u + bX_v, cdU(X_u) + ddU(X_v) > \\
=& < aX_u + bX_v, dU(cX_u + dX_v) > \\
=& < v, dU(w) > .
\end{aligned}
$$

□

Definition 3.4.7 At a point $p \in M$ the **second fundamental form**

$$II_p : T_pM \to \mathbb{R}$$

is defined by $II_p(w) = - < dU(w), w >$ for $w \in T_pM$ for a surface M in $\mathbf{E}^3$ or $\mathbf{L}^3$.

Given $w \in T_pM$, on a coordinate patch $X(u, v)$ containing a neighborhood of p, there is a curve $\alpha : (-\epsilon, \epsilon) \to M$ such that $\alpha(t) = X(u(t), v(t))$, $\alpha(0) = p$ and $\alpha'(0) = w$. Then, differentiating $< (U \circ \alpha)(t), \alpha'(t) >= 0$ we obtain

$$
\begin{aligned}
II_p(w) &= - < dU(w), w > \\
&= - < (U \circ \alpha)'(0), \alpha'(0) >=< (U \circ \alpha)(0), \alpha''(0) > .
\end{aligned}
$$

Since $\alpha''(t) = \dfrac{d^2X}{dt^2} = X_{uu}(u'(t))^2 + 2X_{uv}u'(t)v'(t) + X_{vv}(v'(t))^2 + X_u u''(t) + X_v v''(t)$, the second fundamental form $II_{\alpha(t)}$ along $\alpha(t)$ is

$$\left\langle U \circ \alpha(t), X_{uu}\left(\frac{du}{dt}\right)^2 + 2X_{uv}\frac{du}{dt}\frac{dv}{dt} + X_{vv}\left(\frac{dv}{dt}\right)^2 \right\rangle$$

as $< U, X_u >=< U, X_v >= 0$. Therefore, the second fundamental form II on $X(u, v)$ which is independent of any curve is given by

$$
\begin{aligned}
II &=< X_{uu}(du)^2 + 2X_{uv}dudv + X_{vv}(dv)^2, U > \\
&= Ldu^2 + 2Mdudv + Ndv^2
\end{aligned}
$$

where $L =< X_{uu}, U >$, $M =< X_{uv}, U >$, $N =< X_{vv}, U >$ are called the **coefficient of the 2nd fundamental form**. Gauss himself introduced the notations L, M, N for these quantities.

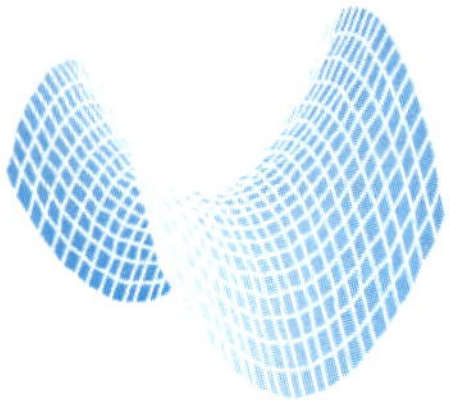

Figure3.4-43

Example 3.4.8 Find the second fundamental form II for a coordinate patch $X(u,v) = (u, v, u^2 - v^2)$ for the hyperbolic paraboloid in $\mathbf{E}^3$ defined on $D = \{(u,v) \in E^2 \mid -2\pi < u < 2\pi,\ -2\pi < u < 2\pi\}$(Figure3.4-43).

Solution. Since $X_u = (1, 0, 2u)$, $X_v = (0, 1, -2v)$ and $X_u \times X_v = (-2u, 2v, 1)$, we have

$$U = \frac{(-2u, 2v, 1)}{\sqrt{4u^2 + 4v^2 + 1}}.$$

Since $X_{uu} = (0, 0, 2)$, $X_{uv} = (0, 0, 0)$ and $X_{vv} = (0, 0, -2)$, we have

$$L =< X_{uu}, U >= \frac{2}{\sqrt{4u^2 + 4v^2 + 1}},$$
$$M =< X_{uv}, U >= 0,$$
$$N =< X_{vv}, U >= \frac{-2}{\sqrt{4u^2 + 4v^2 + 1}}.$$

Thus, $II = \dfrac{2(du)^2 - 2(dv)^2}{\sqrt{4u^2 + 4v^2 + 1}}$. □

Figure3.4-44

Example 3.4.9 Find the second fundamental form for a coordinate patch $X(u,v)$ $= (v\cos u, v\sin u, au)$ for the helicoid in $\mathbf{L}^3$ defined on $D = \{(u,v) \in E^2 \mid 0 < u < 2\pi,\ -\infty < v < \infty,\ v^2 > a^2\}$(Figure3.4-44).

Solution. Since $X_u = (-v\sin u, v\cos u, a)$ and $X_v = (\cos u, \sin u, 0)$, we have $X_u \times_g X_v = (-a\sin u, a\cos u, v)$. Since $g(X_u \times_g X_v, X_u \times_g X_v) = a^2 - v^2 < 0$ and $|X_u \times_g X_v|_g = \sqrt{v^2 - a^2}$, we obtain the timelike unit normal vector field

$$U = \frac{1}{\sqrt{v^2 - a^2}}(-a\sin u, a\cos u, v).$$

Hence, since

$$X_{uu} = (-v\cos u, -v\sin u, 0),$$
$$X_{uv} = (-\sin u, \cos u, 0),$$
$$X_{vv} = (0, 0, 0),$$

we have

$$L = g(U, X_{uu}) = \frac{1}{\sqrt{v^2 - a^2}}(av\sin u\cos u - av\sin u\cos u) = 0,$$
$$M =< U, X_{uv} >= \frac{1}{\sqrt{v^2 - a^2}}(a\sin^2 u + a\cos^2 u) = \frac{a}{\sqrt{v^2 - a^2}},$$
$$N =< U, X_{vv} >= 0.$$

Thus, $II = \dfrac{2adudv}{\sqrt{v^2 - a^2}}$. □

Representation of the second fundamental form

Suppose $X(u + \Delta u, v + \Delta v)$ is a neighboring point of $X(u, v)$ on the same coordinate patch X in M. As $(u, v) \to (u + \Delta u, v + \Delta v)$, the distance from $X(u, v)$ to the point which is the projection of $X(u+\Delta u, v+\Delta v)$ in the direction of U(Figure3.4-45) is given by

$$< (X(u + \Delta u, v + \Delta v) - X(u, v), U > .$$

By using the Taylor's expansion,

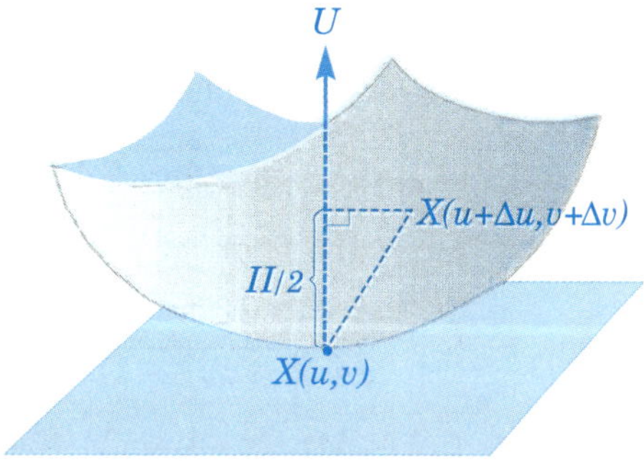

Figure3.4-45

$$\begin{aligned}
&X(u + \Delta u, v + \Delta v) - X(u, v)\\
=&X_u \Delta u + X_v \Delta v + \frac{1}{2}\left[X_{uu}(\Delta u)^2 + 2X_{uv}\Delta u\Delta u + X_{vv}(\Delta v)^2\right]\\
&+ o(\Delta u, \Delta v).
\end{aligned}$$

Since X_u, X_v are perpendicular to U, as $(\Delta u, \Delta v) \to (0, 0)$, $< X(u+\Delta u, v+ \Delta v) - X(u, v), U >$ tends to

$$\frac{1}{2}\left[< X_{uu}du^2 + 2X_{uv}dudv + X_v dv^2, U >\right] = \frac{II}{2}.$$

Therefore, we may classify the points of the given patch $X(u, v)$ as follows.

(1) If $(LN - M^2)(p) > 0$, p is called an **elliptic point**(Figure3.4-46).

(2) If $(LN - M^2)(p) < 0$, p is called a **hyperbolic point**(Figure3.4-47).

(3) If $(LN - M^2)(p) = 0$ and $L^2 + M^2 + N^2 \neq 0$, p is called a **parabolic point**(Figure3.4-48).

(4) If $(LN - M^2)(p) = 0$ and $L = M = N = 0$, p is called a **planar point**(Figure4.2-9).

Example 3.4.10 Show that every point of the paraboloid $X(u, v) = (u, v, u^2 + v^2)$ in $\mathbf{E}^3$ is elliptic.

Figure3.4-46

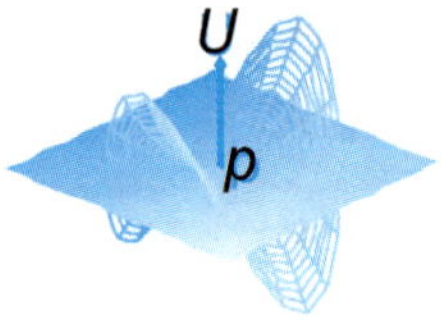

Figure3.4-47

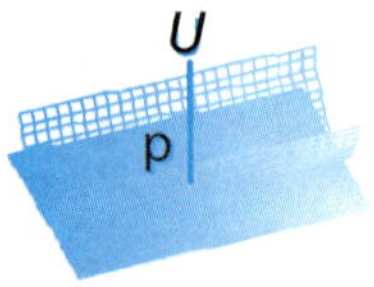

Figure3.4-48

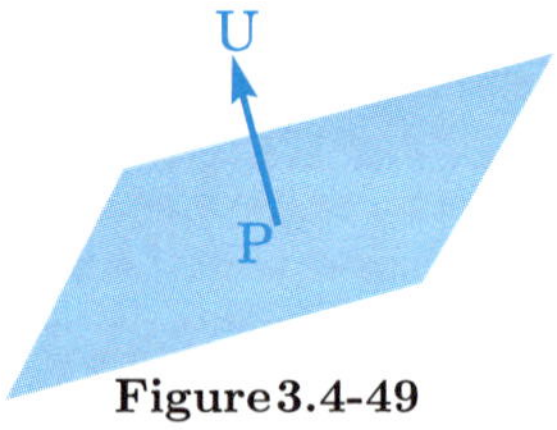

Figure3.4-49

Solution. $X_u = (1, 0, 2u)$, $X_v = (0, 1, 2v)$, $X_u \times X_v = (-2u, -2v, 1)$.

$|X_u \times X_v| = \sqrt{4u^2 + 4v^2 + 1}$, $U = \dfrac{(-2u, -2v, 1)}{\sqrt{4u^2 + 4v^2 + 1}}$.

$X_{uu} = (0, 0, 2), X_{uv} = (0, 0, 0), X_{vv} = (0, 0, 2)$.

$L = \dfrac{2}{\sqrt{4u^2 + 4v^2 + 1}}$, $M = 0$, $N = \dfrac{2}{\sqrt{4u^2 + 4v^2 + 1}}$.

$LN - M^2 = \dfrac{4}{4u^2 + 4v^2 + 1} > 0$. □

Example 3.4.11 Show that the points of the paraboloid $X(u, v) = (u, v, u^2 + v^2)$ in $\mathbf{L}^3$ are elliptic for $u^2 + v^2 \neq \frac{1}{4}$.

Solution. $X_u = (1, 0, 2u)$, $X_v = (0, 1, 2v)$, $X_u \times_g X_v = (-2u, -2v, -1)$.

For $u^2 + v^2 < \dfrac{1}{4}$, it is a spacelike surface since $g(X_u \times_g X_v, X_u \times_g X_v) = 4u^2 + 4v^2 - 1 < 0$.

$|X_u \times_g X_v|_g = \sqrt{1 - 4u^2 - 4v^2}$, $U = \dfrac{(-2u, -2v, -1)}{\sqrt{1 - 4u^2 - 4v^2}}$.

$X_{uu} = (0, 0, 2)$, $X_{uv} = (0, 0, 0)$, $X_{vv} = (0, 0, 2)$.

$L = \dfrac{2}{\sqrt{1 - 4u^2 - 4v^2}}$, $M = 0$, $N = \dfrac{2}{\sqrt{1 - 4u^2 - 4v^2}}$.

$LN - M^2 = \dfrac{4}{1 - 4u^2 - 4v^2} > 0$.

On the other hand, for $u^2 + v^2 > \dfrac{1}{4}$, it is a timelike surface since $g(X_u \times_g X_v, X_u \times_g X_v) = 4u^2 + 4v^2 - 1 > 0$.

$|X_u \times_g X_v|_g = \sqrt{4u^2 + 4v^2 - 1}$, $U = \dfrac{(-2u, -2v, -1)}{\sqrt{4u^2 + 4v^2 - 1}}$.

$X_{uu} = (0, 0, 2)$, $X_{uv} = (0, 0, 0)$, $X_{vv} = (0, 0, 2)$.

$L = \dfrac{2}{\sqrt{4u^2 + 4v^2 - 1}}$, $M = 0$, $N = \dfrac{2}{\sqrt{4u^2 + 4v^2 - 1}}$.

$LN - M^2 = \dfrac{4}{4u^2 + 4v^2 - 1} > 0$. □

Example 3.4.12 Classify the elliptic, hyperbolic and parabolic point on the torus given by

$$X(u,v) = ((R + r\cos u)\cos v, (R + r\cos u)\sin v, r\sin u)$$

for $0 < r < R$ in $\mathbf{E}^3$.

Solution. $X_u = (-r\sin u\cos v, -r\sin u\sin v, r\cos u)$,

$X_v = (-(R + r\cos u)\sin v, (R + r\cos u)\cos v, 0)$,

$$\begin{aligned} X_u \times X_v &= (-r(R + r\cos u)\cos u\cos v, \\ &\quad - r(R + r\cos u)\cos u\sin v, -r(R + r\cos u)\sin u) \\ &= r(R + r\cos u)(-\cos u\cos v, -\cos u\sin v, -\sin u). \end{aligned}$$

$|X_u \times X_v| = r(R + r\cos u)$, $U = (-\cos u\cos v, -\cos u\sin v, -\sin u)$.

Now, $X_{uu} = (-r\cos u\cos v, -r\cos u\sin v, -r\sin u)$,

$X_{uv} = (r\sin u\sin v, -r\sin u\cos v, 0)$,

$X_{vv} = (-(R + r\cos u)\cos v, -(R + r\cos u)\sin v, 0)$.

Hence, $L = r$, $M = 0$, $N = (R + r\cos u)\cos u$.

Thus, $LN - M^2 = r(R + r\cos u)\cos u$.

Since $r(R+r\cos u) > 0$, the sign of $LN - M^2$ is equal to the sign of $\cos u$(Figure 3.4-50). Therefore,

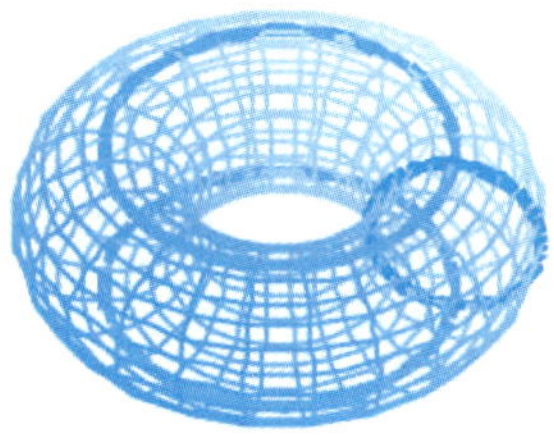

Figure3.4-50

$$LN - M^2 > 0 \Leftrightarrow 0 < u < \frac{\pi}{2}, \text{ or } \frac{3\pi}{2} < u < 2\pi.$$

$$LN - M^2 < 0 \Leftrightarrow \frac{\pi}{2} < u < \frac{3\pi}{2}.$$

$$LN - M^2 = 0 \Leftrightarrow u = \frac{\pi}{2}, \text{ or } u = \frac{3\pi}{2}.$$ □

For further study, see the following references.

1. McCleary, J., Geometry from a Differentiable Viewpoint, Cambridge University Press, 1994.

2. Millman, R. S. and G. D. Parker,Elements of Differential Geometry, Prentice-Hall, Englewood Cliffs, NJ, 1977.

3. O'Neill, B., Elementary Differential Geometry, Academic Press, New York, 1966.

4. Pressley, Andrew, Elementary Differential Geometry, Springer Undergraduate Mathematics Series, 2002.

Chapter 4
Geometry of Surfaces

4.1 The Normal Curvature and Shape Operators

Normal Vectors N and U

Let M be a surface in $\mathbf{E}^3$ or a surface in $\mathbf{L}^3$ with everywhere non-lightlike normal vector field and γ be a unit speed curve with the Frenet-Serret apparatus $\{\kappa, \tau, \mathbf{T}, \mathbf{N}, \mathbf{B}\}$ in M. Sometimes students confuse the unit normal vector field $\mathbf{N}$ along γ with the unit normal vector field U to the surface M. We begin with the following example to clarify the two normal vector fields(Figure4.1-1).

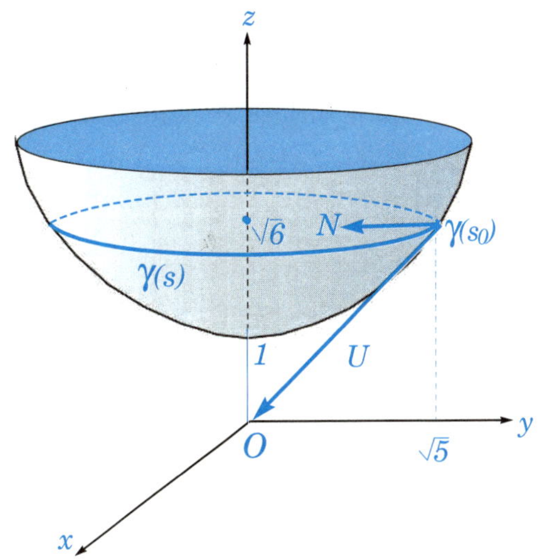

Figure4.1-1

Example 4.1.1 Let M be a upper sheet of hyperboloid $x^2+y^2-z^2=-1$ in $\mathbf{L}^3$ and $\gamma(s)=\left(\sqrt{5}\cos\frac{s}{\sqrt{5}}, \sqrt{5}\sin\frac{s}{\sqrt{5}}, \sqrt{6}\right)$ be a curve on M. Then we have

$$\gamma'(s)=T(s)=\left(-\sin\frac{s}{\sqrt{5}}, \cos\frac{s}{\sqrt{5}}, 0\right),$$

$$T'(s)=\left(-\frac{1}{\sqrt{5}}\cos\frac{s}{\sqrt{5}}, -\frac{1}{\sqrt{5}}\sin\frac{s}{\sqrt{5}}, 0\right),$$

and

$$N(s) = \left(-\cos\frac{s}{\sqrt{5}}, -\sin\frac{s}{\sqrt{5}}, 0\right).$$

For $s_0 = \dfrac{\sqrt{5}\pi}{2}$, $\gamma(s_0) = \left(0, \sqrt{5}, \sqrt{6}\right)$ and $N(s_0) = (0, -1, 0)$.

Now, for a coordinate patch $X(u,v) = (\sinh u \cos v, \sinh u \sin v, \cosh u)$, we have (for $u \neq 0$)

$$X_u = (\cosh u \cos v, \cosh u \sin v, \sinh u),$$
$$X_v = (-\sinh u \sin v, \sinh u \cos v, 0),$$
$$X_u \times_g X_v = (-\sinh^2 u \cos v, -\sinh^2 u \sin v, -\cosh u \sinh u),$$
$$|X_u \times_g X_v| = |\sinh u|,$$

and

$$U = (-\sinh u \cos v, -\sinh u \sin v, -\cosh u).$$

Thus, $U = -X(u,v)$, which implies that $U(\gamma(s)) = -\gamma(s)$ and

$$U\left(\gamma(s_0)\right) = \left(0, -\sqrt{5}, -\sqrt{6}\right).$$

Normal Curvatures

Definition 4.1.2 Let $X(u,v)$ be a proper patch containing a neighborhood of p on a surface M and U a unit normal vector field to M in $\mathbf{E}^3$ (or $\mathbf{L}^3$). For a regular curve $\alpha(t) = X(u(t), v(t))$, with everywhere nonlightlike tangent vector in the case of $\mathbf{L}^3$, the **normal curvature** k_n of $\alpha(t)$ is defined by reparametrizing $\alpha(t)$ by arc length as $\beta(s) = \alpha(t(s))$ and setting

$$k_n\left(\beta'(s)\right) =< T'(s), U\left(\beta(s)\right) >=< \beta''(s), U\left(\beta(s)\right) > .$$

Since $\beta''(s) = \dfrac{d\beta}{ds} = \dfrac{d}{ds}(T(t)) = \dfrac{T'(t)}{|\alpha'(t)|}$, we obtain

$$k_n(\beta'(t)) = \frac{1}{|\alpha'(t)|} < T'(t), U(\alpha(t)) >$$

for which we conclude that

$$k_n(\alpha'(t)) = \frac{1}{|\alpha'(t)|} < T'(t), U(\alpha(t)) > .$$

Accordingly, for a (nonlightlike) tangent vector $w \in T_pM$, we may define

$$k_n(w) = k_n(\alpha'(0)) = \frac{1}{|w|} < T'(0), U(p) >$$

where where $\alpha(t)$ is any regular curve in M with $\alpha(0) = p$ and $\alpha'(0) = w$.

Lemma 4.1.3 *For a nonlightlike $w \in T_pM$,*

$$k_n(w) = \frac{II(w)}{|w|^2}$$

so that if w is Euclidean or spacelike, then

$$k_n(w) = \frac{II(w)}{I(w)}$$

and if w is timelike, then

$$k_n(w) = -\frac{II(w)}{I(w)}$$

where I and II denote the first and second fundamental forms respectively.

Proof. Differentiating $< T(t), U(\alpha(t)) >= 0$, we have

$$\begin{aligned} k_n(w) = k_n(\alpha'(0)) &= \frac{1}{|\alpha'(0)|} < T'(0), U(\alpha(0)) > \\ &= -\frac{1}{|\alpha'(0)|} < T(0), (U \circ \alpha)'(0) > \\ &= -\frac{1}{|\alpha'(0)|^2} < \alpha'(0), dU(\alpha'(0)) > \\ &= \frac{II(w)}{|w|^2}. \end{aligned}$$

□

Remark 4.1.4 Because the terms in the right hand side of the formula for $k_n(w)$ in Lemma 4.1.3, depend only on w and the coefficients of I and II at p, the normal curvature $k_n(w)$ is independent of the choice of curve $\alpha(t)$ in M with $\alpha'(0) = w$. The result also implies that $k_n(w) = k_n(-w)$ for any $w \in T_pM$.

Example 4.1.5 Let $X(u, v)$ be a coordinate patch for the sphere S^2 of radius r in $\mathbf{E}^3$ given by $X(u, v) = (r \cos u \cos v, r \sin u \cos v, r \sin v)$, $-\pi \leq u \leq \pi$, $-\frac{\pi}{2} \leq v \leq \frac{\pi}{2}$. Then the normal curvature of S^2 is $-\frac{1}{r}$.

Solution. Since

$$X_u = (-r \sin u \cos v, r \cos u \cos v, 0)$$

and

$$X_v = (-r \cos u \sin v, -r \sin u \sin v, r \cos v),$$

we have

$$X_u \times X_v = \left(r^2 \cos u \cos^2 v, r^2 \sin u \cos^2 v, r^2 \sin v \cos v\right).$$

Hence

$$|X_u \times X_v| = r^2\sqrt{\cos^4 v + \cos^2 v \sin^2 v} = r^2 \cos v$$

and $U = (\cos u \cos v, \sin u \cos v, \sin v)$.

Now, since

$$X_{uu} = (-r\cos u \cos v, -r \sin u \cos v, 0),$$

$$X_{uv} = (r \sin u \sin v, -r\cos u \sin v, 0)$$

and

$$X_{vv} = (-r\cos u\cos v, -r\sin u \cos v, -r\sin v),$$

we also have

$$E = r^2\cos^2 v,\ \ F = 0,\ \ G = r^2,$$

and

$$L = -r\cos^2 v,\ \ M = 0,\ \ N = -r.$$

Thus,

$$k_n = \frac{-r\cos^2 v du^2 - r dv^2}{r^2\cos^2 v du^2 + r^2 dv^2} = -\frac{1}{r}.$$

□

Example 4.1.6 Let $X(u,v)$ be a coordinate patch for the hyperboloid H^2 in $\mathbf{L}^3$ given by $X(u,v) = (r\sinh u \cos v, r\sinh u \sin v, r\cosh u)$, $0 < u < \infty$, $0 < v < 2\pi$. Then H^2 is a spacelike hyperboloid on which the normal curvature is $\frac{1}{r}$. If $-\infty < u < 0$, $0 < v < 2\pi$, then H^2 is a timelike hyperboloid on which the normal curvature is $\frac{1}{r}$.

Solution. Since

$$X_u = (r\cosh u \cos v, r\cosh u \sin v, r\sinh u)$$

and

$$X_v = (-r\sinh u \sin v, r\sinh u \cos v, 0),$$

we have

$$X_u \times_g X_v = (-r^2\sinh^2 u \cos v, -r^2\sinh^2 u \sin v, -r^2\cosh u \sinh u),$$

and

$$\begin{aligned}|X_u \times_g X_v|^2 &= r^4\sinh^4 u - r^4\cosh^2 u \sinh^2 u\\ &= r^4\sinh^2 u(\sinh^2 u - \cosh^2 u)\\ &= -r^4\sinh^2 u < 0.\end{aligned}$$

Hence

$$|X_u \times_g X_v| = r^2 \sinh u,$$

and

$$U = (-\sinh u \cos v, -\sinh u \sin v, -\cosh u).$$

Now, since

$$X_{uu} = (r \sinh u \cos v, r \sinh u \sin v, r \cosh u),$$

$$X_{uv} = (-r \cosh u \sin v, r \cosh u \cos u, 0)$$

and since

$$X_{vv} = (-r \sinh u \cos v, -r \sinh u \sin v, 0),$$

we have

$$E = r^2, \ F = 0, \ G = r^2 \sinh^2 u$$

and

$$L = r, \ M = 0, \ N = r \sinh^2 u.$$

Thus,

$$k_n = \frac{r du^2 + r \sinh^2 u dv^2}{r^2 du^2 + r^2 \sinh^2 u dv^2} = \frac{1}{r}$$

for $u > 0$ as the surface is spacelike and all tangent vectors are spacelike. □

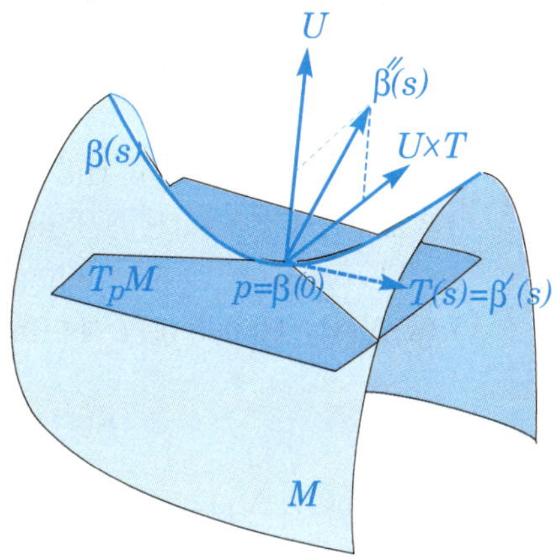

Figure4.1-2

Remark 4.1.7 First, let M be a Euclidean surface in $\mathbf{E}^3$. If $\beta(s)$ is a reparametrization by arc length of $\alpha(t)$, then $k_n =< \beta''(s), U >$. Since $\beta''(s)$ is perpendicular to T and since U and $U \times T$ are perpendicular to T, $\beta''(s)$ is a linear combination of U and $U \times T$ as follows:

$$\beta''(s) = k_n U + k_g U \times T$$

where k_n is the normal curvature of β and $k_g =< \beta''(s), U \times T >$ is called the **geodesic curvature** of β. Hence, the normal curvature k_n determines the normal component of $\beta''(s)$ to M(Figure 4.1-2). Now, assume further for $\beta(s)$ that the curvature $\kappa(s) > 0$, so that the Frenet-Serret apparatus my be constructed at all points of β. Since $\beta''(s) = \kappa N$, we have the following equation from calculating $< \beta''(s), \beta''(s) >= (\kappa(s))^2 < N, N >$:

$$\kappa^2 = k_n^2 + k_g^2.$$

Second, let M be a surface in $\mathbf{L}^3$. Consider first the case that M is a spacelike surface, hence U is unit timelike. Now any unit speed curve $\beta(s)$ on M has $\beta'(s)$, hence $T = \beta'(s)$, unit spacelike since $< U, T >= 0$, by Theorem 1.3.12-(1), also

$$< U \times_g T, U \times_g T >= - < U, U >< T, T >= 1$$

and $U \times_g T$ is unit spacelike. Again, as U, T and $U \times_g T$ are mutually orthogonal vectors and $\beta''(s) = T'(s)$ is orthogonal to T, $\beta''(s)$ is a linear combination of U and $U \times_g T$. As in the Euclidean case, $k_n =< \beta''(s), U >$ and we again define the geodesic curvature k_g of $\beta(s)$ as

$$k_g = k_g(\beta(s)) =< \beta''(s), U \times_g T > .$$

In this case, we then find that

$$\beta''(s) = -k_n U + k_g(U \times_g T).$$

Suppose further that the curve $\beta(s)$ has the property that its curvature $\kappa(s)$ in $\mathbf{L}^3$ is strictly positive. Then the Frenet-Serret apparatus may be constructed as in Theorem 2.1.9 and whether $N(s)$ is unit timelike or unit spacelike, $\beta''(s) = \kappa(s)N(s)$. Hence as U is unit timelike and $U \times_g T$ is unit spacelike, we obtain

$$(\kappa(s))^2 g(N(s), N(s)) = -(k_n)^2 + (k_g)^2$$

where $g(N, N) = -1$ or $+1$ depending on which case of Theorem 2.1.9 applies.

The case that M is a timelike surface in $\mathbf{L}^3$ (hence U is unit spacelike) and $\beta(s)$ is unit speed curve in M is more complicated, because $U \times_g T$ may be either timelike or spacelike. Indeed, Theorem 1.3.12-(1) shows that $U \times_g T$ is timelike (resp., spacelike) if and only if $\beta'(s)$ is spacelike (resp., timelike). [For a lightlike curve $\alpha(t)$ on M, the unit tangent vector field T can not be formed since $|\alpha'(t)| = 0$].

In the case that $\beta'(s)$ is unit spacelike, defining k_n and $k_g =< \beta''(s), U \times_g T >$ as above, one obtains

$$\beta''(s) = k_n U - k_g(U \times_g T)$$

whereas for $\beta'(s)$ unit timelike one has

$$\beta''(s) = k_n U + k_g(U \times_g T).$$

Suppose $\kappa(s) > 0$ for $\beta(s)$ for all s. Hence, $\beta''(s) = \kappa(s)N$ and calculating the inner product $< \beta''(s), \beta''(s) >$ yields

$$(\kappa(s))^2 g(N,N) = (k_n)^2 + (k_g)^2$$

for $\beta'(s)$ unit spacelike (and $g(N,N) = -1$ or $+1$) and yields

$$(\kappa(s))^2 = (k_n)^2 + (k_g)^2$$

if $\beta'(s)$ is unit timelike.

The geometric interpretation of the normal curvature $k_n(w)$ in terms of the so called **normal section** of the surface in direction w traces back to Jean Baptiste Meusnier in in 1776.

Definition 4.1.8 Let M be a regular surface in $\mathbf{E}^3$ and $p \in M$. For a unit vector $u \in T_pM$, the **normal section** of M at p in the direction u is the curve (near p) cut from the surface M by the 2-plane in $\mathbf{E}^3$ determined by the vectors $U(p)$ and u.

The normal section is important because its curvature at p (as a curve in $\mathbf{E}^3$) calculates the normal curvature $k_n(u)$ of the surface in the direction u.

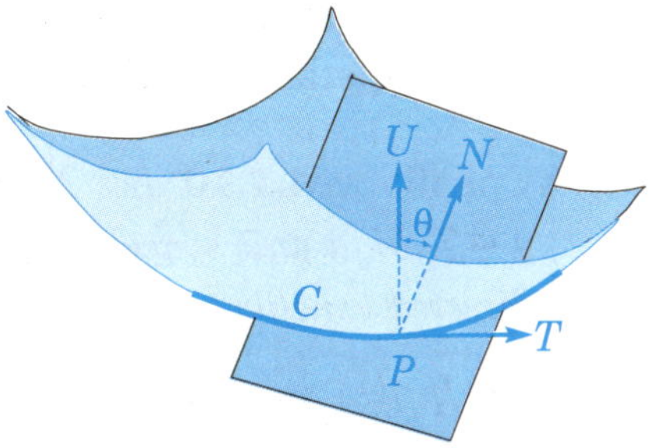

Figure4.1-3

Theorem 4.1.9 *(a) If $C : \gamma(s)$ is any unit speed curve in M with $\gamma(0) = p, \gamma'(0) = u$ and curvature $\kappa(s)$ in $\mathbf{E}^3$ with $\kappa(s) > 0$, then*

$$k_n(u) = \kappa(s)\cos\theta$$

where θ is the angle between $N(0)$ and $U(p)$, (cf. Figure 4.1-3).

(b) If the unit speed parametrization $\beta(s)$ of the normal section of M at p in the direction $u \in T_pM$ has positive curvature $\kappa(0) > 0$ in $\mathbf{E}^3$ at p, then

$$k_n(u) = \pm\kappa(0).$$

Proof. (a) Since γ is unit speed and has positive curvature near $s = 0$, $N(s)$ exists and $\gamma''(0) = \kappa(0)N(0)$. On the other hand, from Definition 4.1.2 we have

$$\begin{aligned} k_n(u) &= < \gamma''(0), U(p) > \\ &= < \kappa(0)N(0), U(p) > \\ &= \kappa(0) < N(0), U(p) > \\ &= \kappa(0)\cos\theta \end{aligned}$$

as $|N(0)| = |U(p)| = 1$.

(b) Given the formula in part (a), it suffices to show that $\theta = 0$ or π for the normal section $\beta(s)$ at $s = 0$. Let P denote the plane in $\mathbf{E}^3$ spanned by $U(p)$ and u. By construction of normal section, $T(s) = \beta'(s)$ lies in P for all small s. Hence $T'(0) = \beta''(0)$ also lies in P. Since $\kappa(0) > 0$, again $N(0)$ exists and also lies in P since $\beta''(0) = \kappa(0)N(0)$. Thus $N(0)$ lies in the 2-plane P and is also orthogonal to u, as is $U(p)$. Hence $N(0) = \pm U(p)$. □

Similar definitions and calculations may be made for a spacelike surface M in $\mathbf{L}^3$, assuming that $U(p)$ has been chosen to be future directed and supposing that the spacelike curve $\gamma(t)$ in M with $\gamma(0) = p$ and $\gamma'(0) = u$ has future directed timelike normal $N(0)$ in $\mathbf{L}^3$ (recall Theorem 1.3.6-(5) and the discussion of the hyperbolic angle between future timelike vectors).

Corollary 4.1.10 *Let M be a regular spacelike surface in $\mathbf{L}^3$ and $p \in M$, $u \in T_pM$ a unit spacelike vector, and choose the unit normal $U(p)$ to be future directed timelike.*

(a) If $\gamma(s)$ is a differentiable spacelike curve on M with $\gamma(0) = p$, $\gamma'(0) = u$ and $N(0)$ future directed timelike, then

$$k_n(u) = -\kappa(0)\cosh\theta$$

where θ is the hyperbolic angle between $U(p)$ and $N(0)$, and $\kappa(0)$ is the curvature of $\gamma(s)$ in $\mathbf{L}^3$ at $s = 0$.

(b) If the normal section $\beta(s)$ has the future directed timelike unit normal $N(0)$, then $k_n(u) = -\kappa(0)$ where $\kappa(0)$ is the curvature of $\beta(s)$ in $\mathbf{L}^3$ at $s = 0$.

Proof. (a) As in the Euclidean case, $\gamma''(0) = \kappa(0)N(0)$ by Theorem 2.1.9.

Again

$$\begin{aligned} k_n(u) &= g(\gamma''(0), U(p)) \\ &= g(\kappa(0)N(0), U(p)) \\ &= \kappa(0)g(N(0), U(p)) \\ &= -\kappa(0)|N(0)||U(p)|\cosh\theta \\ &= -\kappa(0)\cosh\theta. \end{aligned}$$

(b) In the case of the hypotheses on the normal section, arguing along the lines of Theorem 4.1.9-(b) we obtain $N(0) = U(p)$. Thus $g(N(0), U(p)) = -1$ (or $\theta = 0$) and the result follows from the formula in part (a). □

Shape Operators

A second technique for calculating the normal curvature employs the shape operator associated to a surface. This operator provides a measure of how the surface is bending in different tangent directions in $\mathbf{E}^3$ or in $\mathbf{L}^3$.

We first recall some properties of the covariant derivative operator on $\mathbf{E}^3$ or on $\mathbf{L}^3$. Let E_1, E_2, E_3 be the standard frame field for $\mathbb{R}^3$ as in Section 1.2. If U is the vector field on an open subset of $\mathbf{E}^3$ or $\mathbf{L}^3$, and one expands

$$U(p) = \sum_{i=1}^{3} u_i(p)E_i(p),$$

then for any tangent vector v at p, the covariant derivative $\nabla_v U$ in either $\mathbf{E}^3$ or $\mathbf{L}^3$ is given by

$$\nabla_v U = \sum_{i=1}^{3} v[u_i]E_i(p).$$

Then it may be established for either the Euclidean metric on $\mathbf{E}^3$ or the Minkowskian metric on $\mathbf{L}^3$ (both denoted by $<\ ,\ >$) and vector fields V, W defined on an open neighborhood of p that

$$u(< V, W >) =< \nabla_u V, W(p) > + < V(p), \nabla_u W > .$$

Now let M be a surface in $\mathbf{E}^3$ or let M be a surface in $\mathbf{L}^3$ with everywhere unit timelike or everywhere unit spacelike normal field. In either of these three cases, let U be a choice of normal field for M, so that $< U, U >= \pm 1$. For any $v \in T_pM$, the above inequality implies

$$v(< U, U >) = 2 < \nabla_v U, U(p) >= v(\pm 1) = 0$$

so that $\nabla_v U$ is orthogonal to the normal vector $U(p)$. Hence, $\nabla_v U \in T_pM$ and the following definition makes sense.

Definition 4.1.11 Let M be a surface $\mathbf{E}^3$ with unit normal vector field U or let M be a surface $\mathbf{L}^3$ with everywhere unit timelike or unit spacelike normal field U in a neighborhood of p in M. The **shape operator** $S_p : T_pM \to T_pM$ is defined for any v in T_pM by $S_p(v) = -\nabla_v U$ where ∇ denotes the covariant derivative in $\mathbf{E}^3$ (resp., $\mathbf{L}^3$).

Remark 4.1.12 Since U may be replaced by $-U$, the shape operator S_p is only defined up to sign. From the properties of the covariant derivatives, the map S_p is linear. Recall that in Section 3.4, the differential dU_p of the Gauss map $U : M \to TM$ is also computed as

$$dU_p(v) = \sum_{i=1}^{3} v[u_i]E_i(p)$$

where $U = (u_1, u_2, u_3)$. Hence, $S_p = -dU_p$ on T_pM. By Theorem 3.4.6, S_p is thus a self-adjoint map of T_pM for both surfaces in $\mathbf{E}^3$ and in $\mathbf{L}^3$, i.e., $< S_p(v), w >=< v, S_p(w) >$ for all $v, w \in T_pM$. From previous calculations of the Gauss Map in Section 3.4, for the basic Euclidean surfaces we have $S_p = 0$ on a plane and $S_p = -Id$ on the unit sphere. From Example 3.4.4, $S_p = \begin{pmatrix} -1 & 0 \\ 0 & 0 \end{pmatrix}$ on the timelike cylinder $\Gamma = \{(x, y, z) | x^2 + y^2 = 1\}$ in $\mathbf{L}^3$, and a similar result is valid for the Euclidean cylinder in $\mathbf{E}^3$.

Lemma 4.1.13 *If α is a differentiable curve on a surface M in $\mathbf{E}^3$ or in a surface M in $\mathbf{L}^3$ with everywhere timelike or everywhere spacelike normal, then*

$$< \alpha'', U \circ \alpha >=< S_\alpha(\alpha'), \alpha' > .$$

Proof. Since $< \alpha', U >= 0$, by differentiating both sides $< \alpha'', U > + < \alpha', U' >= 0$. Since $S_\alpha(\alpha') = -\nabla_{\alpha'} U = -\sum_{i=1}^{3} \alpha'[u_i]E_i = -\frac{d}{dt}(U \circ \alpha)(t) = -U'$, we have $< \alpha'', U >=< \alpha', -U' >=< \alpha', S_\alpha(\alpha') >$. □

With Lemma 4.1.13 in hand, it may now be shown that the normal curvatures as given in Definition 4.1.2 may be alternatively calculated by the shape operator for surfaces either in $\mathbf{E}^3$ or $\mathbf{L}^3$. This is, in effect, a restatement of the result in Lemma 4.1.3 that the normal curvatures may be calculated from the second fundamental form.

Remark 4.1.14 For a unit speed curve $\beta(s)$ in a surface M, $k_n =< T'(s), U >=< \beta''(s), U >=< S_\beta(\beta'), \beta' >=< S_\beta(T(s)), T(s) >$. Thus, for the unit vector u

in T_pM we have the normal curvature $k_n(u)$ of M in the u direction given by $k_n(u) =< S_p(u), u >$.

Principal Curvatures

In the first portion of this subsection, let M be a surface in $\mathbf{E}^3$ or a spacelike surface in $\mathbf{L}^3$. Then $K = \{v \in T_pM; |v| = 1\}$ is a compact set in T_pM in both cases. Because the normal curvature function $k_n : K \to \mathbb{R}$ is continuous, k_n assumes maximum and minimum values k_1 and k_2 at e_1 and e_2 in K respectively. That is,

$$k_n(e_1) = k_1, k_n(e_2) = k_2, \text{ and } k_2 \leq k(u) \leq k_1$$

for all u in K. Because $k_n(-u) = k_n(u)$ for all $u \in K$, if k_n takes on a maximum at e_1, then k_n also takes on a maximum at $-e_1$, and similarly for e_2 and $-e_2$. Thus while k_1 and k_2 are called the **principal curvatures** of M at p, the tangent vectors e_1 and e_2 are often called the **principal directions** rather than the **principal vectors** of the surface M at p(Figure 4.1-4).

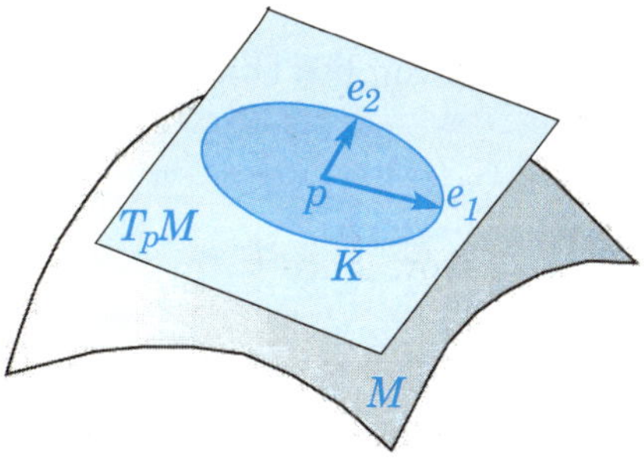

Figure4.1-4

Definition 4.1.15 A point p of the surface M in $\mathbf{E}^3$ or of a spacelike surface in $\mathbf{L}^3$ is said to be an **umbilic** if the normal curvature k_n is constant on $K = \{v \in T_pM; |v| = 1\}$.

For examples, all points of the planes and spheres in $\mathbf{E}^3$ are umbilics.

Theorem 4.1.16 *let M be an Euclidean surface in $\mathbf{E}^3$ or a spacelike surface in $\mathbf{L}^3$. If $p \in M$ is not an umbilic, there are exactly two orthogonal principal directions for which the principal curvatures are eigenvalues of the shape operator S_p.*

Proof. Let the normal curvature k_n have the maximum value k_1 at e_1. Then by Remark 4.1.14,

$$k_1 = k_n(e_1) =< S_p(e_1), e_1 > .$$

Now, take a unit tangent vector e_2 in T_pM which is orthogonal to e_1. For any unit vector u in T_pM,

$$u = u(\theta) = \cos\theta e_1 + \sin\theta e_2.$$

If $S_{ij} =< S_p(e_i), e_j >$, $1 \leq i,\ j \leq 2$, then $S_{ii} = k_i$ and $S_{ij} = S_{ji}$. Hence, the normal curvature in the direction u is

$$\begin{aligned} k_n(\theta) &= k_n(u(\theta)) \\ &=< S_p(u(\theta)), u(\theta) > \\ &=< S_p(\cos\theta e_1 + \sin\theta e_2), \cos\theta e_1 + \sin\theta e_2 > \\ &= S_{11}\cos^2\theta + 2S_{12}\cos\theta\sin\theta + S_{22}\sin^2\theta. \end{aligned}$$

Thus,

$$\frac{dk_n}{d\theta} = 2(S_{22} - S_{11})\cos\theta\sin\theta + 2S_{12}(\cos^2\theta - \sin^2\theta).$$

If $\theta = 0$, then $u(0) = e_1$. Since $k_n(u(0)) = k_n(e_1) = k_1$ is a relative maximum value at $\theta = 0$,

$$\frac{dk_n}{d\theta}(0) = 0,$$

which implies that $S_{12} = 0$. Therefore,

$$k_n(\theta) = k_n(u(\theta)) = S_{11}\cos^2\theta + S_{22}\sin^2\theta.$$

Now since $k_n(e_1) = k_1 = k_n(0)$ we have $S_{11} = k_1$ and

$$k_n(\theta) = k_1\cos(\theta)^2 + S_{22}\sin(\theta)^2$$

for all θ. Since k_1 is the maximum of k_n on K and $S_{22} = k_n(\frac{\pi}{2})$, we have $S_{22} \leq k_1$. If $S_{22} = k_1$, it follows from the above formula that $k_n(\theta) = k_1$ for all θ and p is an umbilic, in contradiction to the hypothesis. Hence, $S_{22} < k_1$ and it can be deduced from the formula $k_n(\theta) = k_1\cos(\theta)^2 + S_{22}\sin(\theta)^2$ that $k_n(\theta)$ takes on its minimum S_{22} when $\theta = \pm\frac{\pi}{2}$ and $u = \pm e_2$. Hence $k_2 = S_{22}$ and the above vector e_2 taken to be orthogonal to e_1 is the second principal direction. Moreover, we now obtain the **Euler's formula**

$$k_n(\theta) = k_1\cos(\theta)^2 + k_2\sin(\theta)^2.$$

This formula indeed implies that when $k_2 < k_1$ the principal directions are unique and that $k_n(\theta)$ takes on its maximum at $\pm e_1$ and its minimum at $\pm e_2$. It remains to show that $S_p(e_j) = k_j e_j$ so that the principal directions are eigenvectors of the shape operator. But because $\{e_1,\ e_2\}$ as above is an orthonormal basis for T_pM, we have

$$\begin{aligned} S_p(e_1) &=< S_p(e_1), e_1 > e_1 + < S_p(e_1), e_2 > e_2 \\ &= S_{11}e_1 + S_{12}e_2 \\ &= k_1 e_1 \end{aligned}$$

and a similar argument shows that $S_p(e_2) = k_2e_2$. □

Remark 4.1.17 From the proof of Theorem 4.1.16, it follows that for a surface $M \in \mathbf{E}^3$ or a spacelike surface in $\mathbf{L}^3$ that the eigenvalues k_1, k_2 for the shape operator $S_p : T_pM \to T_pM$ are given by

$$k_1 = \max\{< S_p(u), u > \mid u \in K\}$$

and

$$k_2 = \min\{< S_p(u), u > \mid u \in K\}.$$

Also from the above Theorem for either a surface in $\mathbf{E}^3$ or a spacelike surface in $\mathbf{L}^3$, the matrix which represents the shape operator is

$$S_p = \begin{pmatrix} k_1 & 0 \\ 0 & k_2 \end{pmatrix}.$$

[If p is an umbilic, then $S_p(v) = kv$ for all v so that the representation with $k_1 = k_2 = k$ is immediate.]

Theorem 4.1.16 and Remark 4.1.17 suggest considering the eigenvectors of the shape operator S_p as well as studying $< S_p(u), u >$ on $K = \{v \in T_pM; |v| = 1\}$ when M is a timelike surface in $\mathbf{L}^3$. Now, however, K is a closed subset of T_pM consisting of the unit timelike and the unit spacelike tangent vectors, but K is not compact. Also, while S_p is still self-adjoint with respect to the Lorentz metric on T_pM, this metric is not positive definite, Hence, S_p can not always be diagonalized unlike the Euclidean or spacelike cases in Theorem 4.1.16.

Example 4.1.18 Consider again the cylinder $M = \{(x, y, z) \in L^3 \mid x^2+y^2 = 1\}$ which is a timelike surface in $\mathbf{L}^3$ and was previously studied in Example 3.4.4. First using the definition of shape operator S_p as given in Definition 4.1.11, we verify that $S_p = -dU_p$ as used in Remark 4.1.12. As in Example 3.4.4, take $U(x, y, z) = (x, y, 0)$. Hence If $v = (a, b, c)$ in T_pM then

$$\nabla_v U = (v(x), v(y), v(0)) = (a, b, 0).$$

Thus $S_p((a, b, c)) = -(a, b, 0)$.

Consider first the eigenvalue problem $S_p(v) = kv$ for $k \in \mathbb{R}$ and $v = (a, b, c)$ nonzero. Then $kc = 0$. If $k = 0$, then $S_p(v) = (0, 0, 0) = -(a, b, 0)$ and we obtain the timelike eigendirection $e_1 = (0, 0, 1)$ corresponding to the eigenvalue $k_1 = 0$. If $c = 0$, then $v = (a, b, 0)$ with one of a or b nonzero, and considering $S_p(v) = -(a, b, 0) = k(a, b, 0)$, we obtain eigenvalue $k_2 = -1$. If $p = (x_0, y_0, z_0)$

in M and $v = (a, b, 0)$ is in T_PM, it may be seen that v is a multiple of $e_2 = (-y_0, x_0, 0)$. Hence this eigenvalue/eigenvector calculation is consistent with the results in Example 3.4.4 for which $dU_p(X_u) = X_u$ and $dU_p(X_v) = (0, 0, 0)$.

Now consider extremizing $< S_p(u), u >$ over the set K of unit vectors in any tangent space to this timelike cylinder M. For $u = (a, b, c)$ unit timelike, we have $a^2 + b^2 - c^2 = -1$. Hence

$$< S_p(u), u >=< -(a, b, 0), (a, b, c) >= -a^2 - b^2 = 1 - c^2,$$

with $|c| \geq 1$. Since c maybe arbitrarily large, $< S_p(u), u >$ takes on no minimum value but takes on a maximum value $k = 0$ when $c = \pm 1$ and hence in the direction $e_1 = (0, 0, 1)$. In this case, e_1 is an eigen-direction of S_p and the value $k = 0$ is the corresponding eigenvalue since $S_p(e_1) = (0, 0, 0) = 0e_1$.
For the unit spacelike vectors, $a^2 + b^2 - c^2 = 1$ so that

$$< S_p(u), u >= -a^2 - b^2 = -1 - c^2.$$

Hence $< S_p(u), u >$ takes on maximum, value -1 and no minimum on the unit spacelike vectors. The maximum value is taken on when $c = 0$ and if $p = (x_0, y_0, z_0)$, the direction u achieving this maximum may be seen to be $e_2 = (-y_0, x_0, 0)$ corresponding to the above eigenvalue calculation. In this case, the maximum value of $< S_p(u), u >$ over the spacelike vectors does agree with the eigenvalue obtained in our above calculation.

Remark 4.1.19 If M is a timelike surface in $\mathbf{L}^3$ and $p \in M$, then the shape operator $S_p : T_pM \to T_pM$ is self-adjoint with respect to the Lorentz inner product $I_p =< , >$ induced on T_pM from the restriction of the Lorentz metric $g =< , >$ on $\mathbf{L}^3$, i.e., $< S_p(v), w >=< v, S_p(w) >$ for all $v, w \in T_pM$. However, since I_p is not positive definite in T_pM, it is not necessarily the case that S_p is diagonalizable with respect to I_p(as is always true for a self-adjoint linear transformation and a positive definite inner product, cf. Theorem 4.1.16). Here S_p may be said to be **diagonalizable** at p if the exist linearly independent vectors $e_1, e_2 \in T_pM$ and $k_1, k_2 \in \mathbb{R}$ with $S_p(e_j) = k_je_j$ for $j = 1, 2$. Timelike surfaces may be constructed for which S_p has only a single lightlike direction in T_pM as eigendirection.

Nonetheless, a partial parallel to Theorem 4.1.16 may be obtained for timelike surfaces.

Theorem 4.1.20 *Let M be a timelike surface in $\mathbf{L}^3$ and suppose for $p \in M$, that $S_p : T_pM \to T_pM$ has two different real eigenvalues $k_1 \neq k_2$. Then S_p has two unique eigendirections, one represented by a unit timelike eigenvector (say e_1) and the other by a unit spacelike eigenvector (say e_2).*

Proof. By definition of eigenvectors, given distinct eigenvalues k_1, k_2, there exist nonzero $e_1, e_2 \in T_pM$ with $S_p(e_j) = k_j e_j$ for $j = 1, 2$. By linear algebraic considerations, $k_1 \neq k_2$ implies that e_1 and e_2 are linearly independent and also that the eigendirections are unique. Now

$$\begin{aligned} k_1 < e_1, e_2 > &=< k_1 e_1, e_2 >=< S_p(e_1), e_2 > \\ &=< e_1, S_p(e_2) >= k_2 < e_1, e_2 > . \end{aligned}$$

Hence as $k_1 \neq k_2$, we must have $< e_1, e_2 >= 0$. Since e_1 and e_2 are linearly independent, they can not both be nonspacelike(cf. Corollary 3.3.9). Also, since $\dim(T_pM) = 2$, it may be shown directly that the orthogonal vectors e_1, e_2 can not both be spacelike or be spacelike and lightlike respectively (or consider the Lorentz metric $-g$ on T_pM to deduce this). Hence, one of the vectors, say e_1, must be timelike and the other, say e_2, spacelike. □

If $S_p(e_1) = ke_1$ and $S_p(e_2) = ke_2$ for $\{e_1, e_2\}$ linearly independent, then $S_p(v) = kv$ for all $v \in T_pM$ by linear algebra. Hence, p is an umbilic point and $\{e_1, e_2\}$ may be selected to be an orthogonal unit timelike vector and unit spacelike vector. In both of these cases, if p is a point of a timelike surface M in $\mathbf{L}^3$ at which the shape operator is diagonalizable, then S_p is represented by a matrix

$$S_p = \begin{pmatrix} k_1 & 0 \\ 0 & k_2 \end{pmatrix}.$$

For further study, see the following references.

1. McCleary, J., Geometry from a Differentiable Viewpoint, Cambridge University Press, 1994.

2. Millman, R. S. and G. D. Parker,Elements of Differential Geometry, Prentice-Hall, Englewood Cliffs, NJ, 1977.

3. O'Neill, B., Elementary Differential Geometry, Academic Press, New York, 1966.

4. Pressley, Andrew, Elementary Differential Geometry, Springer Undergraduate Mathematics Series, 2002.

4.2 Gaussian and Mean Curvatures

The result in this section for Euclidean surfaces stem from Gauss's 1827 paper " General investigations of curved surfaces" already cited in Section 3.3. Gauss defined what is not now called the Gaussian Curvature K as a limit of area ratios akin to Theorem 4.2.4 of this section. Then Gauss showed that this curvature quantity $K = \dfrac{LN - M^2}{EG - F^2}$ may thus be calculated from the coefficients of the first and second fundamental forms. He also showed that this curvature related to the principal curvatures earlier introduced by Euler(cf. Definition 4.2.1) Gauss also derived formulas which showed that the Gaussian curvature of a surface only depends on the coefficients E, F, and G of the first fundamental form and their partial derivatives, despite the apparent dependence on the extrinsic geometry (how the surface sits as a submanifold of $\mathbf{E}^3$) in the numerator $LN - M^2$ of the Gaussian curvature. Since E, F and G are functions only of the surface patch $X : D \to \mathbf{E}^3$, the Gaussian curvature is only a function of this paramerization and not of how the surface is embedded in $\mathbf{E}^3$. Gauss term this discovery the "theorema egregium", Latin for a "most excellent theorem." The derivation given in this section employs the Christoffel symbols, which were defined by Elwin Bruno Christoffel(1829-1900) in a paper published in 1869.

Let M be a surface in $\mathbf{E}^3$, a spacelike surface in $\mathbf{L}^3$, or a timelike surface in $\mathbf{L}^3$ for which p is a diagonalizable point for S_p as discussed in Theorem 4.1.20.

Recall that the matrix which represents the shape operator S_p in Section 4.1 was given as follows:

$$\begin{pmatrix} k_1 & 0 \\ 0 & k_2 \end{pmatrix}$$

where k_1 and k_2 are the principal curvatures of the surface M at a point p.

Definition 4.2.1 The **Gaussian curvature** of M at p is given by

$$K = \det S_p = k_1 k_2.$$

The **mean curvature** of M at p is given by

$$H = \mathrm{tr} S_p = \frac{k_1 + k_2}{2}.$$

If $H = 0$ on M in $\mathbf{E}^3$(or for a spacelike surface in $\mathbf{L}^3$), then M is called a **minimal surface**(or a **maximal surface**).

For a spacelike surface M in $\mathbf{L}^3$, some authors like Kühnel in the reference cited in Section 3.3 take the Gaussian curvature to be defined by $K = -k_1 k_2$. At a point p on the timelike surface at which the shape operator S_p has orthogonal timelike and spacelike principal directions, Kühnel would take the Gaussian curvature to be $K = k_1 k_2$.

If U is the unit timelike normal field to a spacelike surface, Kühnel would take the mean curvature vector H to be

$$H = -\left[\frac{k_1 + k_2}{2}\right] U$$

and for a timelike surface with unit spacelike normal field U and at a point p at which the shape operator S_p has orthogonal timelike and spacelike principal directions, Kühnel would take the mean curvature vector H to be

$$H = \left[\frac{k_1 + k_2}{2}\right] U.$$

However, the mean curvature is only defined up to sign since U may be replaced with $-U$. Hence one could take $H = \dfrac{k_1 + k_2}{2}$ in both cases.

Computation of Curvatures

However, it is generally not easy to find the principal curvatures k_1 and k_2 or the matrix representing the shape operator S_p. Hence, we need to find the formulas to compute the Gaussian curvature K and the mean curvature H by using the coefficients of the first fundamental form and the second fundamental form of the proper patch $X(u, v)$ containing a neighborhood of p.

Let M be a surface and $p \in M$ and let $S_p : T_pM \to T_pM$ be a shape operator at p on M. Since $\{X_u,\ X_v\}$ is a basis for the tangent space T_pM, $S(X_u)$ and $S(X_v)$ are linear combinations of X_u and X_v as follows:

$$(4.2\text{-}1) \qquad \begin{cases} S(X_u) = a_{11}X_u + a_{12}X_v \\ S(X_v) = a_{21}X_u + a_{22}X_v. \end{cases}$$

By definition of the shape operator S_p,

$$(4.2\text{-}2) \qquad \begin{cases} S(X_u) = -\nabla_{X_u}U = -dU(X_u) = -U_u \\ S(X_v) = -\nabla_{X_v}U = -dU(X_v) = -U_v. \end{cases}$$

Using (4.2-1) and (4.2-2), the coefficients of the second fundamental form are given by

$$(4.2\text{-}3) \qquad \begin{cases} L =< U, X_{uu} >= - < U_u, X_u >= a_{11}E + a_{12}F \\ M =< U, X_{vu} >= - < U_u, X_v >= a_{11}F + a_{12}G \\ M =< U, X_{uv} >= - < U_v, X_u >= a_{21}E + a_{22}F \\ N =< U, X_{vv} >= - < U_v, X_v >= a_{21}F + a_{22}G. \end{cases}$$

Hence,

$$\begin{pmatrix} L & M \\ M & N \end{pmatrix} = \begin{pmatrix} a_{11} & a_{12} \\ a_{21} & a_{22} \end{pmatrix} \begin{pmatrix} E & F \\ F & G \end{pmatrix},$$

and $EG - F^2 \neq 0$ since I_p is assumed to be nondegenerate. Thus,

$$(4.2\text{-}4) \qquad S_p = \begin{pmatrix} a_{11} & a_{12} \\ a_{21} & a_{22} \end{pmatrix} = \begin{pmatrix} L & M \\ M & N \end{pmatrix} \begin{pmatrix} E & F \\ F & G \end{pmatrix}^{-1}$$

$$= \begin{pmatrix} L & M \\ M & N \end{pmatrix} \frac{1}{EG - F^2} \begin{pmatrix} G & -F \\ -F & E \end{pmatrix}$$

$$(4.2\text{-}5) \qquad = \frac{1}{EG - F^2} \begin{pmatrix} GL - FM & -FL + EM \\ -FN + GM & EN - FM \end{pmatrix}.$$

To obtain the eigenvalues of the matrix $\begin{pmatrix} a_{11} & a_{12} \\ a_{21} & a_{22} \end{pmatrix}$ for the shape operator with respect to $\{X_u, X_v\}$, we have to find the solution k of the eigenfunction $S - kI = 0$.

For, $(a_{11}-k)(a_{22}-k)-a_{12}a_{21} = 0$ implies $k^2-(a_{11}+a_{22})k+a_{11}a_{22}-a_{12}a_{21} = 0$, so that

$$(4.2\text{-}6) \qquad k^2 - \operatorname{tr}(S)k + \det(S) = 0,$$

where we take $\operatorname{tr}(S) = a_{11} + a_{22}$ and $\det(S) = a_{11}a_{22} - a_{12}a_{21}$ in all three cases. Since the solution of the above equation (4.2-6) are the eigenvalues k_1 and k_2 of $\begin{pmatrix} a_{11} & a_{12} \\ a_{21} & a_{22} \end{pmatrix}$, we have the two equations $k_1 + k_2 = \operatorname{tr}(S)$ and $k_1 k_2 = \det(S)$.

Since $\det(S) = \det \begin{pmatrix} a_{11} & a_{12} \\ a_{21} & a_{22} \end{pmatrix}$ and since $\operatorname{tr}(S) = \operatorname{tr} \begin{pmatrix} a_{11} & a_{12} \\ a_{21} & a_{22} \end{pmatrix}$, from (4.2-5) we have

$$K = \det \begin{pmatrix} a_{11} & a_{12} \\ a_{21} & a_{22} \end{pmatrix} = \frac{LN - M^2}{EG - F^2}$$

and

$$H = \frac{1}{2}\operatorname{tr} \begin{pmatrix} a_{11} & a_{12} \\ a_{21} & a_{22} \end{pmatrix} = \frac{EN + GL - 2FM}{2(EG - F^2)}.$$

Further, from (4.2-6), $k^2 - 2Hk + K = 0$. Hence, the principal curvatures are

$$(4.2\text{-}7) \qquad k = H \pm \sqrt{H^2 - K}.$$

Example 4.2.2 Let M be a helicoid given by the patch $X(u,v) = (u\cos v, u\sin v, bv)$, $b \neq 0$ in $\mathbf{E}^3$(Figure4.2-5). Then M is a minimal surface with Gaussian curvature $-\frac{1}{b^2} \leq K < 0$.

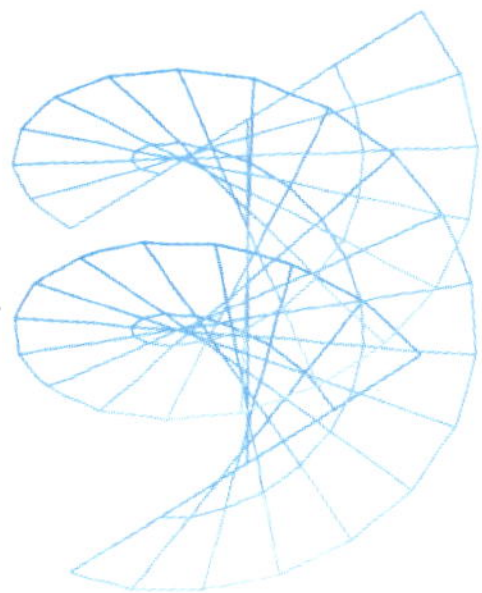

Figure4.2-5

Solution. Since $X_u = (\cos v, \sin v, 0)$ and $X_v = (-u\sin v, u\cos v, b)$, the coefficients of the 1st fundamental form are

$$E = 1,\ F = 0,\ G = u^2 + b^2.$$

Moreover, since $X_u \times X_v = (b\sin v, -b\cos v, u)$ and since $|X_u \times X_v| = \sqrt{u^2+b^2}$, the unit normal curvature to M is

$$U = \frac{(b\sin v, -b\cos v, u)}{\sqrt{u^2+b^2}}.$$

Now, since $X_{uu} = (0,0,0)$, $X_{uv} = (-\sin v, \cos v, 0)$, and since $X_{vv} = (-u\cos v, -u\sin v, 0)$, the coefficients of the 2nd fundamental form are

$$L =< X_{uu}, U >= 0, M =< X_{uv}, U >= -\frac{b}{\sqrt{u^2+b^2}}, N =< X_{vv}, U >= 0.$$

Therefore, the Gaussian curvature K and the mean curvature H are

$$K = \frac{LN - M^2}{EG - F^2} = -\frac{b^2}{(u^2+b^2)^2},\quad H = \frac{EN + GL - 2FM}{2(EG - F^2)} = 0.$$

When $u \to 0$, the Gaussian curvature K has the minimum value $\frac{-1}{b^2}$. And if $|u| \to \infty$, $K \to 0$. □

Example 4.2.3 Let M be a helicoid given by the patch $X(u,v) = (u\cos v, u\sin v, bv)$, $u^2 > b^2$, $b \neq 0$ in $\mathbf{L}^3$(Figure4.2-5). Then M is a spacelike maximal surface with Gaussian curvature $K < 0$.

Solution. Since $X_u = (\cos v, \sin v, 0)$ and $X_v = (-u\sin v, u\cos v, b)$, the coefficients of the 1st fundamental form are

$$E = 1,\ F = 0,\ G = u^2 - b^2.$$

Moreover, since $X_u \times_g X_v = (b\sin v, -b\cos v, -u)$ and since $|X_u \times_g X_v| = \sqrt{u^2 - b^2}$, the unit normal curvature to M is

$$U = \frac{(b\sin v, -b\cos v, -u)}{\sqrt{u^2 - b^2}}$$

which is timelike. Thus M is a spacelike surface for $u^2 > b^2$. Now, since $X_{uu} = (0,0,0)$, $X_{uv} = (-\sin v, \cos v, 0)$, and $X_{vv} = -(u\cos v, u\sin v, 0)$, the coefficients of the 2nd fundamental form are

$$L =< X_{uu}, U >= 0, M =< X_{uv}, U >= -\frac{b}{\sqrt{u^2 - b^2}}, N =< X_{vv}, U >= 0.$$

Therefore, the Gaussian curvature K and the mean curvature H are

$$K = \frac{LN - M^2}{EG - F^2} = -\frac{b^2}{(u^2 - b^2)^2} \text{ for } u^2 > b^2,\ H = \frac{EN + GL - 2FM}{2(EG - F^2)} = 0.$$

Thus, $K < 0$. For $0 < u < b$, the surface M is timelike and we obtain $H = 0$ and $K = \dfrac{b^2}{(b^2 - u^2)^2}$ so that $\dfrac{1}{b^2} \leq K < +\infty$. □

Gauss Map and Curvature

Let M be an orientable surface and let S^2 be the unit sphere in $\mathbf{E}^3$ centered at the origin. Recall that the Gauss map is a map $U : M \to S^2$ given by associating to each point p in M the unit normal vector U_p at p. For a proper patch $X : D \to M$ containing a neighborhood of p let $\mathcal{A}_X(D)$ be the area of $X(D)$ and $\mathcal{A}_U(D)$ the area of $U(D) = U(X(D))$ in Figure3.4-41.

Theorem 4.2.4 *Let $D_\delta(p) = \{(u,v) \in D | (u - u_0)^2 + (v - v_0)^2 \leq \delta^2\}$ be the disc of radius δ with center (u_0, v_0) where $p = X(u_0, v_0)$. Then, at p*

$$\lim_{\delta \to 0} \frac{\mathcal{A}_U(D_\delta(p))}{\mathcal{A}_X(D_\delta(p))} = |K(p)|.$$

Proof. From (4.2-1) and (4.2-2),

$$\begin{aligned} U_u \times U_v &= (a_{11}X_u + a_{12}X_v) \times (a_{21}X_u + a_{22}X_v) \\ &= \det \begin{pmatrix} a_{11} & a_{12} \\ a_{21} & a_{22} \end{pmatrix} X_u \times X_v. \end{aligned}$$

From (4.2-4),

$$\begin{aligned} \begin{vmatrix} a_{11} & a_{12} \\ a_{21} & a_{22} \end{vmatrix} X_u \times X_v &= \begin{vmatrix} L & M \\ M & N \end{vmatrix} \begin{vmatrix} E & F \\ F & G \end{vmatrix}^{-1} X_u \times X_v \\ &= \frac{LN - M^2}{EG - F^2} X_u \times X_v \\ &= K X_u \times X_v. \end{aligned}$$

Hence,

$$|U_u \times U_v| = |K||X_u \times X_v|.$$

Since K is continuous at p, by the Mean Value Theorem for double integrals, there is a point $q(u,v)$ with $(u,v) \in D_\delta(p)$ such that

$$\iint_{D_\delta(p)} |U_u \times U_v| dudv = |K(q)| \iint_{D_\delta(p)} |X_u \times X_v| dudv.$$

Thus,

$$\mathcal{A}_U(D_\delta(p)) = |K(q)| \mathcal{A}_X(D_\delta(p)).$$

Since $q \to p$ as $\delta \to 0$, we are done. □

The Sign of the Curvature

Since $EG - F^2 > 0$, if M is a surface in $\mathbf{E}^3$ (or a spacelike surface $\mathbf{L}^3$), the sign of $K(p)$ is the same as the sign of $(LN - M^2)(p)$.

(1) If $K(p) > 0$, p is an elliptic point.

Example 4.2.5 Let M be the surface in $\mathbf{E}^3$ given by the patch

$$X(u,v) = (u, v, -u^2 - 2v^2).$$

Then $0 < K \leq 8$ on M. Thus, all points of M are elliptic. Further, the principal curvatures at $p = (0,0,0)$ are $k_1 = -2$ and $k_2 = -4$(Figure4.2-6).

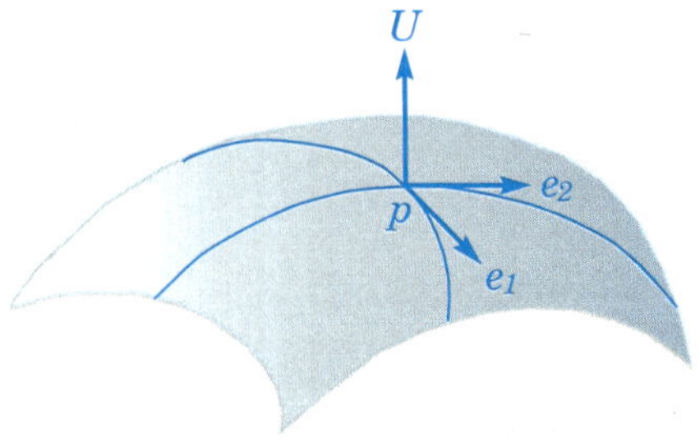

Figure4.2-6

Solution. Since $X_u = (1, 0, -2u)$ and $X_v = (0, 1, -4v)$, the coefficients of the 1st fundamental form are

$$E = 1 + 4u^2,\ F = 8uv,\ G = 1 + 16v^2.$$

Moreover, since $X_u \times X_v = (2u, 4v, 1)$ and $|X_u \times X_v| = \sqrt{4u^2 + 16v^2 + 1}$, the unit normal vector to M is

$$U = \frac{(2u, 4v, 1)}{\sqrt{4u^2 + 16v^2 + 1}}.$$

Now,

$$X_{uu} = (0, 0, -2),\ X_{uv} = (0, 0, 0),\ X_{vv} = (0, 0, -4).$$

Thus,

$$L = -\frac{2}{\sqrt{4u^2 + 16v^2 + 1}},\ M = 0,\ N = -\frac{4}{\sqrt{4u^2 + 16v^2 + 1}}.$$

Therefore,

$$K = \frac{LN - M^2}{EG - F^2} = \frac{8}{(4u^2 + 16v^2 + 1)^2} > 0.$$

Further, at $p = (0, 0, 0)$, $E = 1,\ F = 0,\ G = 1$, and $L = -2,\ M = 0,\ N = -4$. Hence, $K = 8,\ H = -3$. From (4.2-7), $k = -3 \pm \sqrt{9 - 8} = -3 \pm 1$. $k_1 = -2$ and $k_2 = -4$. □

(2) If $K(p) < 0$, p is a hyperbolic point.

Example 4.2.6 Let M be the surface in $\mathbf{E}^3$ given by the patch

$$X(u, v) = (u, v, uv).$$

Then $-1 \leq K < 0$ on M. Thus, all points of M are hyperbolic. Further, the principal curvatures at $p = (0, 0, 0)$ are $k_1 = 1$ and $k_2 = -1$(Figure4.2-7).

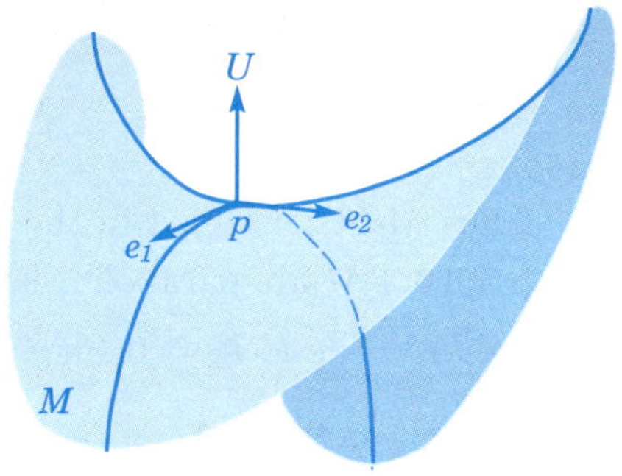

Figure4.2-7

Solution. Since $X_u = (1,0,v)$ and $X_v = (0,1,u)$, the coefficients of the 1st fundamental form are

$$E = 1+v^2,\ F = uv,\ G = 1+u^2.$$

Moreover, since $X_u \times X_v = (-v,-u,1)$ and $|X_u \times X_v| = \sqrt{u^2+v^2+1}$, the unit normal vector to M is

$$U = \frac{(-v,-u,1)}{\sqrt{u^2+v^2+1}}.$$

Now,

$$X_{uu} = (0,0,0),\ X_{uv} = (0,0,1),\ X_{vv} = (0,0,0).$$

Thus,

$$L = 0,\ M = \frac{1}{\sqrt{u^2+v^2+1}},\ N = 0.$$

Therefore,

$$K = \frac{LN - M^2}{EG - F^2} = -\frac{1}{(u^2+v^2+1)^2} < 0.$$

Further, at $p = (0,0,0)$, $E = 1,\ F = 0,\ G = 1$, and $L = 0,\ M = 1,\ N = 0$. Hence, $K = -1,\ H = 0$. From (4.2-7), $k = \pm 1$. □

(3) If $K(p) = 0$ and if $L^2 + M^2 + N^2 \neq 0$, p is a parabolic point.

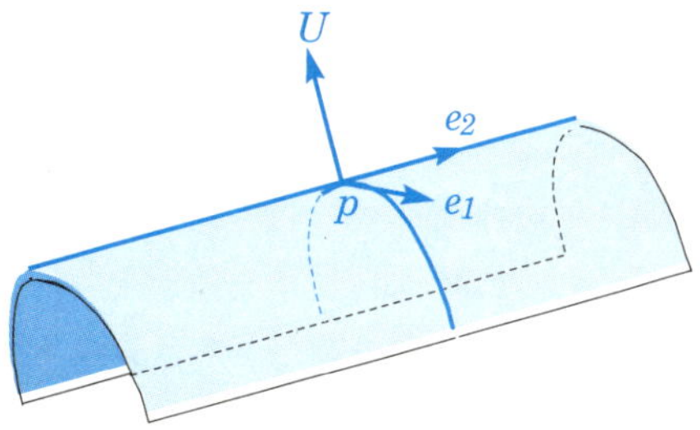

Figure4.2-8

Example 4.2.7 Let M be a surface given in $\mathbf{E}^3$ by a patch

$$X(u,v) = (r\cos u, v, r\sin u).$$

Thus M is contained in the cylinder $x^2 + z^2 = r^2$. Then $K = 0$ on M and all points of M are parabolic. Further, the principal curvatures at any point $p \in M$ are $k_1 = \dfrac{1}{r}$ and $k_2 = 0$(Figure 4.2-8).

Solution. Since $X_u = (-r\sin u, 0, r\cos u)$ and $X_v = (0,1,0)$, the coefficients of the 1st fundamental form are

$$E = r^2, \;\; F = 0, \;\; G = 1.$$

Moreover, since $X_u \times X_v = (-r\cos u, 0, -r\sin u)$ and since $|X_u \times X_v| = r$, the unit normal vector to M is

$$U = (-\cos u, 0, -\sin u).$$

Now,

$$X_{uu} = (-r\cos u, 0, -r\sin u), \;\; X_{uv} = (0,0,0), \;\; X_{vv} = (0,0,0).$$

Thus,

$$L = r, \;\; M = 0, \;\; N = 0.$$

Therefore,

$$K = \frac{LN - M^2}{EG - F^2} = \frac{0}{r^2} = 0.$$

Further, at any point $p \in M$, $H = \dfrac{r}{2r^2} = \dfrac{1}{2r}$. From (4.2-7), $k = \dfrac{1}{2r} \pm \dfrac{1}{2r}$. Thus, $k_1 = \dfrac{1}{r}$ and $k_2 = 0$. □

(4) If $K(p) = 0$ and if $L = M = N = 0$, p is a planar point.

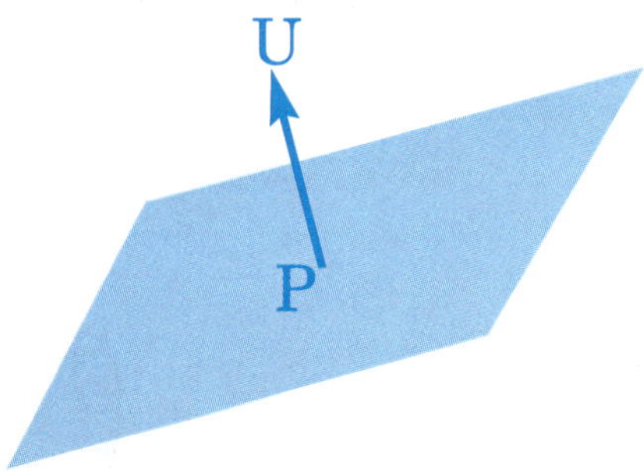

Figure4.2-9

Example 4.2.8 Consider the plane M in $\mathbf{E}^3$ given by $X(u,v) = (u,v,r)$ (by rotation and translation if necessary). Then $K = 0$ on M and all points of M are planar. Further, all normal curvatures at any point $p \in M$ are $k = 0$(Figure 4.2-9).

Solution. Since $X_u = (1,0,0)$ and $X_v = (0,1,0)$, the coefficients of the 1st fundamental form are

$$E = 1, \ F = 0, \ G = 1.$$

Moreover, since $X_u \times X_v = (0,0,1)$, the unit normal vector to M is

$$U = (0,0,1).$$

Now,

$$X_{uu} = (0,0,0), \ X_{uv} = (0,0,0), \ X_{vv} = (0,0,0).$$

Thus,

$$L = 0, \ M = 0, \ N = 0.$$

Therefore, $K = 0$.

Further, at any point $p \in M$, $H = 0$. From (4.2-7), $k = 0$. □

Alternatively, since U is a constant vector field, $S_p(v) = -\nabla_v U = (0,0,0)$ for any v. Thus $k_n(u) =< S_p(u), u >= 0$ recalling Remark 4.1.14.

The Christoffel Symbols

Since $\{X_u, X_v, U\}$ is a basis for the tangent space of $\mathbf{E}^3$ (or $\mathbf{L}^3$), at all points p of M,

$$\text{(4.2-8)} \quad X_{uu} = \Gamma_{11}^1 X_u + \Gamma_{11}^2 X_v + L_1 U,$$
$$\text{(4.2-9)} \quad X_{uv} = \Gamma_{12}^1 X_u + \Gamma_{12}^2 X_v + L_2 U,$$
$$\text{(4.2-10)} \quad X_{vu} = \Gamma_{21}^1 X_u + \Gamma_{21}^2 X_v + L_3 U,$$
$$\text{(4.2-11)} \quad X_{vv} = \Gamma_{22}^1 X_u + \Gamma_{22}^2 X_v + L_4 U.$$

Since $X_{uv} = X_{vu}$, $\Gamma_{12}^k = \Gamma_{21}^k$, $k = 1,2$.

Moreover, in $\mathbf{E}^3$ or for a timelike surface in $\mathbf{L}^3$,

$$L =< X_{uu}, U >= L_1,$$
$$M =< X_{uv}, U >= L_2,$$
$$M =< X_{vu}, U >= L_3,$$
$$N =< X_{vv}, U >= L_4.$$

Now note that $E_u = \dfrac{\partial}{\partial u} < X_u, X_u >= 2 < X_{uu}, X_u >$. Thus taking the inner product of X_u, X_v with (4.2-8) respectively,

$$\Gamma_{11}^1 E + \Gamma_{11}^2 F =< X_{uu}, X_u >= \frac{1}{2} E_u,$$

$$\Gamma_{11}^1 F + \Gamma_{11}^2 G =< X_{uu}, X_v >= F_u - < X_u, X_{vu} >= F_u - \frac{1}{2}E_v,$$

Taking the inner product of X_u, X_v to (4.2-9) or (4.2-10) respectively,

$$\Gamma_{12}^1 E + \Gamma_{12}^2 F =< X_{uv}, X_u >= \frac{1}{2}E_v,$$

$$\Gamma_{12}^1 F + \Gamma_{12}^2 G =< X_{uv}, X_v >= \frac{1}{2}G_u,$$

Taking the inner product of X_u, X_v to (4.2-11) respectively,

$$\Gamma_{22}^1 E + \Gamma_{22}^2 F =< X_{vv}, X_u >= F_v - < X_v, X_{uv} >= F_v - \frac{1}{2}G_u,$$

$$\Gamma_{22}^1 F + \Gamma_{22}^2 G =< X_{vv}, X_v >= \frac{1}{2}G_v.$$

Thus, using Cramer's Rule,

$$\Gamma_{11}^1 = \frac{E_u G - 2F_u F + E_v F}{2(EG - F^2)}, \quad \Gamma_{11}^2 = -\frac{E_u F - 2F_u E + E_v E}{2(EG - F^2)},$$

$$\Gamma_{12}^1 = \frac{E_v G - G_u F}{2(EG - F^2)}, \quad \Gamma_{12}^2 = \frac{G_u E - E_v F}{2(EG - F^2)},$$

$$\Gamma_{22}^1 = -\frac{G_u G - 2F_v G + G_v F}{2(EG - F^2)}, \quad \Gamma_{22}^2 = \frac{G_u F - 2F_v F + G_v E}{2(EG - F^2)}.$$

For smooth vector fields $F(u,v)$ and $G(u,v)$ in $\mathbf{L}^3$ paramerized on the domain D of a surface patch, it may be verified that $\frac{\partial}{\partial u} < F, G >=< F_u, G >$ $+ < F, G_u >$ and $\frac{\partial}{\partial v} < F, G >=< F_v, G > + < F, G_v >$ where $< \, , \, >$ denotes the Lorentz metric on $\mathbf{L}^3$. Hence, since the above derivation does not rely on $< U, U >= +1$ or $< U, U >= -1$, the formulas derived above for the Christoffel symbols remain valid for a timelike or spacelike surface in $\mathbf{L}^3$. In the timelike case, $EG - F^2 < 0$ by Theorem 3.3.11 and the denominators in these expressions are well defined.

The Gauss Equation

Since the Gaussian curvature K and the mean curvature H are given by the coefficients of the first and second fundamental forms, they seem to be the invariants of the extrinsic geometry (how the surface M sits in $\mathbf{E}^3$ or in $\mathbf{L}^3$).

However, the Gaussian curvature K only depends on the intrinsic geometry. In other words, K tells us the geometry of the surface M on M itself but not in $\mathbf{E}^3$ or $\mathbf{L}^3$.

Theorem 4.2.9 (Gauss's Theorema Egregium) *K can be represented by only the coefficients of the first fundamental form of the surface M in $\mathbf{E}^3$ (or of a timelike surface in $\mathbf{L}^3$).*

Proof.

$$\begin{aligned}
0 =&(X_{uu})_v - (X_{uv})_u \\
=&(\Gamma_{11}^1 X_u + \Gamma_{11}^2 X_v + LU)_v - (\Gamma_{12}^1 X_u + \Gamma_{12}^2 X_v + MU)_u \\
=&\Gamma_{11}^1 X_{uv} + \Gamma_{11}^2 X_{vv} + LU_v - \Gamma_{12}^1 X_{uu} - \Gamma_{12}^2 X_{vu} - MU_u \\
&+ (\Gamma_{11}^1)_v X_u + (\Gamma_{11}^2)_v X_v + L_v U - (\Gamma_{12}^1)_u X_u - (\Gamma_{12}^2)_u X_v - M_u U \\
=&\Gamma_{11}^1(\Gamma_{12}^1 X_u + \Gamma_{12}^2 X_v + MU) + \Gamma_{11}^2(\Gamma_{22}^1 X_u + \Gamma_{22}^2 X_v + NU) \\
&+ L(-a_{21}X_u - a_{22}X_v) - \Gamma_{12}^1(\Gamma_{11}^1 X_u + \Gamma_{11}^2 X_v + LU) \\
&- \Gamma_{12}^2(\Gamma_{12}^1 X_u + \Gamma_{12}^2 X_v + MU) - M(-a_{11}X_u - a_{12}X_v) \\
&+ (\Gamma_{11}^1)_v X_u + (\Gamma_{11}^2)_v X_v + L_v U - (\Gamma_{12}^1)_u X_u - (\Gamma_{12}^2)_u X_v - M_u U.
\end{aligned}$$

Since X_u, X_v and U are linearly independent, taking the coefficient of X_v, we obtain

$$\Gamma_{11}^1\Gamma_{12}^2 + \Gamma_{11}^2\Gamma_{22}^2 + a_{22}L - \Gamma_{12}^1\Gamma_{11}^2 - \Gamma_{12}^2\Gamma_{12}^2 - a_{12}M + (\Gamma_{11}^2)_v - (\Gamma_{12}^2)_u = 0.$$

Hence,

$$\begin{aligned}
&(\Gamma_{11}^2)_v - (\Gamma_{12}^2)_u + \Gamma_{11}^1\Gamma_{12}^2 + \Gamma_{11}^2\Gamma_{22}^2 - \Gamma_{12}^1\Gamma_{11}^2 - \Gamma_{12}^2\Gamma_{12}^2 \\
=&a_{22}L - a_{12}M \\
=&L\frac{EN - FM}{EG - F^2} - M\frac{EM - FL}{EG - F^2} \\
=&E\frac{LN - M^2}{EG - F^2} \\
=&EK.
\end{aligned}$$

Thus, we have the equation called the **Gauss equation** given by

$$K = \frac{1}{E}\left\{(\Gamma_{11}^2)_v - (\Gamma_{12}^2)_u + \Gamma_{11}^1\Gamma_{12}^2 + \Gamma_{11}^2\Gamma_{22}^2 - \Gamma_{12}^1\Gamma_{11}^2 - \Gamma_{12}^2\Gamma_{12}^2\right\}. \qquad \square$$

With our convention for the sign of the Gaussian Curvature, exactly the same derivation is valid for the timelike surface in $\mathbf{L}^3$.

Remark 4.2.10 If U is the unit timelike normal vector field on a spacelike surface M in $\mathbf{L}^3$ we may check that

$$\begin{aligned}
L &=< X_{uu}, U >= -L_1, \\
M &=< X_{uv}, U >= -L_2, \\
M &=< X_{vu}, U >= -L_3, \\
N &=< X_{vv}, U >= -L_4.
\end{aligned}$$

Thus, we have the following Gauss's Theorema Egregium for a spacelike surface M with U unit timelike normal in $\mathbf{L}^3$

Corollary 4.2.11 *K can be represented by only the coefficients of the first fundamental form of a spacelike surface M in $\mathbf{L}^3$.*

Proof.

$$\begin{aligned}
0 =&(X_{uu})_v - (X_{uv})_u \\
=&(\Gamma^1_{11}X_u + \Gamma^2_{11}X_v - LU)_v - (\Gamma^1_{12}X_u + \Gamma^2_{12}X_v - MU)_u \\
=&\Gamma^1_{11}X_{uv} + \Gamma^2_{11}X_{vv} - LU_v - \Gamma^1_{12}X_{uu} - \Gamma^2_{12}X_{vu} + MU_u \\
&+ (\Gamma^1_{11})_v X_u + (\Gamma^2_{11})_v X_v - L_v U - (\Gamma^1_{12})_u X_u - (\Gamma^2_{12})_u X_v + M_u U \\
=&\Gamma^1_{11}(\Gamma^1_{12}X_u + \Gamma^2_{12}X_v - MU) + \Gamma^2_{11}(\Gamma^1_{22}X_u + \Gamma^2_{22}X_v - NU) \\
&+ L(a_{21}X_u + a_{22}X_v) - \Gamma^1_{12}(\Gamma^1_{11}X_u + \Gamma^2_{11}X_v - LU) \\
&- \Gamma^2_{12}(\Gamma^1_{12}X_u + \Gamma^2_{12}X_v - MU) - M(a_{11}X_u + a_{12}X_v) + (\Gamma^1_{11})_v X_u \\
&+ (\Gamma^2_{11})_v X_v - L_v U - (\Gamma^1_{12})_u X_u - (\Gamma^2_{12})_u X_v + M_u U.
\end{aligned}$$

Taking the coefficient of X_v,

$$\Gamma^1_{11}\Gamma^2_{12} + \Gamma^2_{11}\Gamma^2_{22} + a_{22}L - \Gamma^1_{12}\Gamma^2_{11} - \Gamma^2_{12}\Gamma^2_{12} - a_{12}M + (\Gamma^2_{11})_v - (\Gamma^2_{12})_u = 0.$$

Hence,

$$\begin{aligned}
&(\Gamma^2_{11})_v - (\Gamma^2_{12})_u + \Gamma^1_{11}\Gamma^2_{12} + \Gamma^2_{11}\Gamma^2_{22} - \Gamma^1_{12}\Gamma^2_{11} - \Gamma^2_{12}\Gamma^2_{12} \\
=& - a_{22}L + a_{12}M \\
=& - L\frac{EN - FM}{EG - F^2} + M\frac{EM - FL}{EG - F^2} \\
=& - E\frac{LN - M^2}{EG - F^2} \\
=& - EK.
\end{aligned}$$

Thus,

$$K = -\frac{1}{E}\left\{(\Gamma^2_{11})_v - (\Gamma^2_{12})_u + \Gamma^1_{11}\Gamma^2_{12} + \Gamma^2_{11}\Gamma^2_{22} - \Gamma^1_{12}\Gamma^2_{11} - \Gamma^2_{12}\Gamma^2_{12}\right\}. \qquad \square$$

For further study, see the following references.

1. do Carmo, Manfredo Perdigan, Differential Geometry of Curves and Surfaces. Englewood Cliffs, NJ: Prentice-Hall, 1976.

2. Kuhnel, Wolfgang. Differential Geometry: Curves-Surfaces-Manifolds. Student mathematical library, vol. 16. Providence, RI: American Mathematical Society, 2002.

3. Pressley, Andrew, Elementary Differential Geometry, Springer Undergraduate Mathematics Series, 2002.

4.3 Isometries and the Fundamental Theorem for Surfaces

In this section, M and N will either denote a pair of Euclidean surfaces in $\mathbf{E}^3$, a pair of timelike surfaces in $\mathbf{L}^3$, a pair of spacelike surfaces in $\mathbf{L}^3$, or a spacelike surface in $\mathbf{L}^3$ and a Euclidean surface in $\mathbf{E}^3$.

Distances Between Two Points

Let p, q be two points of a surface M in $\mathbf{E}^3$ or $\mathbf{L}^3$ and let $\alpha : [a, b] \to M$ be a piecewise smooth curve from p to q. Recall that the length of the curve α is given by

$$L(\alpha) = \int_a^b |\alpha'(t)| dt.$$

Definition 4.3.1 The **distance** $\rho_E(p, q)$ on M in $\mathbf{E}^3$ from p to q given by

$$\rho_E(p, q) = \inf \left\{ L(\alpha) | \alpha \in \Omega^E_{p,q} \right\}$$

where $\Omega_{p,q}$ is the collection of all piecewise smooth curves on M from p to q.

In fact, $\rho_E(p, q)$ is the intrinsic distance on M differing from the extrinsic distance $d(p, q)$ in $\mathbf{E}^3$.

Example 4.3.2 Let p be the south pole and q be the north pole on the unit sphere S^2. The $\rho_E(p, q) = \pi$ while $d(p, q) = 2$(Figure4.3-10).

However, in $\mathbf{L}^3$, motivated by relativity theory, we restrict the curves to the future-directed nonspacelike curves on a timelike surface M to define the Lorentz distance on M from p to q as follows.

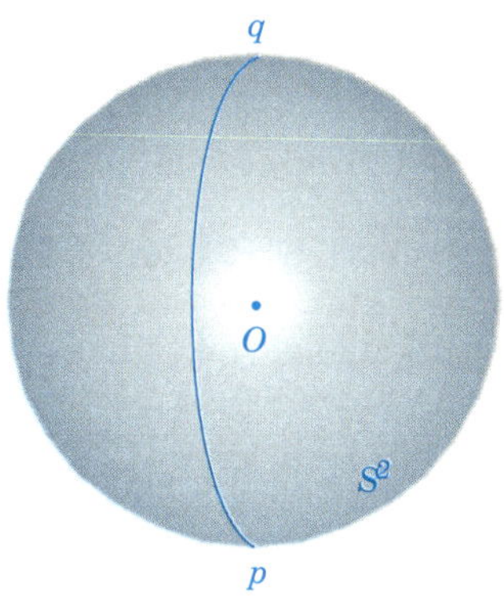

Figure4.3-10

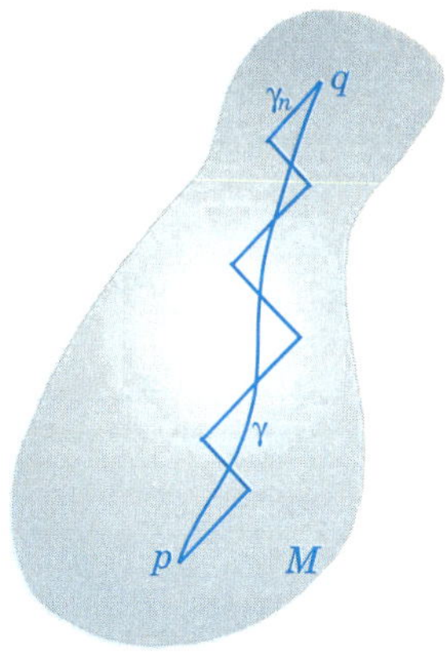

Figure4.3-11

Definition 4.3.3 Let $\Omega_{p,q}^L$ be the collection of all future-directed nonspacelike curve from p to q in a timelike surface M in $\mathbf{L}^3$. Then the **Lorentz distance** on M from p to q is given by

$$\rho_L(p,q) = \sup\left\{L(\alpha)|\alpha \in \Omega_{p,q}^L\right\}$$

and $\rho_L(p,q) = 0$ if there is no future-directed nonspacelike curve from p to q.

Remark 4.3.4 If α is a future-directed lightlike curve from p to q, then $L(\alpha) = 0$. Moreover, if we took the infimum of $\left\{L(\alpha)|\alpha \in \Omega_{p,q}^L\right\}$, then $\rho_L(p,q)$ would always be zero since a future-directed timelike curve γ may be approximated by a sequence of curves $\{\gamma_n\}$ such that $\gamma_n \to \gamma$ and $L(\gamma_n) \to 0$(Figure 4.3-11).

Isometries and Local Isometries

First, let M (resp., N) be either a surface in $\mathbf{E}^3$ or a spacelike surface in $\mathbf{L}^3$.

Definition 4.3.5 An **isometry** is a diffeomorphism $f : M \to N$ such that for any smooth curve $\gamma : [a,b] \to M$ f preserves the length of γ for the Euclidean or Minkowski inner product, i.e., $L(\gamma) = L(f \circ \gamma)$. M and N are **isometric** if there is an isometry from M to N.

Remark 4.3.6 If $f : M \to N$ is an isometry, then, since for any $t \in [a,b]$

$$\int_a^t |\gamma'(\sigma)|d\sigma = \int_a^t |(f \circ \gamma)'(\sigma)|d\sigma,$$

we have $|\gamma'(t)| = |(f \circ \gamma)'(t)|$. Now, for any $w \in T_pM$ there is a curve γ such that $\gamma(0) = p$ and $\gamma'(0) = w$. Thus $|w| = |df_p(w)|$. Therefore, since the inner products restricted to M and N are positive definite, by polarization

$$< df_p(v), df_p(w) >=< v, w >$$

for any $v, w \in T_pM$, and $p \in M$.

Turning to the case of timelike surfaces in $\mathbf{L}^3$, let

$$M = N = \{(x_1, 0, x_3) \in \mathbf{L}^3; x_1, x_3 \in \mathbb{R}\}.$$

Then the diffeomorphism $f : M \to N$ given by $f((x_1,0,x_3)) = (x_3,0,x_1)$ satisfies $< df_p(w), df_p(w) >= - < w, w >$ for any w tangent to M (and where $< , >$ denotes the Lorentz metric on $\mathbf{L}^3$). Hence, f preserves the norm $|w| = \sqrt{|< w,w >|}$ and thus the lengths of piecewise smooth curves. But f takes timelike (resp., spacelike) curves in M to spacelike (resp., timelike) curves in N. Consequently, f would not usually be considered to be an isometry. Thus, for timelike surfaces, the following definition of isometry is given, which is also used in place of Definition 4.3.5 in many texts. Because $(f \circ \gamma)'(t) = df(\gamma'(t))$, it is immediate that an isometry given as in Definition 4.3.7 preserves the lengths of curves.

Definition 4.3.7 Let M and N be timelike surfaces in $\mathbf{L}^3$. An **isometry** $f : M \to N$ is a diffeomorphism of M and N such that

$$< df_p(v), df_p(w) >=< v, w >$$

for any $v, w \in T_pM$, and $p \in M$.

Since isometries preserve the length of curves, it may be checked that if $f : M \to N$ is an isometry of surfaces in $\mathbf{E}^3$ (resp., timelike surfaces in $\mathbf{L}^3$), then

$$\rho_E(f(p), f(q)) = \rho_E(p, q)$$

(resp., $\rho_L(f(p), f(q)) = \rho_L(p, q)$) for all $p, q \in M$.

Definition 4.3.8 A map $f : M \to N$ is a **local isometry** about $p \in M$ if there are two open sets $U \subset M$, $V \subset N$ with $p \in U$ such that $f : U \to V$ is an isometry. Two surfaces M and N are **locally isometric** if for each $p \in M$ there is a local isometry at every point in M.

Note that an isometry is a local isometry. In fact, if $f : M \to N$ is a isometry, for a coordinate patch $X(u, v)$ about $p \in M$, $\bar{X}(u, v) = f(X(u, v))$ for $(u, v) \in D$ is a coordinate patch about $f(p)$ such that $f : X(D) \to \bar{X}(D)$ is an isometry.

Theorem 4.3.9 *$f : M \to N$ is a local isometry if and only if for each $p \in M$ there are two coordinate patches $X : D \to M$ about p and $\bar{X} : D \to N$ about $f(p)$ such that $E = \bar{E}, F = \bar{F}$, and $G = \bar{G}$ where $\{E, F, G\}$ and $\{\bar{E}, \bar{F}, \bar{G}\}$ are the coefficients of the first fundamental forms of M and N respectively.*

Proof. Suppose $f : M \to N$ is a local isometry and $p \in M$. Then there are two open sets $U \subset M$, $V \subset N$ with $p \in U$ such that $f : U \to V$ is an isometry. Take a coordinate patch $X(D)$ about p in U so that $\bar{X} = f \circ X : D \to V$ is a coordinate patch about $f(p)$. Then $\bar{X}_u = df_p(X_u)$ and $\bar{X}_v = df_p(X_v)$. By Remark 4.3.6 or Definition 4.3.7, $E = \bar{E}, F = \bar{F}$, and $G = \bar{G}$.

Conversely, consider a map $f : X(D) \to \bar{X}(D)$ defined by $f(X(u, v)) = \bar{X}(u, v)$ for $(u, v) \in D$. Then f is a diffeomorphism. For a curve $\gamma : [a, b] \to$

$X(D)$ we represent $\gamma(t) = X(u(t), v(t)),\ t \in [a, b]$. Hence

$$\begin{aligned} L(\gamma) &= \int_a^b |\gamma'(t)| dt \\ &= \int_a^b \sqrt{|E(u')^2 + 2Fu'v' + G(v')^2|} dt \\ &= \int_a^b \sqrt{|\bar{E}(u')^2 + 2\bar{F}u'v' + \bar{G}(v')^2|} dt \\ &= \int_a^b |(f \circ \gamma)'(t)| dt \\ &= L(f \circ \gamma). \end{aligned}$$

Therefore, $f : X(D) \to \bar{X}(D)$ is an isometry in the Euclidean or spacelike case since f preserves length of curves as required in Definition 4.3.5. In the case M and N are timelike surfaces in $\mathbf{L}^3$, since $E = \bar{E}, F = \bar{F}$, and $G = \bar{G}$ and any tangent vector to M may be represented as $w = aX_u + bX_v$, the differential df preserves the inner product as required in Definition 4.3.7. □

Example 4.3.10 Let M be the catenoid in $\mathbf{E}^3$ given by $X(u, v) = (\cosh u \cos v, \cosh u \sin v, u)$ for $-\infty < u < \infty$ and $0 < v < 2\pi$(Figure4.3-12) and let N be the helicoid in $\mathbf{E}^3$ given by $Y(u, v) = (u \cos v, u \sin v, v)$ for $-\infty < u < \infty$ and $0 < v < 2\pi$(Figure4.3-13). Then the map $f : M \to N$ defined by $f(X(u, v)) = Y(\sinh u, v)$ is a local isometry.

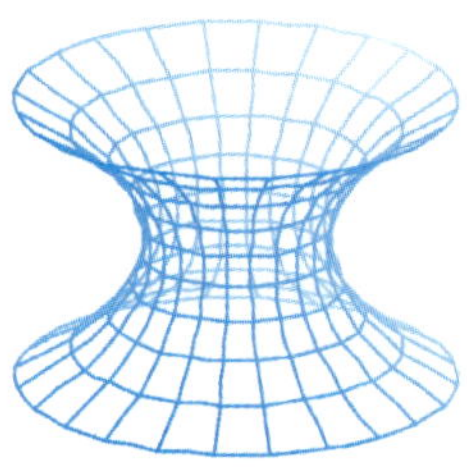

Figure4.3-12

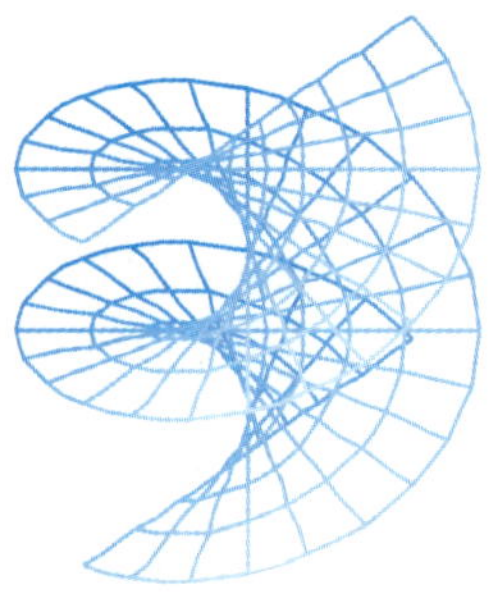

Figure4.3-13

Solution. Since

$$X_u = (\sinh u \cos v, \sinh u \sin v, 1),$$
$$X_v = (-\cosh u \sin v, \cosh u \cos v, 0),$$

we have

$$E = \sinh^2 u + 1 = \cosh^2 u, \ F = 0, \ G = \cosh^2 u.$$

Since the corresponding coordinate patch $\bar{X}(u,v)$ on N of $X(u,v)$ is given by

$$\bar{X}(u,v) = Y(\sinh u, v) = (\sinh u \cos v, \sinh u \sin v, v),$$

we also have

$$\bar{X}_u = (\cosh u \cos v, \cosh u \sin v, 0),$$

$$\bar{X}_v = (-\sinh u \sin v, \sinh u \cos v, 1),$$

and

$$\bar{E} = \cosh^2 u, \bar{F} = 0, \bar{G} = \sinh^2 u + 1 = \cosh^2 u.$$

Therefore,

$$E = \bar{E}, F = \bar{F}, G = \bar{G}.$$

By Theorem 4.3.9, f is a local isometry. □

Local Isometries and Gaussian Curvature

Theorem 4.2.9(**Gauss's Theorema Egregium**) established that the Gaussian curvature K can be computed from only the coefficients of the first fundamental form. Using this Theorem, we have also the following corollary.

Corollary 4.3.11 *If $f : M \to N$ is a local isometry of surfaces in $\mathbf{E}^3$ or a local isometry of timelike surfaces in $\mathbf{L}^3$, then their Gaussian curvatures at corresponding points are equal, i.e., $K_p = K_{f(p)}$ for $p \in M$.*

Proof. Since M and N are locally isometric, for each $p \in M$ there are open sets $U \subset M$ and $V \subset N$ of p and $f(p)$ respectively such that $f : U \to V$ is an isometry. If $X(u,v)$ is a coordinate patch about p, then $\bar{X}(u,v) = f(X(u,v))$ is a coordinate patch about $f(p)$. By Theorem 4.2.9, the Gauss formula for M in $\mathbf{E}^3$ or a timelike surface M in $\mathbf{L}^3$ is

$$K_p = \frac{1}{E}\left\{(\Gamma^2_{11})_v - (\Gamma^2_{12})_u + \Gamma^1_{11}\Gamma^2_{12} + \Gamma^2_{11}\Gamma^2_{22} - \Gamma^1_{12}\Gamma^2_{11} - \Gamma^2_{12}\Gamma^2_{12}\right\}$$

and for a spacelike surface in M in $\mathbf{L}^3$ is

$$K = -\frac{1}{E}\left\{(\Gamma^2_{11})_v - (\Gamma^2_{12})_u + \Gamma^1_{11}\Gamma^2_{12} + \Gamma^2_{11}\Gamma^2_{22} - \Gamma^1_{12}\Gamma^2_{11} - \Gamma^2_{12}\Gamma^2_{12}\right\}.$$

Since the Christoffel symbols are represented only by E, F, G and their derivatives as in Section 4.2, by Theorem 4.3.9,

$$\Gamma^k_{ij} = \bar{\Gamma}^k_{ij}, \ 1 \leq i, j, k \leq 2.$$

Thus, $K_p = K_{f(p)}$ if M and N are Euclidean surfaces in $\mathbf{E}^3$ or timelike surfaces in $\mathbf{L}^3$. □

Remark 4.3.12 If one of M and N is a spacelike surface in $\mathbf{L}^3$ and the other is a Euclidean surface in $\mathbf{E}^3$ which are locally isometric, then with the sign conventions for Gaussian curvature in $\mathbf{L}^3$ used in this book,

$$K_p = -K_{f(p)}$$

for all $p \in M$.

However, a local isometry does not necessarily preserve the mean curvature at corresponding points. It may easily be checked that a plane and a cylinder have the same coefficients of the first fundamental form but different mean curvatures as in the following Example 4.3.13. Moreover, the converse of the above Corollary 4.3.11 does not hold as the following Example 4.3.14 shows.

Example 4.3.13 Let M be a circular cylinder $\mathbf{E}^3$ given by a patch $X(u, v) = (\cos u, \sin u, v)$ where $0 < u < 2\pi$, $-\infty < v < \infty$. Define an isometry $f : M \to N$ defined by $\bar{X}(u, v) = f(X(u, v)) := (u, v, 0)$. Then f does not preserve the mean curvatures at the corresponding points.

Solution. For the coordinate patch $X(u, v) = (\cos u, \sin u, v)$ on M, we have

$X_u = (-\sin u, \cos u, 0)$,

$X_v = (0, 0, 1)$, and

$E = 1,\ F = 0,\ G = 1.$

Since the principal curvatures of M are $k_1 = 1,\ k_2 = 0$,

$H = \dfrac{1}{2}$ on M.

For the coordinate patch $\bar{X}(u, v) = (u, v, 0)$ on N,

$\bar{X}_u = (1, 0, 0)$,

$\bar{X}_v = (0, 1, 0)$, and

$\bar{E} = 1,\ \bar{F} = 0,\ \bar{G} = 1.$

Since the principal curvatures of N are $k_1 = 0,\ k_2 = 0$,

$H = 0$ on N. □

Example 4.3.14 Let M and N be two surfaces in $\mathbf{E}^3$ given by $X(u, v) = (u \cos v, u \sin v, \ln u)$ for $0 < u,\ 0 < v < 2\pi$ and $\bar{X}(u, v) = (u \cos v, u \sin v, v)$ for $0 < u,\ 0 < v < 2\pi$. Then M and N are not locally isometric but the Gaussian curvature of both surfaces at (u, v) is $K = -\dfrac{1}{(1 + u^2)^2}$.

Solution. For the coordinate patch $X(u, v) = (u \cos v, u \sin v, \ln u)$, we have

$X_u = \left(\cos v, \sin v, \dfrac{1}{u}\right)$,

$$X_v = (-u\sin v, u\cos v, 0),$$
$$E = 1 + \frac{1}{u^2},\ F = 0,\ G = u^2.$$

To find the unit normal vector field on M,

$$X_u \times X_v = (-\cos v, -\sin v, u),$$
$$|X_u \times X_v| = \sqrt{1+u^2},$$
$$U = \left(-\frac{\cos v}{\sqrt{1+u^2}}, -\frac{\sin v}{\sqrt{1+u^2}}, \frac{u}{\sqrt{1+u^2}}\right).$$

Hence we compute the second fundamental form on M.

$$X_{uu} = \left(0, 0, -\frac{1}{u^2}\right),$$
$$X_{uv} = (-\sin v, \cos v, 0),$$
$$X_{vv} = (-u\cos v, -u\sin v, 0),$$
$$L = -\frac{1}{u\sqrt{1+u^2}},\ M = 0,\ N = \frac{u}{\sqrt{1+u^2}}.$$

Thus, we have the Gaussian curvature K on M:

$$K = -\frac{1}{(1+u^2)^2}.$$

For the coordinate patch $\bar{X}(u,v) = (u\cos v, u\sin v, v)$ we have

$$\bar{X}_u = (\cos v, \sin v, 0),$$
$$\bar{X}_v = (-u\sin v, u\cos v, 1).$$

To find the unit normal vector field on N,

$$\bar{E} = 1,\ \bar{F} = 0,\ \bar{G} = u^2 + 1,$$
$$\bar{X}_u \times \bar{X}_v = (\sin v, -\cos v, u),$$
$$|\bar{X}_u \times \bar{X}_v| = \sqrt{1+u^2},$$
$$\bar{U} = \left(\frac{\sin v}{\sqrt{1+u^2}}, -\frac{\cos v}{\sqrt{1+u^2}}, \frac{u}{\sqrt{1+u^2}}\right).$$

Hence we compute the second fundamental form on N.

$$\bar{X}_{uu} = (0, 0, 0),$$
$$\bar{X}_{uv} = (-\sin v, \cos v, 0),$$
$$\bar{X}_{vv} = (-u\cos v, -u\sin v, 0),$$
$$\bar{L} = 0,\ \bar{M} = -\frac{1}{\sqrt{1+u^2}},\ \bar{N} = 0.$$

Thus, we have the Gaussian curvature $\bar{K}$ on N:

$$\bar{K} = -\frac{1}{(1+u^2)^2}.$$

□

Rigid Motion

In Corollary 4.3.11 we observed that a isometry preserves the Gaussian curvature but not the mean curvature at corresponding points. Now, what class of maps preserves both curvatures?

Definition 4.3.15 A **rigid motion of $\mathbf{E}^2$** is a map $F : \mathbf{E}^2 \to \mathbf{E}^2$ given by $F = T_C \circ T_\theta$ where T_θ is a rotation by θ

$$\begin{pmatrix} \cos\theta & -\sin\theta \\ \sin\theta & \cos\theta \end{pmatrix}$$

and T_C is a translation by C.
In $\mathbf{L}^2$, a rigid motion $F : \mathbf{L}^2 \to \mathbf{L}^2$ is given by $F = T_C \circ T_\theta$ where T_θ is a (hyperbolic) rotation by θ

$$\begin{pmatrix} \cosh\theta & \sinh\theta \\ \sinh\theta & \cosh\theta \end{pmatrix}$$

and T_C is a translation by C.

Now we define a rotation for $\mathbf{E}^3$ to be a 3×3 real matrix A with $A^{-1} = {}^tA$ and $\det A = 1$. We call such a matrix a **special orthogonal matrix**. If $A = \begin{pmatrix} a & b \\ c & d \end{pmatrix}$ is such a 2×2 real matrix, then

$$A^{-1} = \frac{1}{\det A}\begin{pmatrix} d & -b \\ -c & a \end{pmatrix} = \begin{pmatrix} a & c \\ b & d \end{pmatrix}$$

with $\det A = 1$. Hence, $a = d,\ b = -c$. Since $1 = \det A = ad - bc = a^2 + b^2$, we may write $a = \cos\theta,\ b = -\sin\theta$ for $0 \le \theta \le 2\pi$. Thus,

$$A = \begin{pmatrix} \cos\theta & -\sin\theta \\ \sin\theta & \cos\theta \end{pmatrix}.$$

Definition 4.3.16 $F : \mathbf{E}^3 \to \mathbf{E}^3$ is a **rigid motion of $\mathbf{E}^3$** if there is a rotation A of $\mathbf{E}^3$ and a vector $C \in \mathbf{E}^3$ such that $F(w) = Aw + C$.

Revisited Gauss Equations

In the proof of Theorem 4.2.9, taking the coefficient of X_v from the equation $(X_{uu})_v - (X_{uv})_u = 0$, we obtained a Gauss equation:

$$\text{(4.3-1)} \qquad EK = (\Gamma^2_{11})_v - (\Gamma^2_{12})_u + \Gamma^1_{11}\Gamma^2_{12} + \Gamma^2_{11}\Gamma^2_{22} - \Gamma^1_{12}\Gamma^2_{11} - \Gamma^2_{12}\Gamma^2_{12}.$$

Taking the coefficient of X_u from the equation $(X_{uu})_v - (X_{uv})_u = 0$, we may similarly obtain the equation:

(4.3-2) $\quad FK = (\Gamma^1_{12})_u - (\Gamma^1_{11})_v + \Gamma^2_{12}\Gamma^1_{12} - \Gamma^2_{11}\Gamma^1_{22}.$

Also, taking the coefficient of X_v from the equation $(X_{uv})_v - (X_{vv})_u = 0$, we similarly obtained two Gauss equations:

(4.3-3) $\quad GK = (\Gamma^1_{22})_u - (\Gamma^1_{12})_v + \Gamma^1_{22}\Gamma^1_{11} + \Gamma^2_{22}\Gamma^1_{12} - \Gamma^1_{12}\Gamma^1_{12} - \Gamma^2_{12}\Gamma^1_{22}.$

(4.3-4) $\quad FK = -(\Gamma^2_{12})_v + (\Gamma^2_{22})_u + \Gamma^1_{12}\Gamma^2_{12} - \Gamma^1_{22}\Gamma^2_{11}.$

Now, we take the coefficients of U from the equations $(X_{uu})_v - (X_{uv})_u = 0$ and $(X_{uv})_v - (X_{vv})_u = 0$ respectively called the **Mainardi-Codazzi equations**:

(4.3-a) $\quad L_v - M_u = L\Gamma^1_{12} + M(\Gamma^2_{12} - \Gamma^1_{11}) - N\Gamma^2_{11}.$

(4.3-b) $\quad M_v - N_u = L\Gamma^1_{22} + M(\Gamma^2_{22} - \Gamma^1_{12}) - N\Gamma^2_{12}.$

Theorem 4.3.17 (fundamental theorem for surfaces in $\mathbf{E}^3$) *Let $\{E, F, G, L, M, N\}$ be differentiable functions on an open set $V \subset \mathbb{R}^2$ with $E > 0$ and $G > 0$. Suppose that $\{E, F, G, L, M, N\}$ satisfy the Gauss equations (4.3-1) through (4.3-4) and the Mainardi-Codazzi equations (4.3-a b), and suppose $EG - F^2 > 0$. Then at each $p \in V$ there is a neighborhood U of p with $U \subset V$ and a diffeomorphism $X : U \to X(U) \subset \mathbf{E}^3$ such that the coefficients of the first and second fundamental forms for $X(U)$ are equal to $\{E, F, G\}$ and $\{L, M, N\}$ respectively. If $\bar{X} : U \to \bar{X}(U) \subset \mathbf{E}^3$ is another such diffeomorphism, then there is a rigid motion $f : \mathbf{E}^3 \to \mathbf{E}^3$ satisfying $f(X(U)) = \bar{X}(U)$.*

The proof was given by Ossian Bonnet in 1867.

For further study, see the following references.

1. do Carmo, Manfredo Perdigan, Differential Geometry of Curves and Surfaces. Englewood Cliffs, NJ: Prentice-Hall, 1976.

2. McCleary, J., Geometry from a Differentiable Viewpoint, Cambridge University Press, 1994.

3. Millman, R. S. and G. D. Parker,Elements of Differential Geometry, Prentice-Hall, Englewood Cliffs, NJ, 1977.

4. O'Neill, B., Elementary Differential Geometry, Academic Press, New York, 1966.

5. Pressley, Andrew, Elementary Differential Geometry, Springer Undergraduate Mathematics Series, 2002.

4.4 Parallel Translation and Geodesics

Parallel Vector Fields

First, let $\alpha : I \to \mathbb{R}^3$ be a differentiable curve in $\mathbf{E}^3$ or $\mathbf{L}^3$ and let $\{E_1, E_2, E_3\}$ be the standard frame field for $\mathbb{R}^3$. For a vector field $W = (w_1, w_2, w_3)$ and tangent vector v at p, the covariant derivative of W in the direction v has been given for $\mathbf{E}^3$ or $\mathbf{L}^3$ by

$$\nabla_v W = \sum_{i=1}^{3} v[w_i]E_i(p).$$

Now we want to extend the covariant differentiation to vector fields along the curve α. A **vector field** $Y(t)$ **along** $\alpha(t)$assigns a vector in $\mathbf{E}^3$ (or $\mathbf{L}^3$) at each point $\alpha(t)$ of the curve α. Thus

$$Y(t) = (y_1(t), y_2(t), y_3(t)) = \sum_{i=1}^{3} y_i(t)E_i(\alpha(t))$$

and the covariant derivative Y' of Y along α is defined by

$$Y'(t) = \sum_{i=1}^{3} y_i'(t)E_i(\alpha(t))$$

for either $\mathbf{E}^3$ or $\mathbf{L}^3$. AS a special case, if W is a vector field defined in a neighborhood of α, then $Y = W \circ \alpha$ (i.e., $Y(t) = W(\alpha(t))$) is a vector field along α. In this case, for $W = (w_1, w_2, w_3)$, we have $y_i(t) = w_i(\alpha(t))$ and hence

$$y_i'(t) = \frac{d}{dt}[w_i(\alpha(t))] = \alpha'(t)[w_i]$$

for $i = 1, 2, 3$. A differentiable vector field Y along the curve α in $\mathbf{E}^3$ or $\mathbf{L}^3$

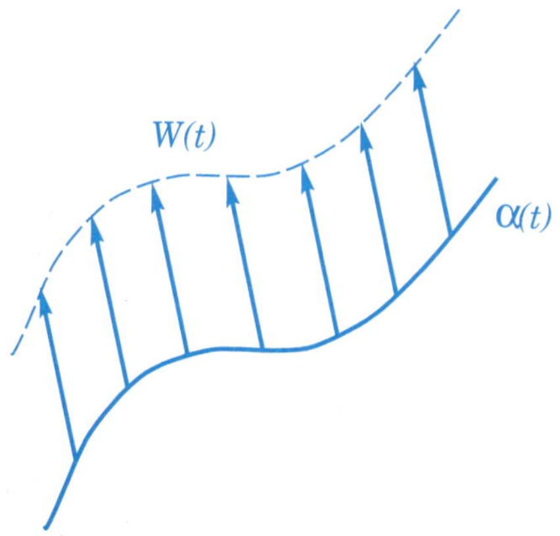

Figure4.4-14 A parallel-field in $\mathbf{E}^3$

is said to be **parallel** in $\mathbf{E}^3$ or $\mathbf{L}^3$ if $Y'(t) = 0$ for all $t \in I$. In this case, the differentiable function $y_i(t)$ satisfy $y_i'(t) = 0$ for $i = 1, 2, 3$, and thus the functions $y_i(t)$ are constant for $i = 1, 2, 3$ and the vector $Y(t)$ is parallel in $\mathbf{E}^3$ or $\mathbf{L}^3$ in the usual sense, cf. Figure4.4-14.

Now let M be a surface in $\mathbf{E}^3$ with unit normal vector field U and let $\alpha : I \to M$ be a differentiable curve in M. Given a vector field W on M, by restriction we obtain the vector field $W(t) = W(\alpha(t))$ along the curve α which also has the property that $W(t) \in T_{\alpha(t)}M$ for each $t \in I$. As in the above paragraph, we may calculate the covariant derivative $W'(t)$ in $\mathbf{E}^3$ for all $t \in I$. Even though $W(t)$ is tangent to M for all t, it is *not* necessarily the case that $W'(t)$ is tangent to M, cf. Figure4.4-15. Rather, the induced covariant derivative $\nabla_{\alpha'(t)}W$ for the surface M is implicitly defined by the equation

$$W'(t) = \nabla_{\alpha'(t)}W + a(t)U$$

where $a(t) =< W'(t), U(\alpha(t)) >$ and $\nabla_{\alpha'(t)}W$ is the (unique) projection of $W'(t)$ as a vector in $\mathbf{E}^3$ at $\alpha(t)$ onto the subspace $T_{\alpha(t)}M$.

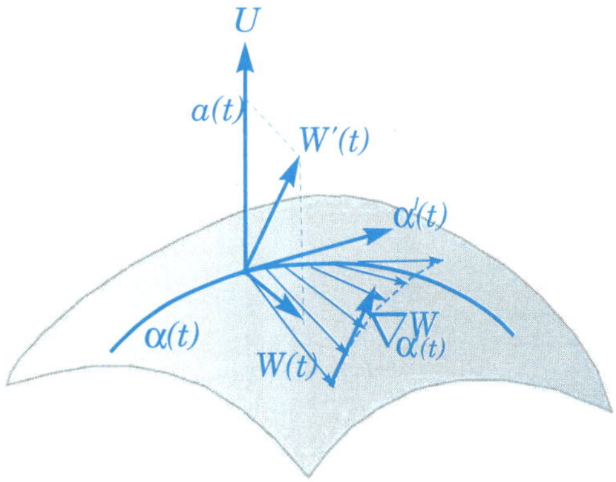

Figure4.4-15

Second, let M be a surface in $\mathbf{L}^3$ with unit normal field U which is either everywhere timelike or everywhere spacelike and let $\alpha : I \to M$ be a differentiable curve in M. Since the Lorentz metric g for $\mathbf{L}^3$ restricted to any tangent space $T_{\alpha(t)}M$ is nondegenerate, it is still possible to define a unique projection of $T_{\alpha(t)}\mathbf{L}^3$ onto $T_{\alpha(t)}M$. Hence , we may again decompose the covariant derivative $W'(t)$ in $\mathbf{L}^3$ of $W(t) = W(\alpha(t))$ as

$$W'(t) = \nabla_{\alpha'(t)}W + a(t)U$$

where $a(t) = \begin{cases} < W'(t), U(\alpha(t)) > & \text{if } U \text{ is unit spacelike} \\ - < W'(t), U(\alpha(t)) > & \text{if } U \text{ is unit timelike.} \end{cases}$

[If U were allowed to be lightlike and at some points of M, then the function $a(t)$ and projection onto $T_{\alpha(t)}M$ would not be well defined.]

Definition 4.4.1 Let M be a surface in $\mathbf{E}^3$ or a surface in $\mathbf{L}^3$ with everywhere timelike or everywhere spacelike normal vector field. A differentiable vector field W along the differentiable curve $\alpha : I \to M$ is said to be a **parallel field** along α if $\nabla_{\alpha'(t)} W = 0$ for all $t \in I$.

Unlike the situation in Figure 4.4-14, a parallel field W to a curve in a surface is *not* necessarily a constant vector field in $\mathbf{E}^3$ or $\mathbf{L}^3$. On the other hand, a vector field V in M along α which also happen to be parallel in $\mathbf{E}^3$ (or $\mathbf{L}^3$), i.e., $V'(t) = 0$, is automatically parallel in M (since $\nabla_{\alpha'(t)} V = 0$ as the projection of the zero vector field onto the tangent space to the surface). Note also that if W is parallel along α in M, then $W'(t)$ is normal to M at all points of $\alpha(t)$.

Example 4.4.2 In Example 4.1.1, $\gamma(t) = \left(\sqrt{5}\cos\frac{t}{\sqrt{5}}, \sqrt{5}\sin\frac{t}{\sqrt{5}}, \sqrt{6}\right)$ is a unit spacelike curve on the upper sheet of the hyperboloid M in $\mathbf{L}^3$. From Example 3.1.19, $U(x, y, z) = (x, y, z)$ is a unit normal to M. Now $X(t) = (0, 0, -1)$ is a parallel field in $\mathbf{L}^3$ along γ, but not a parallel field in M since $X(t)$ is not tangent to M. According to the above discussion,

$$Y(t) = X(t) + < X(t), U(\alpha(t)) > U(\alpha(t))$$

is a parallel field along α in M (but *not* in $\mathbf{L}^3$) which is the projection of $X(t)$ onto $T_{\alpha(t)}M$. A calculation shows that

$$Y(t) = \left(\sqrt{30}\cos\frac{t}{\sqrt{5}}, \sqrt{30}\sin\frac{t}{\sqrt{5}}, 5\right)$$

and also $< Y(t), U(\alpha(t)) >= 5\sqrt{6} - 5\sqrt{6} = 0$(Figure4.4-16).

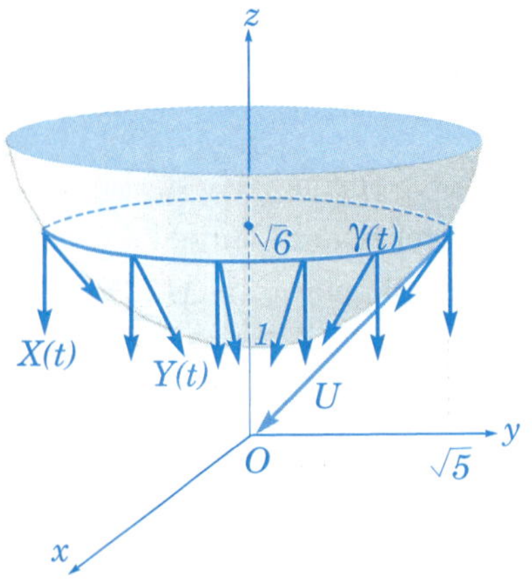

Figure4.4-16

A fundamental property of parallel fields to a differentiable curve in a surface M is the following.

Theorem 4.4.3 *Let M be a surface in $\mathbf{E}^3$ or $\mathbf{L}^3$ with everywhere timelike or everywhere spacelike unit normal field U and let $\alpha : I \to M$ be a differentiable curve. If V and W are parallel fields along α in M, then $< V, W >$ is constant. Thus, $|V|$ and $|W|$ are also constant.*

Proof. Since $\nabla_{\alpha'(t)}V = \nabla_{\alpha'(t)}W = 0$ for all t, we have $V'(t) = a_1(t)U(\alpha(t))$ and $W'(t) = a_2(t)U(\alpha(t))$. By the basic property of covariant differentiation in $\mathbf{E}^3$ or $\mathbf{L}^3$ (which is immediate from the formula for the derivative of a product),

$$\begin{aligned}\frac{d}{dt}(< V(t), W(t) >) &=< V'(t), W(t) > + < V(t), W'(t) > \\ &=< a_1(t)U(\alpha(t)), W(t) > + < V(t), a_2(t)U(\alpha(t)) > \\ &= 0 + 0 = 0,\end{aligned}$$

as $V(t)$ and $W(t)$ are tangent to M for all t. $\square$

Uniqueness and Existence of Parallel Fields

Theorem 4.4.4 *Let M be a regular surface in $\mathbf{E}^3$ and let $\alpha : I \to M$ be a differentiable curve and $t_0 \in I$. Then for any $w_0 \in T_{\alpha(t_0)}M$, there exists a unique differentiable parallel vector field W along α in M such that $W(t_0) = w_0$ and $\nabla_{\alpha'(t)}W = 0$ for all $t \in I$.*

Proof. For a proper a patch $X(u, v)$ containing a neighborhood of $\alpha(t_0)$, represent α by $\alpha(t) = X(u(t), v(t))$ and W by $W(t) = w_1(t)X_u + w_2(t)X_v$ along $\alpha(t)$. Hence, recalling the Christoffel symbols (4.2-8) through (4.2-11) of Section 4.2,

$$\begin{aligned}W'(t) &= w_1'(t)X_u + w_2'(t)X_v \\ &\quad + w_1(t)(X_{uu}u'(t) + X_{uv}v'(t)) + w_2(t)(X_{vu}u'(t) + X_{vv}v'(t)) \\ &= w_1'(t)X_u + w_2'(t)X_v \\ &\quad + w_1(t)[(\Gamma^1_{11}X_u + \Gamma^2_{11}X_v + LU)u'(t) + (\Gamma^1_{12}X_u + \Gamma^2_{12}X_v + MU)v'(t)] \\ &\quad + w_2(t)[(\Gamma^1_{12}X_u + \Gamma^2_{12}X_v + MU)u'(t) + (\Gamma^1_{22}X_u + \Gamma^2_{22}X_v + NU)v'(t)].\end{aligned}$$

Since $\nabla_{\alpha'(t)}W$ is the tangential part of $W'(t)$, and X_u and X_v are orthogonal to U,

$$\begin{aligned}&\nabla_{\alpha'(t)}W \\ &=[w_1'(t) + (\Gamma^1_{11}u'(t) + \Gamma^1_{12}v'(t))w_1(t) + (\Gamma^1_{12}u'(t) + \Gamma^1_{22}v'(t))w_2(t)]X_u \\ &\quad + [w_2'(t) + (\Gamma^2_{11}u'(t) + \Gamma^2_{12}v'(t))w_1(t) + (\Gamma^2_{12}u'(t) + \Gamma^2_{22}v'(t))w_2(t)]X_v.\end{aligned}$$

Using the initial conditions and and the definition of the parallel vector field, (as X_u and X_v are assumed to be linearly independent),

$$w_k'(t) + (\Gamma_{11}^k u'(t) + \Gamma_{12}^k v'(t))w_1(t) + (\Gamma_{12}^k u'(t) + \Gamma_{22}^k v'(t))w_2(t) = 0,$$

$k = 1, 2$ with $W(t_0) = (w_1(t_0), w_2(t_0)) = w_0$.

Now the function $u'(t), v'(t)$ and $\Gamma_{ij}^k(\alpha(t)) = \Gamma_{ij}^k(u(t), v(t))$ are all continuous in I. Hence by the uniqueness and existence of solutions for any system of first order linear differential equations, we have a unique solution $W(t) = w_1(t)X_u + w_2(t)X_v$ along $\alpha(t)$ on the entire interval I. □

Remark 4.4.5 If M is a timelike or spacelike surface in $\mathbf{L}^3$, in Section 4.2(cf. Remark 4.2.10) the second derivatives X_{uu}, X_{uv} and X_{vv} have identical expansions as in the Euclidean case in equations (4.2-8) through (4.2-11) except for sign changes in the coefficient of U. Hence exactly the same set of first order linear differential equations for parallel fields for such surfaces in $\mathbf{L}^3$ is obtained as given in the proof of Theorem 4.4.4. Therefore, if M is a timelike surface or spacelike surface in $\mathbf{L}^3$ and $\alpha : I \to M$ is a differentiable curve, exactly the same existence and uniqueness result for parallel fields holds as in Theorem 4.4.4.

Parallel Translation

With the existence of parallel fields with arbitrary initial tangent vector $v \in T_{\alpha(t_0)}M$ in hand, we may now define parallel translation along an arbitrary differentiable curve $\alpha : I \to M$ for a Euclidean surface M in $\mathbf{E}^3$ or a spacelike or timelike surface M in $\mathbf{L}^3$.

Definition 4.4.6 Given $v \in T_{\alpha(t_0)}M$, for any $t_1 \in I$ we obtain a second vector $V(t_1)$ in $T_{\alpha(t_1)}M$ where V is the unique parallel vector field along α with $V(t_0) = v$. Thus a map called the **parallel translation**

$$P_\alpha : T_{\alpha(t_0)} \to T_{\alpha(t_1)}$$

along $\alpha(t)$ is obtained, given by $P_\alpha(v) = V(t_1)$.

Corresponding to each possible choice of t_0, t_1 in I, a different map P_α is determined.

Example 4.4.7 Let α be a circle on $M = S^2$ in $\mathbf{E}^3$ given by

$$\alpha(t) = \left(\frac{1}{\sqrt{2}}\cos t, \frac{1}{\sqrt{2}}\sin t, \frac{1}{\sqrt{2}}\right)$$

for the coordinate patch $X(u,v) = (\cos u \sin v, \sin u \sin v, \cos v)$ of S^2. Let $w = X_v\left(0, \frac{\pi}{4}\right) = \left(\frac{1}{\sqrt{2}}, 0, -\frac{1}{\sqrt{2}}\right)$. Then the parallel vector field W along α with $W(0) = w$ is

$$W(t) = -\sqrt{2}\sin\left(\frac{t}{\sqrt{2}}\right) X_u + \cos\left(\frac{t}{\sqrt{2}}\right) X_v.$$

Solution. $X_u = (-\sin u \sin v, \cos u \sin v, 0)$,

$X_v = (\cos u \cos v, \sin u \cos v, -\sin v)$,

$E = \sin^2 v,\ F = 0,\ G = 1.$

Then, the Christoffel symbols given in Section 4.2 are

$$\Gamma^1_{11} = \frac{E_u G - 2F_u F + E_v F}{2(EG - F^2)} = 0,$$

$$\Gamma^2_{11} = -\frac{E_u F - 2F_u E + E_v E}{2(EG - F^2)} = -\sin v \cos v,$$

$$\Gamma^1_{12} = \frac{E_v G - G_u F}{2(EG - F^2)} = \cot v,$$

$$\Gamma^2_{12} = \frac{G_u E - E_v F}{2(EG - F^2)} = 0,$$

$$\Gamma^1_{22} = -\frac{G_u G - 2F_v G + G_v F}{2(EG - F^2)} = 0,$$

$$\Gamma^2_{22} = \frac{G_u F - 2F_v F + G_v E}{2(EG - F^2)} = 0.$$

Since $\alpha(t) = X(u(t), v(t)) = X\left(t, \frac{\pi}{4}\right)$, we have $u(t) = t,\ v(t) = \frac{\pi}{4}$.

Now let's compute the following differential equations to find the parallel vector field W with $W(0) = w$.

$$\begin{cases} w_1'(t) + \cot\left(\frac{\pi}{4}\right) w_2(t) = 0, \\ w_2'(t) - \sin\left(\frac{\pi}{4}\right)\cos\left(\frac{\pi}{4}\right) w_1(t) = 0, \end{cases}$$

with $w_1(0) = 0,\ w_2(0) = 1$(since $W\left(0, \frac{\pi}{4}\right) = 0 \cdot X_u + 1 \cdot X_v$.).

That is,

$$\begin{cases} w_1'(t) + w_2(t) = 0, \\ w_2'(t) - \frac{1}{2} w_1(t) = 0, \end{cases}$$

with $w_1(0) = 0,\ w_2(0) = 1$.

Therefore, the solution of the above differential equation system is $W(t) = -\sqrt{2}\sin\left(\frac{t}{\sqrt{2}}\right) X_u + \cos\left(\frac{t}{\sqrt{2}}\right) X_v$.

Let $t_0 = 0$ and $t_1 = 2\pi$, so that $\alpha(0) = \alpha(2\pi) = \left(\frac{1}{\sqrt{2}}, 0, \frac{1}{\sqrt{2}}\right)$. Consider the parallel translation of the above initial vector $w = X_v\left(0, \frac{\pi}{4}\right)$ around the curve α in S^2 back to $\alpha(2\pi) = \alpha(0)$. According to the above calculation,

$$\begin{aligned} P_\alpha(w) =& W(2\pi) = -\sqrt{2}\sin\left(\frac{2\pi}{\sqrt{2}}\right) X_u\left(2\pi, \frac{\pi}{4}\right) \\ &+ \cos\left(\frac{2\pi}{\sqrt{2}}\right) X_v\left(2\pi, \frac{\pi}{4}\right) \\ =& -\sqrt{2}\sin\left(\frac{2\pi}{\sqrt{2}}\right) X_u\left(0, \frac{\pi}{4}\right) + \cos\left(\frac{2\pi}{\sqrt{2}}\right) X_v\left(0, \frac{\pi}{4}\right). \end{aligned}$$

Thus, $P_\alpha(w) \neq w$ even though $\alpha(2\pi) = \alpha(0)$. □

Theorem 4.4.8 *Let M be a surface in $\mathbf{E}^3$ or a timelike or spacelike surface in $\mathbf{L}^3$ and $\alpha : I \to M$ a differentiable curve. Then for any t_0, t_1 in I,*

$$P_\alpha : T_{\alpha(t_0)}M \to T_{\alpha(t_1)}M$$

is a linear isomorphism which preserves the Euclidean (resp., Lorentz) inner product.

Proof. (i) P_α is linear.

Given $v,\ w \in T_{\alpha(t_0)}M$, there are two parallel vector fields $V,\ W$ such that $V(t_0) = v,\ W(t_0) = w$. For $v + w \in T_{\alpha(t_0)}$ there is a parallel vector field U such that $U(t_0) = v + w$. Hence, $U(t_0) = v + w = V(t_0) + W(t_0) = (V + W)(t_0)$. Since $V + W$ is a parallel vector field along α such that $(V + W)(t_0) = v + w$, by uniqueness of the parallel vector field for the initial vector, $U = V + W$. Thus, $P_\alpha(v + w) = U(t_1) = (V + W)(t_1) = V(t_1) + W(t_1) = P_\alpha(v) + P_\alpha(w)$.

Now, for a real number $a \in \mathbb{R}$ and given $av \in T_{\alpha(t_0)}M$ there is a parallel vector field U along α such that $U(t_0) = av$. By uniqueness, $U = aV$ with V as above. Hence, $P_\alpha(av) = U(t_1) = (aV)(t_1) = aV(t_1) = aP_\alpha(v)$.

(ii) P_α is an one to one map.

Suppose $P_\alpha(v) = 0$. Then for the parallel vector field V such that $V(t_0) = v$, $V(t_1) = P_\alpha(v) = 0$. O is also a parallel vector field along α such that $O(t_1) = 0$. By the uniqueness of the parallel vector field with the same initial conditions $V(t_1) = 0 = O(t_1)$, $V = O$. Therefore, $v = V(t_0) = O(t_0) = 0$.

(iii) P_α is an onto map.

We have a unique parallel vector field W with the initial condition $w \in T_{\alpha(t_1)}M$ along α. Hence, there is $W(t_0) \in T_{\alpha(t_0)}M$ such that $P_\alpha(W(t_0)) = W(t_1) = w$.

(iv) P_α preserves the inner product.

For any $v, w \in T_{\alpha(t_0)}M$, by Theorem 4.4.3 we have

$$\begin{aligned} < P_\alpha(v), P_\alpha(w) > &=< V(t_1), W(t_1) > \\ &=< V(t_0), W(t_0) >=< v, w > \end{aligned}$$

□

Geodesics

Definition 4.4.9 A differentiable unit speed curve $\beta(s) : I \to M$ on a surface M in $\mathbf{E}^3$ or $\mathbf{L}^3$ is said to be a **geodesic** if $\beta''(s)$ is perpendicular to M for all $s \in I$.

As $Y(s) = \beta'(s)$ is itself a vector field along β, the covariant derivative in M of β', denoted by $\nabla_{\beta'(s)}\beta'$ or by $\nabla_{\beta'(s)}\beta'(s)$, may be defined as discussed for general Y along $\beta(s)$ at the beginning of this section. Hence, for a surface M in $\mathbf{E}^3$, or a spacelike or a timelike surface M in $\mathbf{L}^3$, Definition 4.4.9 is equivalent to requiring that $\nabla_{\beta'(s)}\beta'(s) = 0$ in M for all $s \in I$. Lightlike geodesics on a surface M in $\mathbf{L}^3$ may also be considered, but since there is no concept of a unit speed parametrization or a Frenet-Serret apparatus corresponding exactly to Theorem 2.1.9, we shall not treat this kind of geodesic in this section. A lightlike geodesic $\beta(s)$would satisfy $< \beta'(s), \beta'(s) >= 0$ and $\nabla_{\beta'(s)}\beta'(s) = 0$ in M for all $s \in I$.

Proposition 4.4.10 *(1) Any geodesic $\beta(s)$ on M has constant speed in $\mathbf{E}^3$ (or $\mathbf{L}^3$).*
(2) Let $\beta : I \to M$ be a differentiable unit speed curve on a surface M in $\mathbf{E}^3$ or a spacelike or timelike surface M in $\mathbf{L}^3$. If the curve β is a normal section of $\beta'(s)$ for all $s \in I$ (recall Definition 4.1.8), then β is a geodesic M.
(3) If $\beta(s)$ is a unit speed geodesic on M in $\mathbf{E}^3$ with positive curvature in $\mathbf{E}^3$, the shape operator along $\beta(s)$ is

$$S(\beta') = S(T) = \begin{cases} \kappa T - \tau B & \text{if } U(\beta) = N \\ -\kappa T + \tau B & \text{if } U(\beta) = -N. \end{cases}$$

If M is a spacelike or timelike surface in $\mathbf{L}^3$ and $\beta : I \to M$ is a differentiable curve with positive curvature in $\mathbf{L}^3$ and $U(\beta) = N$, then

$$S(\beta') = S(T) = \begin{cases} -\kappa T - \tau B & \text{if } \beta \text{ is timelike, or spacelike curve on} \\ & \text{a spacelike surface} \\ \kappa T + \tau B & \text{if } \beta \text{ is spacelike on a timelike surface.} \end{cases}$$

Proof. Let $\beta(s)$ be a geodesic on a surface M. Then
(1) $\frac{d}{ds} < \beta'(s), \beta'(s) >= 2 < \beta''(s), \beta'(s) >= 0$ by the basic properties of differentiation in $\mathbf{E}^3$ or $\mathbf{L}^3$.

(2) By definition of the normal section, $T'(s) = \beta''(s)$ is parallel to U and hence normal to M.

(3) If $U = N$, since $N = \frac{\beta''(s)}{\kappa}$ is parallel to U, $S(T) = -U'(s) = -N'(s) = \kappa T - \tau B$. By the same manner, the other case may be proved. In $\mathbf{L}^3$, from Theorem 2.1.9

$$N'(s) = \begin{cases} \kappa T + \tau B & \text{if } \beta \text{ is timelike, or spacelike with } N \text{ timelike} \\ -\kappa T - \tau B & \text{if } \beta \text{ is spacelike with } N \text{ spacelike.} \end{cases}$$

Thus we obtain the result as $S(\beta') = -N'(s)$. □

Example 4.4.11 (1) Any straight line on a surface is a geodesic(Figure 4.4-17).

(2) The geodesics of $S^2(r)$ are the great circles of constant speed(Figure 4.4-18).

(3) The geodesics of the circular cylinder $\Gamma : x^2 + y^2 = r^2$ are all curves of the form(Figure 4.4-19)

$$\alpha(t) = (r\cos(at + b), r\sin(at + b), ct + d).$$

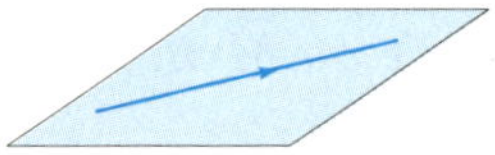

Figure4.4-17

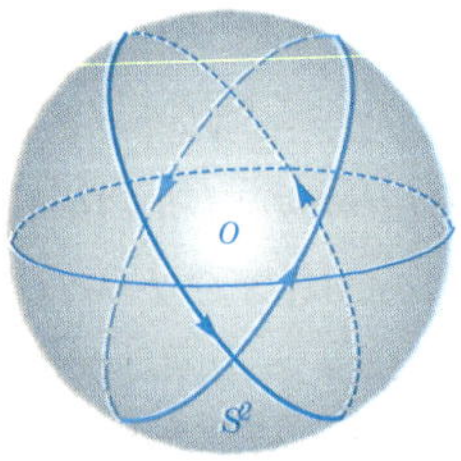

Figure4.4-18

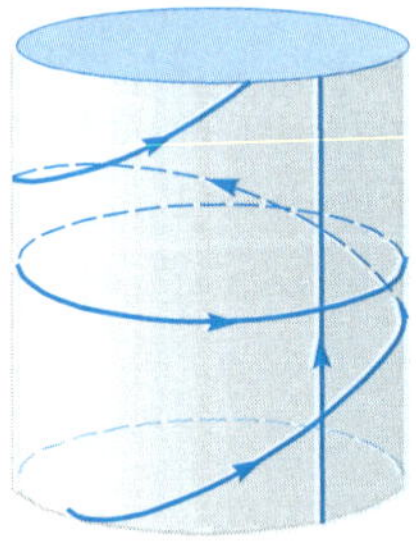

Figure4.4-19

Solution. (1) A straight line may be given by $\beta(t) = A + tB$ where A and B are constant vectors. Clearly, $\beta''(t) = 0$.

(2) Let β be a unit speed geodesic on $S^2(r)$. By Proposition 4.4.10, taking $U = -N$ we have the shape operator $S(T) = -\kappa T + \tau B$. On the other hand, since $U = \frac{1}{r}\sum_{i=1}^{3} x_i E_i$ and for any $v \in T_pM$ $S_p(v) = -dU(v) = -\frac{1}{r}\sum_{i=1}^{3} v[x_i]E_i(p) = -\frac{v}{r}$, we have $S(T) = -\frac{1}{r}T$. Hence, $\kappa = \frac{1}{r}$ and $\tau = 0$, which implies that β is a circle of radius r. Thus, β is a great circle of $S^2(r)$.

Conversely, if α is a great circle of constant speed on $S^2(r)$, the $\alpha''(t)$ has the direction pointing toward the center of $S^2(r)$. Hence $\alpha''(t)$ is normal to $S^2(r)$ and thus α is a geodesic on $S^2(r)$.

(3) We may write any curve $c(t)$ on the circular cylinder as

$$c(t) = (r\cos\theta(t), r\sin\theta(t), h(t)).$$

Since $c(t)$ was supposed to be a geodesic on Γ, $c''(t)$ must be normal to Γ, which implies that $c''(t)$ is a vector parallel to the xy-plane. Hence $h''(t) = 0$, or $h(t) = ct + d$ where c, d are constant. Now, $c'(t) = (-r\sin\theta(t)\cdot\theta'(t), r\cos\theta(t)\cdot\theta'(t), h'(t))$. Since $c(t)$ has a constant speed, $|c'(t)| = \sqrt{r^2(\theta'(t))^2 + (h'(t))^2}$ is constant. Thus $\theta'(t)$ must be constant, so that $\theta(t) = at + b$ where c, d are constant. □

In Remark 4.1.7, we may recall whether M is a surface in $\mathbf{E}^3$ or a timelike or spacelike surface in $\mathbf{L}^3$ that

$$\beta''(s) = \pm k_n U \pm k_g U \times T$$

where k_n is the normal curvature of β and $k_g =< \beta''(s), U \times T >$ is the geodesic curvature of β. Note that $k_g =< T'(s), U \times T >=< U, T \times T'(s) >$ by Theorem 1.3.10-(3) and the corresponding result in $\mathbf{E}^3$.

Theorem 4.4.12 *A unit speed curve $\beta(s)$ is a geodesic on M if and only if $k_g(s) = 0$.*

Proof. It is obvious from the equation of Remark 4.1.7 that $\beta''(s) = \pm k_n(s)U(s)$ if and only if $\beta(s)$ is a geodesic, if and only if $k_g = 0$. □

Example 4.4.13 Let M be a surface given by the Monge patch $X(u,v) = (u, v, -u^2 - v^2)$ with $u^2 + v^2 < \dfrac{1}{4}$ in $\mathbf{L}^3$. Then the curve $\alpha(t) = (t, 0, -t^2)$ with $|t| < \dfrac{1}{2}$ may be reparametrized to be a spacelike geodesic on M(Figure4.4-20).

Solution. Suppose $\beta(s) = \alpha(t(s))$ is a reparametrization by arc length of $\alpha(t)$. Then $\beta'(s) = \alpha'(t)\dfrac{dt}{ds}$ and $\beta''(s) = \alpha''(t)\left(\dfrac{dt}{ds}\right)^2 + \alpha'(t)\dfrac{d^2t}{ds^2}$.

Now, $k_g =< U, T \times T'(s) >=< U, \beta'(s) \times \beta''(s) >=< U, \alpha'(t) \times \alpha''(t) > \left(\dfrac{dt}{ds}\right)^3$.

Since, $X_u = (1, 0, -2u)$, $X_v = (0, 1, -2v)$, $X_u \times X_v = (2u, 2v, -1)$ and $U = \dfrac{(2u, 2v, -1)}{\sqrt{1 - 4u^2 - 4v^2}}$, $U(\alpha(t)) = \dfrac{(2t, 0, -1)}{\sqrt{1 - 4t^2}}$.

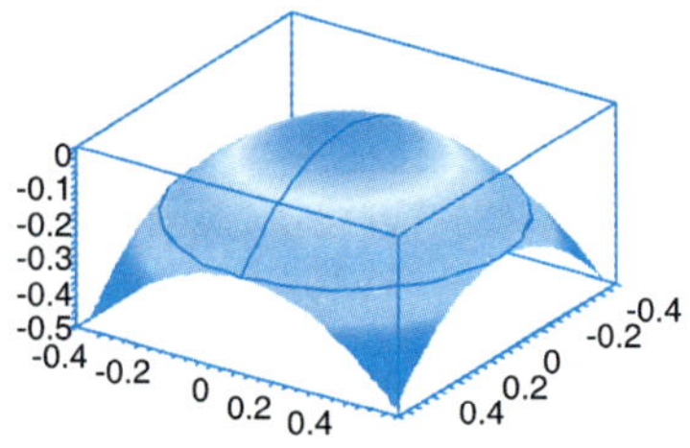

Figure4.4-20

Moreover, since $\alpha'(t) = (1, 0, -2t)$ and $\alpha''(t) = (0, 0, -2)$, $\alpha'(t) \times \alpha''(t) = (0, 2, 0)$. Hence, $< U, \alpha'(t) \times \alpha''(t) >= 0$, and thus $k_g = 0$.
Further, $g(\alpha'(t), \alpha'(t)) = 1 - 4t^2 > 0$ since $u^2 + v^2 < \dfrac{1}{4}$ for $u = t,\ v = 0$, or $|t| < \frac{1}{2}$. Therefore, α is a spacelike curve on M with $k_g = 0$, so may be reparametrized to be a spacelike geodesic. □

Uniqueness and Existence of Geodesics

Theorem 4.4.14 *Let M be a surface in $\mathbf{E}^3$. For any $p \in M$ and unit vector $v \in T_pM$, there is a unique geodesic $\beta : (a, b) \to M$ with $\beta(0) = p$ and $\beta'(0) = v$ for some real numbers a, b with $a < 0 < b$.*

Proof. For a proper patch $X(u, v)$ containing a neighborhood of p in M, we represent the desired geodesic $\beta(s)$ by $\beta(s) = X(u(s), v(s))$. Since $\beta(s)$ is a geodesic if and only if $\nabla_{\beta'(s)}\beta'(s) = 0$ in M, or equivalently, the tangential projection of $\beta''(s) = (\beta')'(s)$ onto $T_{\beta(s)}M$ is zero, we calculate this projection along the lines of the proof of Theorem 4.4.4. First,

$$\beta'(s) = u'(s)X_u + v'(s)X_v$$

and thus

$$\begin{aligned}
\beta''(s) =& \nabla_{\beta'(s)} X_u u'(s) + X_v v'(s) \\
=& \frac{d^2u}{ds^2} X_u + X_{uu} u'(s)^2 + X_{uv} u'(s) v'(s) \\
&+ \frac{d^2v}{ds^2} X_u + X_{vu} u'(s) v'(s) + X_{vv} v'(s)^2 \\
=& \frac{d^2u}{ds^2} X_u + \left(\Gamma^1_{11} X_u + \Gamma^2_{11} X_v + LU\right) u'(s)^2 \\
&+ \left(\Gamma^1_{12} X_u + \Gamma^2_{12} X_v + MU\right) u'(s) v'(s) \\
&+ \frac{d^2v}{ds^2} X_u + \left(\Gamma^1_{12} X_u + \Gamma^2_{12} X_v + MU\right) u'(s) v'(s) \\
&+ \left(\Gamma^1_{22} X_u + \Gamma^2_{22} X_v + NU\right) v'(s)^2.
\end{aligned}$$

Hence,

$$\begin{aligned}
\nabla_{\beta'(s)} \beta'(s) =& \left[\frac{d^2u}{ds^2} + \Gamma^1_{11}\left(\frac{du}{ds}\right)^2 + 2\Gamma^1_{12}\left(\frac{du}{ds}\right)\left(\frac{dv}{ds}\right) + \Gamma^1_{22}\left(\frac{dv}{ds}\right)^2\right] X_u \\
&+ \left[\frac{d^2v}{ds^2} + \Gamma^2_{11}\left(\frac{du}{ds}\right)^2 + 2\Gamma^2_{12}\left(\frac{du}{ds}\right)\left(\frac{dv}{ds}\right) + \Gamma^2_{22}\left(\frac{dv}{ds}\right)^2\right] X_v \\
=& \ 0.
\end{aligned}$$

Thus, we have the 2nd order system of the nonlinear differential equations

$$\begin{aligned}
&\frac{d^2u}{ds^2} + \Gamma^1_{11}\left(\frac{du}{ds}\right)^2 + 2\Gamma^1_{12}\left(\frac{du}{ds}\right)\left(\frac{dv}{ds}\right) + \Gamma^1_{22}\left(\frac{dv}{ds}\right)^2 = 0 \\
&\frac{d^2v}{ds^2} + \Gamma^2_{11}\left(\frac{du}{ds}\right)^2 + 2\Gamma^2_{12}\left(\frac{du}{ds}\right)\left(\frac{dv}{ds}\right) + \Gamma^2_{22}\left(\frac{dv}{ds}\right)^2 = 0
\end{aligned}$$

with $p = X(u(0), v(0))$ and $v = X_u u'(0) + X_v v'(0)$.

By the uniqueness and existence for solutions of systems of 2nd order differential equations with given initial conditions, we obtain a unique solution $u(s), v(s)$ on some interval (a, b) containing 0. □

Remark 4.4.15 By rescaling the initial tangent vector in T_pM, it may be shown for a surface M in $\mathbf{E}^3$ that Theorem 4.4.14 actually implies that for any $v \in T_pM$, there exists a unique geodesic $\gamma(t)$ with $\gamma(0) = p$ and $\gamma'(0) = v$ defined on some initial I containing 0. In this case, $<\gamma'(t), \gamma'(t)> = <v, v>$ for all t, cf. Proposition 4.4.10-(1). (If v is the zero vector in T_pM, then $\gamma(t) = p$ for all t.)

For a spacelike or a timelike surface M in $\mathbf{L}^3$, arguments analogous to those in Theorem 4.4.14 show that given any unit timelike (resp., spacelike) $v \in T_pM$, there exists a unique timelike (resp., spacelike) geodesic $\beta : I \to M$ with

$\beta(0) = p$ and $\beta'(0) = v$. By rescaling, the result extends to arbitrary spacelike or timelike initial vectors and by different arguments, a corresponding result may be obtained for lightlike vectors. Thus it is actually true for a spacelike or timelike surface M in $\mathbf{L}^3$ that any given $v \in T_pM$, there exists a unique geodesic $\gamma : I \to M$, $\gamma = \gamma(t)$, with $\gamma(0) = p$ and $\gamma'(0) = v$ for some interval I containing 0.

For further study, see the following references.

1. do Carmo, Manfredo Perdigan, Differential Geometry of Curves and Surfaces. Englewood Cliffs, NJ: Prentice-Hall, 1976.

2. McCleary, J., Geometry from a Differentiable Viewpoint, Cambridge University Press, 1994.

3. Millman, R. S. and G. D. Parker,Elements of Differential Geometry, Prentice-Hall, Englewood Cliffs, NJ, 1977.

4. O'Neill, B., Elementary Differential Geometry, Academic Press, New York, 1966.

5. Pressley, Andrew, Elementary Differential Geometry, Springer Undergraduate Mathematics Series, 2002.

Chapter 5
Global Theory of Surfaces

5.1 The Local Gauss-Bonnet Theorem

The Gauss-Bonnet theorem is the most important theorem in Mathematics, which shows that the geometrical property "Gaussian curvature" is related to the topological property "Euler number".

The n-Polygonal Curves

Definition 5.1.1 Let M be a surface in $\mathbf{E}^3$. A regular curve $\gamma : [0, L] \to M$ is called a **simple closed curve** with period L if $\gamma(0) = \gamma(L)$ and $\gamma(t_1) \neq \gamma(t_2)$ for $t_1 \neq t_2 \in [0, L)$. γ is said to be **positively oriented** if the region R enclosed by γ in M is positively oriented.

A region R on M is said to be **simply connected** if every closed curve in R may be shrunk continuously to a point in R. Thus, there are no holes in R.

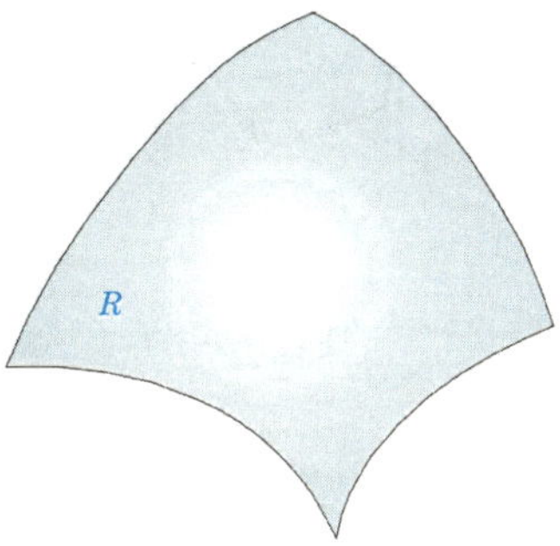

Figure 5.1-1 simply connected region

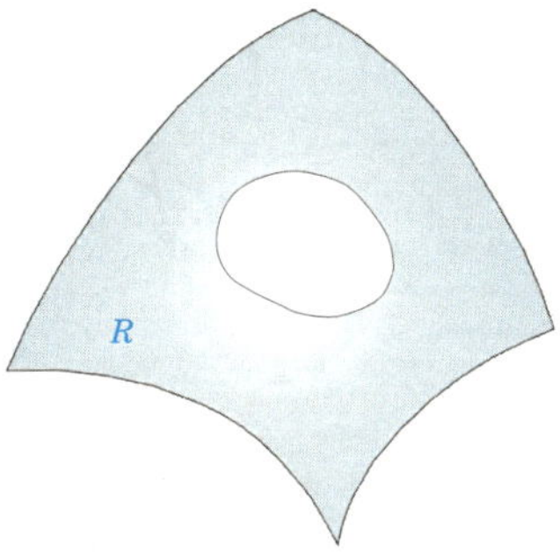

Figure 5.1-2 non simply connected region

Then $c : [a, b] \to M$ is called an n-**polygonal curve**(Figure 5.1-3) if for a partition of $[a, b]$: $a = t_0 < t_1 < t_2 < \cdots < t_n = b$ and for each $i = 1, 2, \ldots\ldots, n$,

(1) $C_i(t) = c(t),\ t \in [t_{i-1}, t_i]$ is an 1-1 regular arc.

(2) $c = C_1 + C_2 + C_3 + \cdots + C_n$, where $C_{i-1} + C_i$ implies that $C_{i-1}(t_{i-1}) = C_i(t_{i-1})$.

(3) c is a simply closed curve with $c(a) = C_1(a) = C_n(b) = c(b)$.

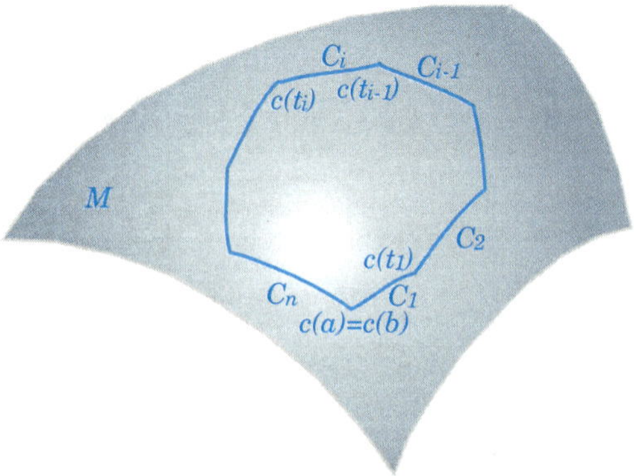

Figure5.1-3

$c(t_i)$ is called the **vertex** of n-polygonal curve c and C_i is called the **edge** of c. if $n = 3$, c is called the **triangle** on M. And if $n = 4$, c is called the **rectangle** on M.

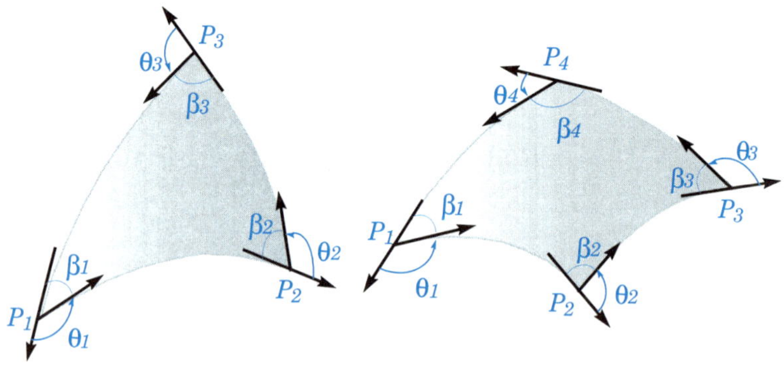

Figure5.1-4

In Figure5.1-4, P_i is the vertex and θ_i and β_i are the **exterior angle** and **interior angle** at the vertex P_i respectively.

Let (X, D) be a coordinate patch compatible with the orientation of the surface M and $c: [a, b] \to X(D) \subset M$ be a simply connected n-polygonal curve in $\mathbf{E}^3$ such that $c = C_1 + C_2 + \cdots + C_n$ where $C_1, C_2, \cdots, C_n$ are unit speed curves, and θ_i be the exterior angle given by $\theta_i = \measuredangle(C'_{i-1}(t_{i-1}), C'_i(t_{i-1}))$ for $1 \leq i \leq n$ with $C'_{n+1} = C'_1$.

Theorem 5.1.2 *Let θ_i be the exterior angle at the vertex $P_i = c(t_i)$ of the n-polygonal curve c on a regular surface in $\mathbf{E}^3$ and $\varphi_i(t)$ is the angle between X_u and $C_i'(t)$. Then*

$$\sum_{i=1}^{n}(\varphi_i(t_i) - \varphi_i(t_{i-1})) + \sum_{i=1}^{n}\theta_i = \pm 2\pi$$

with the sign according to the orientation of c.

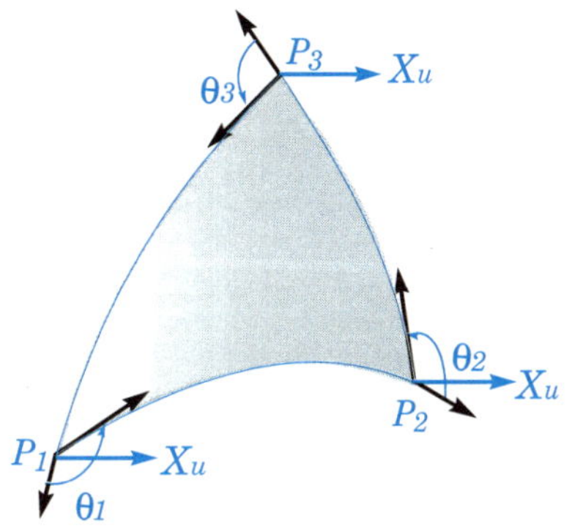

Figure5.1-5

Proof. At first, we show that it holds on a triangle. As in Figure 5.1-5, since the angle $\varphi_i(t)$ between X_u and $C_i'(t)$ jumps the exterior angle θ_i at the vertex, if it turns the edges of the triangle one time counterclockwise,

$$\sum_{i=1}^{3}\int_{t_{i-1}}^{t_i}\varphi_i'(t)dt + \sum_{i=1}^{3}\theta_i = \pm 2\pi$$

and $\displaystyle\int_{t_{i-1}}^{t_i}\varphi_i'(t)dt = \varphi_i(t_i) - \varphi_i(t_{i-1})$. On the n-polygonal curve, it works by the same manner. □

Lemma 5.1.3 *Let M be a surface in $\mathbf{E}^3$ and $\alpha : I \to M$ be a regular curve. Suppose that V and W are unit tangent vector fields and that U is a unit normal vector field along α. If $\overline{V} = U \times V$ and $\overline{W} = U \times W$, then*

$$\left\langle \frac{dW}{dt}, \overline{W} \right\rangle = \left\langle \frac{dV}{dt}, \overline{V} \right\rangle + \frac{d\varphi}{dt}$$

where $\varphi = (V, W)$ denotes an angle between V and W.

Proof. Since $< V, \overline{V} >= 0$ and since $V, \overline{V}$ are unit, $\{V, \overline{V}\}$ is a frame field along α. Hence,

$$W = \cos\varphi V + \sin\varphi \overline{V}$$

and

$$W' = -\sin\varphi \cdot \varphi' V + \cos\varphi V' + \cos\varphi \cdot \varphi' \overline{V} + \sin\varphi \overline{V}'.$$

Thus,

$$\overline{W} = U \times W = \cos\varphi(U \times V) + \sin\varphi(U \times \overline{V}) = \cos\varphi\overline{V} + \sin\varphi(-V).$$

Since $< V', \overline{V} > + < V, \overline{V}' >= 0$ and $< \overline{V}', \overline{V} >=< V', V >$,

$$\begin{aligned}
< W', \overline{W} >= & -\sin\varphi\cos\varphi \cdot \varphi' < V, \overline{V} > + \cos^2\varphi < V', \overline{V} > \\
& + \cos^2\varphi \cdot \varphi' < \overline{V}, \overline{V} > + \sin\varphi\cos\varphi < \overline{V}', \overline{V} > \\
& + \sin^2\varphi \cdot \varphi' < V, V > - \sin\varphi\cos\varphi < V', V > \\
& - \sin\varphi\cos\varphi \cdot \varphi' < \overline{V}, V > - \sin^2\varphi < \overline{V}', V > \\
= & \varphi' + \cos^2\varphi < V', \overline{V} > - \sin^2\varphi < \overline{V}', V > \\
= & \varphi' + < V', \overline{V} > .
\end{aligned}$$

□

Corollary 5.1.4 *Let $V(s)$ is a unit parallel vector field along a curve $\gamma : I \rightarrow M$ in $\mathbf{E}^3$ and let $\varphi = (\gamma'(s), V(s))$. Then*

$$k_g(s) = \frac{d\varphi}{ds}.$$

Proof. Using the fact that $\nabla_{\gamma'(s)} V = 0$ and Lemma 5.1.3,

$$\begin{aligned}
k_g(s) &=< T'(s), U \times T(s) >=< T'(s), \overline{T}(s) > \\
&= \frac{d\varphi}{ds} + < V'(s), \overline{V} >= \frac{d\varphi}{ds} + < \nabla_{\gamma'(s)} V, \overline{V} > \\
&= \frac{d\varphi}{ds}.
\end{aligned}$$

□

The Revisited Hyperbolic Angles

Let M be a everywhere timelike or everywhere spacelike surface in $\mathbf{L}^3$. Then, on the spacelike surface M the Lorentz metric is positive definite, and hence an n polygonal curve on the spacelike surface M is given as on the surface in $\mathbf{E}^3$. On the timelike surface M, we distinguish two kinds of n-polygonal curves: the **timelike n-polygonal curve** given by an n-polygonal curve with C_i timelike arcs for $i = 1, 2, \ldots\ldots, n$(Figure 5.1-6) and a **spacelike n-polygonal curve** with all C_i spacelike arcs for $i = 1, 2, \ldots\ldots, n$(Figure 5.1-6).

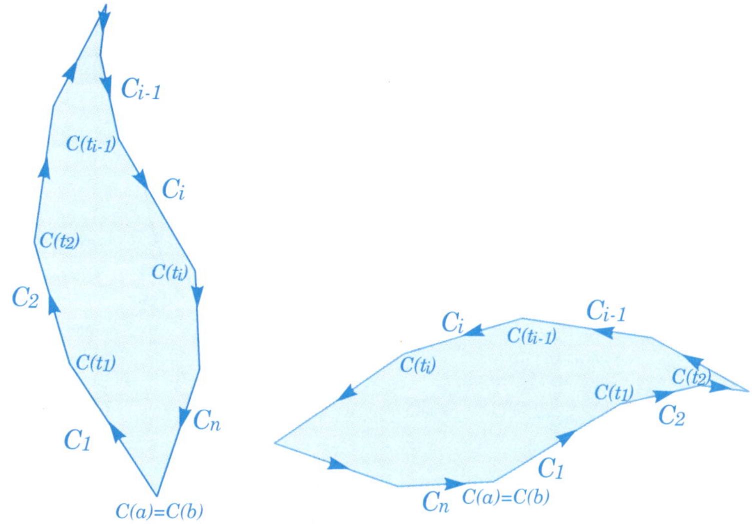

Figure5.1-6 n polygonal curves on the timelike surface

Now, recall the hyperbolic angle θ between two future directed timelike vectors or two spacelike vectors as given in Theorem 1.3.6-(5) and Remark 1.3.7. That is,

$$< x, y >= \begin{cases} -|x||y| \cosh \theta, & x, y \text{ future directed timelike vectors} \\ |x||y| \cosh \theta, & x, y \text{ spacelike vectors.} \end{cases}$$

Let M be the timelike surface in $\mathbf{L}^3$ and let $p \in M$ and $\mathbf{t} \in T_pM$ be a unit timelike vector. Then we may take a unit spacelike vector $\mathbf{t}^\perp$ orthogonal to $\mathbf{t}$ such that $\{\mathbf{t}, \mathbf{t}^\perp\}$ is the right-handed orthogonal basis for T_pM.

As in Section 1.4-Rotation on $\mathbf{L}^2$, the unit hyperbolic circle $\mathbf{H}^1 = \{x \in \mathbf{L}^2 |\ |x| = 1\}$ is separated by

$$\mathbf{H}^1 = \mathbf{H}^1_t \cup \mathbf{H}^1_s$$

where the vectors in $\mathbf{H}^1_t$ are unit timelike and the vectors in $\mathbf{H}^1_s$ are unit spacelike. Moreover, we may also separate $\mathbf{H}^1_t$ and $\mathbf{H}^1_s$ by

$$\mathbf{H}^1_t = \mathbf{H}^{1+}_t \cup \mathbf{H}^{1-}_t \text{ and } \mathbf{H}^1_s = \mathbf{H}^{1R}_s \cup \mathbf{H}^{1L}_s$$

where $\mathbf{H}^{1+}_t$ is a set of unit future directed timelike vectors, $\mathbf{H}^{1-}_t$ of unit past directed timelike vectors, $\mathbf{H}^{1R}_s$ of unit spacelike vectors along the right side curve, and $\mathbf{H}^{1L}_s$ of unit spacelike vectors along the left side curve(Figure5.1-7). Especially, as the hyperbolic angle θ increases, the unit timelike vector moves from left to right along $\mathbf{H}^{1+}_t$ and the unit spacelike vector moves from below to above along $\mathbf{H}^{1R}_s$.

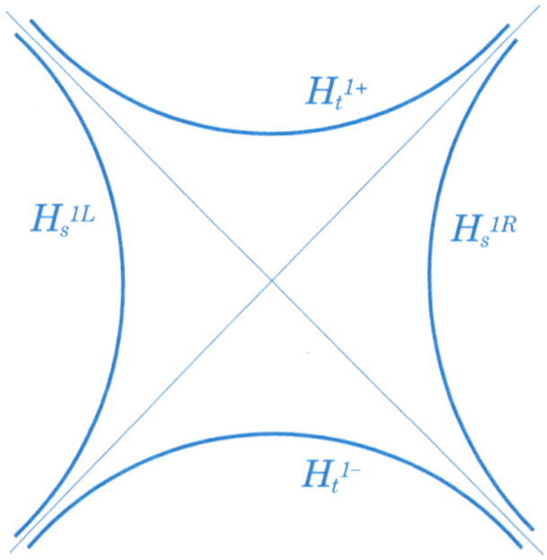

Figure 5.1-7

If x, y are two unit future directed timelike vectors on $\mathbf{H}_t^{1+}$ or unit spacelike vectors on $\mathbf{H}_s^{1R}$, the hyperbolic angle θ from x to y(Figure 5.1-8) is also represented by the rotation on M (cf. Section 4.2-Rotation on $\mathbf{L}^2$). i.e.,

$$\begin{pmatrix} \cosh\theta & \sinh\theta \\ \sinh\theta & \cosh\theta \end{pmatrix}(x) = y.$$

Context, we denote the hyperbolic angle θ from x to y by $\theta = (x, y)$. directed timelike If x is a unit future directed timelike vector and y is a unit past directed

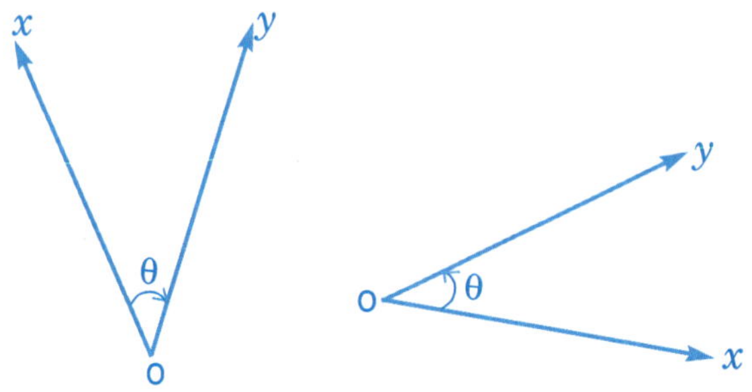

Figure 5.1-8 angle between two unit future directed timelike vectors or two unit spacelike vectors x and y

timelike vector, then the hyperbolic angle from x to y(Figure 5.1-9) is defined by

$$\begin{pmatrix} \cosh\theta & \sinh\theta \\ \sinh\theta & \cosh\theta \end{pmatrix}(-y) = x,$$

$$\begin{pmatrix} \cosh\theta & -\sinh\theta \\ -\sinh\theta & \cosh\theta \end{pmatrix}(x) = -y,$$

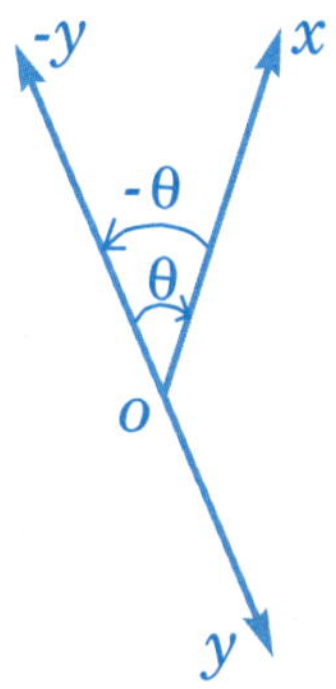

Figure 5.1-9 angle between the future directed timelike vector x and the past directed timelike vector y

or

$$\begin{pmatrix} \cosh(-\theta) & \sinh(-\theta) \\ \sinh(-\theta) & \cosh(-\theta) \end{pmatrix} (x) = \begin{pmatrix} -1 & 0 \\ 0 & -1 \end{pmatrix} y.$$

Thus,

$$\begin{pmatrix} -1 & 0 \\ 0 & -1 \end{pmatrix} \begin{pmatrix} \cosh(-\theta) & \sinh(-\theta) \\ \sinh(-\theta) & \cosh(-\theta) \end{pmatrix} (x) = y.$$

We also obtain the following proposition.

Proposition 5.1.5 *If x, y, z are unit future directed timelike vectors of $\mathbf{H}_t^{1+}$ (or unit spacelike vectors of $\mathbf{H}_s^{1R}$) on a timelike surface, then*

(1) $(x, x) = 0$, $(x, -x) = 0$

(2) $(x, -y) = (x, y)$, $(-x, y) = (x, y)$

(3) $(y, x) = -(x, y)$,

(4) $(x, y) + (y, z) = (x, z)$

Proof. Parts (1), (2), (3) follow easily from the above discussion. For (4), let

$$\begin{pmatrix} \cosh\theta & \sinh\theta \\ \sinh\theta & \cosh\theta \end{pmatrix} (x) = y \text{ and } \begin{pmatrix} \cosh\phi & \sinh\phi \\ \sinh\phi & \cosh\phi \end{pmatrix} (y) = z.$$

Then

$$\begin{aligned} z &= \begin{pmatrix} \cosh\phi & \sinh\phi \\ \sinh\phi & \cosh\phi \end{pmatrix} \begin{pmatrix} \cosh\theta & \sinh\theta \\ \sinh\theta & \cosh\theta \end{pmatrix} (x) \\ &= \begin{pmatrix} \cosh\phi\cosh\theta + \sinh\phi\sinh\theta & \cosh\phi\sinh\theta + \sinh\phi\cosh\theta \\ \cosh\phi\sinh\theta + \sinh\phi\cosh\theta & \cosh\phi\cosh\theta + \sinh\phi\sinh\theta \end{pmatrix} (x) \\ &= \begin{pmatrix} \cosh(\theta+\phi) & \sinh(\theta+\phi) \\ \sinh(\theta+\phi) & \cosh(\theta+\phi) \end{pmatrix} (x). \end{aligned}$$

□

The result of the above Proposition 5.1.5-(4) is the same as Remark 1.4.1. i.e., $(x, y) + (y, z) + (z, x) = 0$. Moreover, we may also obtain the same result of (4) in two more cases on $\mathbf{H}_t^1$.

(a) x, y are unit future directed timelike vectors and z is a unit past directed timelike vector(Figure5.1-10).

(b) x, z are unit future directed timelike vectors and y is a unit past directed timelike vector(Figure5.1-11).

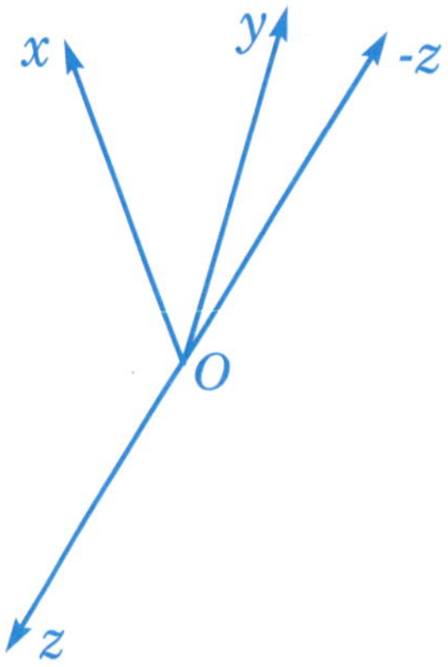

Figure5.1-10 case (a)

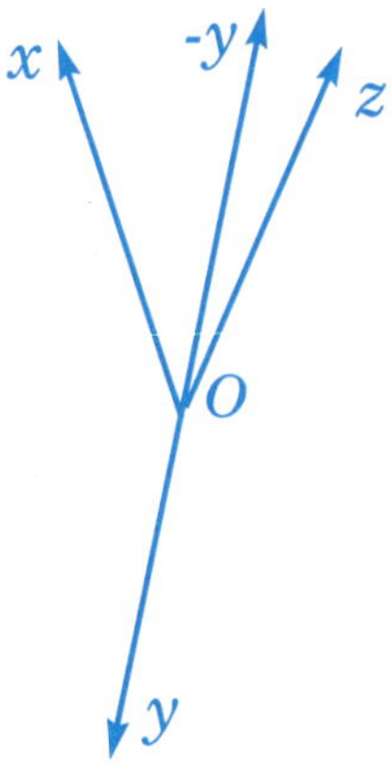

Figure5.1-11 case (b)

Let (X, D) be a coordinate patch compatible with the orientation of the timelike surface M and $c : [a, b] \to X(D) \subset M$ be a simply connected timelike n-polygonal curve (or a unit speed simply connected spacelike n-polygonal curve, resp.) in $\mathbf{L}^3$ such that $c = C_1 + C_2 + \cdots + C_n$ where $C_1, C_2, \cdots, C_n$ are unit speed timelike (or spacelike, resp.) curves, and θ_i be the exterior angle given by $\theta_i = (C'_{i-1}, C'_i)(c(t_{i-1}))$ for $1 \le i \le n$ with $c(t_n) = c(t_0)$ and $C'_{n+1} = C'_1$. Let $\varphi_i(t)$ be the angle from X_v to $C'_i(t)$ denoted by $\varphi_i(t) = (X_v, C'_i)(c(t_i))$ (or X_u to $C'_i(t)$ denoted by $\varphi_i(t) = (X_u, C'_i)(c(t_i)))$ where X_v is a timelike velocity vector field along the v-curve(or X_u is a spacelike velocity vector field along the u-curve).

Theorem 5.1.6 *Let θ_i be the exterior angle at the vertex $P_i = c(t_i)$ of the timelike (or spacelike) n-polygonal curve c on a timelike surface in $\mathbf{L}^3$ and $\varphi_i(t) = (X_v, C'_i)(c(t_i))$ (or $\varphi_i(t) = (X_u, C'_i)(c(t_i))$). Then*

$$\sum_{i=1}^{n}(\varphi_i(t_i) - \varphi_i(t_{i-1})) + \sum_{i=1}^{n} \theta_i = 0.$$

Proof. As in the proof of Theorem 5.1.2, we may show on a timelike(or spacelike) triangle.

$$\begin{aligned}
&\sum_{i=1}^{3}(\varphi_i(t_i) - \varphi_i(t_{i-1})) \\
&= (\varphi_1(t_1) - \varphi_1(t_0)) + (\varphi_2(t_2) - \varphi_2(t_1)) + (\varphi_3(t_3) - \varphi_3(t_2)) \\
&= (X_v, C_1')(c(t_1)) - (X_v, C_1')(c(t_0)) + (X_v, C_2')(c(t_2)) \\
&\quad - (X_v, C_2')(c(t_1)) + (X_v, C_3')(c(t_3)) - (X_v, C_3')(c(t_2))
\end{aligned}$$

Using the Proposition 5.1.5,

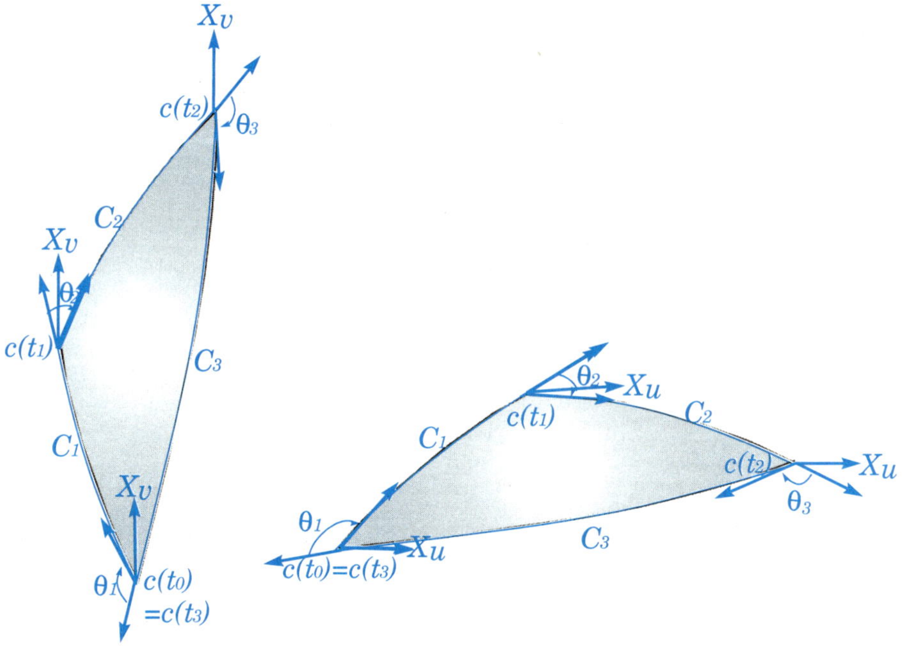

Figure5.1-12

$$\begin{aligned}
&\sum_{i=1}^{3}\theta_i = \theta_1 + \theta_2 + \theta_3 \\
&= (C_0', C_1')(c(t_0)) + (C_1', C_2')(c(t_1)) + (C_2', C_3')(c(t_2)) \\
&= (X_v, C_1')(c(t_0)) - (X_v, C_3')(c(t_0)) + (X_v, C_2')(c(t_1)) \\
&\quad - (X_v, C_1')(c(t_1)) + (X_v, C_3')(c(t_2)) - (X_v, C_2')(c(t_2))
\end{aligned}$$

Hence,

$$\sum_{i=1}^{3}(\varphi_i(t_i) - \varphi_i(t_{i-1})) + \sum_{i=1}^{3}\theta_i = 0.$$

On the n-polygonal curve, it works by the same manner. □

Lemma 5.1.7 *Let M be a timelike surface in $\mathbf{L}^3$ and $\alpha : I \to M$ be a non-lightlike regular curve. Suppose that V and W are unit timelike vector fields tangent to M and that U is a unit spacelike normal vector field along α. If $\overline{V} = U \times V$ and $\overline{W} = U \times W$, then*

$$\left\langle \frac{dW}{dt}, \overline{W} \right\rangle = \left\langle \frac{dV}{dt}, \overline{V} \right\rangle + \frac{d\varphi}{dt}$$

where $\varphi = (V, W)$ denotes the hyperbolic angle from V to W.

Proof. Since $< V, \overline{V} >= 0$ and since $V, \overline{V}$ are unit, $\{V, \overline{V}\}$ is a frame field along α.

Since V, W are timelike, $\overline{V}, \overline{W}$ are spacelike. By using some results of the first two subsections in Section 1.4,

$$W = - < W, V > V + < W, \overline{V} > \overline{V} = \cosh\varphi V + \sinh\varphi \overline{V}$$

and

$$W' = \sinh\varphi \cdot \varphi' V + \cosh\varphi V' + \cosh\varphi \cdot \varphi' \overline{V} + \sinh\varphi \overline{V}'.$$

Since $U \times \overline{V} = -\overline{V} \times U = -(U \times V) \times U = V$,

$$\overline{W} = U \times W = \cosh\varphi (U \times V) + \sinh\varphi (U \times \overline{V}) = \cosh\varphi \overline{V} + \sinh\varphi V.$$

Since $< V', \overline{V} > + < V, \overline{V}' >= 0$,

$$\begin{aligned} < W', \overline{W} >=& \sinh\varphi\cosh\varphi \cdot \varphi' < V, \overline{V} > + \cosh^2\varphi < V', \overline{V} > \\ &+ \cosh^2\varphi \cdot \varphi' < \overline{V}, \overline{V} > + \sinh\varphi\cosh\varphi < \overline{V}', \overline{V} > \\ &+ \sinh^2\varphi \cdot \varphi' < V, V > + \sinh\varphi\cosh\varphi < V', V > \\ &+ \sinh\varphi\cosh\varphi \cdot \varphi' < \overline{V}, V > + \sinh^2\varphi < \overline{V}', V > \\ =& \cosh^2\varphi \cdot \varphi' - \sinh^2\varphi \cdot \varphi' + \cosh^2\varphi < V', \overline{V} > \\ &+ \sinh^2\varphi < \overline{V}', V > \\ =& \varphi' + < V', \overline{V} > . \end{aligned}$$

□

Lemma 5.1.8 *Let M be a timelike surface in $\mathbf{L}^3$ and $\alpha : I \to M$ be a non-lightlike regular curve. Suppose that V and W are unit spacelike vector fields tangent to M and that U is a unit normal vector field along α. If $\overline{V} = U \times V$ and $\overline{W} = U \times W$, then*

$$\left\langle \frac{dW}{dt}, \overline{W} \right\rangle = \left\langle \frac{dV}{dt}, \overline{V} \right\rangle - \frac{d\varphi}{dt}$$

where $\varphi = (V, W)$.

Proof. Since V, W are spacelike, $\overline{V}, \overline{W}$ are timelike.

$$W =< W, V > V - < W, \overline{V} > \overline{V} = \cosh\varphi V + \sinh\varphi \overline{V}$$

and

$$W' = \sinh\varphi \cdot \varphi' V + \cosh\varphi V' + \cosh\varphi \cdot \varphi' \overline{V} + \sinh\varphi \overline{V}'.$$

Since $U \times \overline{V} = V$,

$$\overline{W} = U \times W = \cosh\varphi (U \times V) + \sinh\varphi (U \times \overline{V}) = \cosh\varphi \overline{V} + \sinh\varphi V.$$

Since $< V', \overline{V} > + < V, \overline{V}' >= 0$,

$$\begin{aligned} < W', \overline{W} > &= \sinh\varphi\cosh\varphi \cdot \varphi' < V, \overline{V} > + \cosh^2\varphi < V', \overline{V} > \\ &\quad + \cosh^2\varphi \cdot \varphi' < \overline{V}, \overline{V} > + \sinh\varphi\cosh\varphi < \overline{V}', \overline{V} > \\ &\quad + \sinh^2\varphi \cdot \varphi' < V, V > + \sinh\varphi\cosh\varphi < V', V > \\ &\quad + \sinh\varphi\cosh\varphi \cdot \varphi' < \overline{V}, V > + \sinh^2\varphi < \overline{V}', V > \\ &= -\cosh^2\varphi \cdot \varphi' + \sinh^2\varphi \cdot \varphi' + \cosh^2\varphi < V', \overline{V} > \\ &\quad + \sinh^2\varphi < \overline{V}', V > \\ &= -\varphi' + < V', \overline{V} > . \end{aligned}$$

□

Corollary 5.1.9 *Let $V(s)$ is a unit timelike parallel vector field along a unit speed timelike curve $\gamma : I \to M$ in $\mathbf{L}^3$ and let $\varphi = (\gamma'(s), V(s))$. Then*

$$k_g(s) = \frac{d\varphi}{ds}.$$

Proof. Using the fact that $\nabla_{\gamma'(s)} V = 0$, Lemma 5.1.7, and recalling Remark 4.1.7, we have

$$\begin{aligned} k_g(s) &=< T'(s), U \times T(s) >=< T'(s), \overline{T}(s) > \\ &= \frac{d\varphi}{ds} + < V'(s), \overline{V} >= \frac{d\varphi}{ds} + < \nabla_{\gamma'(s)} V, \overline{V} > \\ &= \frac{d\varphi}{ds}. \end{aligned}$$

□

Corollary 5.1.10 *Let $V(s)$ is a unit spacelike parallel vector field along a unit speed spacelike curve $\gamma : I \to M$ in $\mathbf{L}^3$ and let $\varphi = (\gamma'(s), V(s))$. Then*

$$k_g(s) = -\frac{d\varphi}{ds}.$$

Proof. Using the fact that $\nabla_{\gamma'(s)}V = 0$ and Lemma 5.1.8,

$$\begin{aligned} k_g(s) &=< T'(s), U \times T(s) >=< T'(s), \overline{T}(s) > \\ &= -\frac{d\varphi}{ds} + < V'(s), \overline{V} >= -\frac{d\varphi}{ds} + < \nabla_{\gamma'(s)}V, \overline{V} > \\ &= -\frac{d\varphi}{ds}. \end{aligned}$$

□

The Orthogonal Coordinate Patch

Definition 5.1.11 Let $X : D \to M$ be a proper patch of a surface M in $\mathbf{E}^3$ or $\mathbf{L}^3$ containing a neighborhood of $p \in M$. X is said to be an **orthogonal coordinate patch** if $F =< X_u, X_v >= 0$.

Remark 5.1.12 For any $p \in M$, take a proper patch $\overline{X} : \overline{D} \to M$ containing a neighborhood of p. Then we may define two orthogonal vector fields on $\overline{X}(\overline{D})$ by $V = \overline{X}_{\overline{u}}$ and $W = -\frac{\overline{F}}{\overline{E}}\overline{X}_{\overline{u}} + \overline{X}_{\overline{v}}$ where $\overline{E}, \overline{F}, \overline{G}$ are coefficients of the first fundamental form in $\overline{X}$. Since $< V(q), W(q) >= 0$ for any $q \in \overline{X}(\overline{D})$, by the Theorem in Section 3.4 in "Differential Geometry of curves and Surfaces" written by Manfredo P. do Carmo, Prentice-Hall, Inc., 1976, p.182, there exists an orthogonal proper patch X in a neighborhood of p whose u-curve and v-curve are tangent to the lines determined by $V(q)$ and $W(q)$.

Proposition 5.1.13 *For an orthogonal proper patch $X : D \to M$ on a surface M in $\mathbf{E}^3$, if W is a parallel vector field along a regular curve $\alpha(t) = X(u(t), v(t))$, then*

$$\left\langle \frac{dW}{dt}, U \times W \right\rangle = \frac{1}{2\sqrt{EG}} \left(G_u \frac{dv}{dt} - E_v \frac{du}{dt} \right) + \frac{d\varphi}{dt}$$

where $\varphi = (X_u, W)$ and V is the normal field to M which orients X.

Proof. Given $X : D \to M$ is an orthogonal patch, since $\mathbf{e}_1 = \frac{X_u}{\sqrt{E}}$ and $\mathbf{e}_2 = \frac{X_v}{\sqrt{G}}$ are orthonormal vector fields in the patch $X(u, v)$, by Lemma 5.1.3,

$$\begin{aligned} \left\langle \frac{dW}{dt}, \overline{W} \right\rangle &= \left\langle \frac{d\mathbf{e}_1}{dt}, \overline{\mathbf{e}}_1 \right\rangle + \frac{d\varphi}{dt} \\ &= \frac{d\varphi}{dt} + \left\langle \frac{d\mathbf{e}_1}{dt}, \mathbf{e}_2 \right\rangle \quad = \frac{d\varphi}{dt} + \left\langle (\mathbf{e}_1)_u \frac{du}{dt} + (\mathbf{e}_1)_v \frac{dv}{dt}, \mathbf{e}_2 \right\rangle. \end{aligned}$$

Here

$$\begin{aligned}
< (\mathbf{e}_1)_u, \mathbf{e}_2 > &= \left\langle \left(\frac{X_u}{\sqrt{E}}\right)_u, \frac{X_v}{\sqrt{G}} \right\rangle \\
&= \left\langle \frac{1}{E}\left(X_{uu}\sqrt{E} - X_u \frac{E_u}{2\sqrt{E}}\right), \frac{X_v}{\sqrt{G}} \right\rangle \\
&= \left\langle \frac{2X_{uu}E - X_u E_u}{2E\sqrt{E}}, \frac{X_v}{\sqrt{G}} \right\rangle \\
&= \frac{< X_{uu}, X_v >}{\sqrt{EG}} \\
&= \frac{1}{\sqrt{EG}}\left(\frac{\partial}{\partial u} < X_u, X_v > - < X_u, X_{uv} >\right) \\
&= -\frac{E_v}{2\sqrt{EG}}
\end{aligned}$$

and

$$\begin{aligned}
< (\mathbf{e}_1)_v, \mathbf{e}_2 > &= \left\langle \left(\frac{X_u}{\sqrt{E}}\right)_v, \frac{X_v}{\sqrt{G}} \right\rangle \\
&= \left\langle \frac{2X_{uv}E - X_u E_v}{2E\sqrt{E}}, \frac{X_v}{\sqrt{G}} \right\rangle \\
&= \frac{< X_{uv}, X_v >}{\sqrt{EG}} \\
&= \frac{G_u}{2\sqrt{EG}}.
\end{aligned}$$

Thus,

$$\left\langle \frac{dW}{dt}, U \times W \right\rangle = \frac{d\varphi}{dt} + \frac{1}{2\sqrt{EG}}\left(G_u \frac{dv}{dt} - E_v \frac{du}{dt}\right).$$

□

Proposition 5.1.14 *For an orthogonal proper patch $X : D \to M$ on a timelike surface M in $\mathbf{L}^3$, if W is a unit timelike parallel vector field along a unit speed curve $\alpha(t) = X(u(t), v(t))$, then*

$$\left\langle \frac{dW}{dt}, U \times W \right\rangle = \frac{1}{2\sqrt{-EG}}\left(G_u \frac{dv}{dt} - E_v \frac{du}{dt}\right) + \frac{d\varphi}{dt}$$

where $\varphi = (X_u, W)$.

Proof. We may assume that $X : D \to M$ is an orthogonal patch with X_u timelike. Since $\mathbf{e}_1 = \dfrac{X_u}{\sqrt{-E}}$ and $\mathbf{e}_2 = \dfrac{X_v}{\sqrt{G}}$ are orthonormal vector fields in the patch $X(u,v)$, by Lemma 5.1.7,

$$\left\langle \frac{dW}{dt}, \overline{W} \right\rangle = \left\langle \frac{d\mathbf{e}_1}{dt}, \overline{\mathbf{e}}_1 \right\rangle + \frac{d\varphi}{dt}$$

$$= \frac{d\varphi}{dt} + \left\langle \frac{d\mathbf{e}_1}{dt}, \mathbf{e}_2 \right\rangle \qquad = \frac{d\varphi}{dt} + \left\langle (\mathbf{e}_1)_u \frac{du}{dt} + (\mathbf{e}_1)_v \frac{dv}{dt}, \mathbf{e}_2 \right\rangle.$$

Here

$$\begin{aligned} < (\mathbf{e}_1)_u, \mathbf{e}_2 > &= \left\langle \left(\frac{X_u}{\sqrt{-E}} \right)_u, \frac{X_v}{\sqrt{G}} \right\rangle \\ &= \left\langle \frac{1}{-E} \left(X_{uu}\sqrt{-E} - X_u \frac{-E_u}{2\sqrt{-E}} \right), \frac{X_v}{\sqrt{G}} \right\rangle \\ &= \frac{< X_{uu}, X_v >}{\sqrt{-EG}} \\ &= \frac{1}{\sqrt{-EG}} \left(\frac{\partial}{\partial u} < X_u, X_v > - < X_u, X_{uv} > \right) \\ &= -\frac{E_v}{2\sqrt{-EG}} \end{aligned}$$

and

$$\begin{aligned} < (\mathbf{e}_1)_v, \mathbf{e}_2 > &= \left\langle \left(\frac{X_u}{\sqrt{-E}} \right)_v, \frac{X_v}{\sqrt{G}} \right\rangle \\ &= \left\langle \frac{1}{-E} \left(X_{uv}\sqrt{-E} - X_u \frac{-E_v}{2\sqrt{-E}} \right), \frac{X_v}{\sqrt{G}} \right\rangle \\ &= \frac{< X_{uv}, X_v >}{\sqrt{-EG}} \\ &= \frac{G_u}{2\sqrt{-EG}}. \end{aligned}$$

Thus,

$$\left\langle \frac{dW}{dt}, U \times W \right\rangle = \frac{d\varphi}{dt} + \frac{1}{2\sqrt{-EG}} \left(G_u \frac{dv}{dt} - E_v \frac{du}{dt} \right).$$

□

Corollary 5.1.15 *For a proper patch $X : D \to M$ on a surface M in $\mathbf{L^3}$, if W is a unit spacelike parallel vector field along a unit speed curve $\alpha(t) = X(u(t), v(t))$, then*

$$\left\langle \frac{dW}{dt}, U \times W \right\rangle = \frac{1}{2\sqrt{-EG}} \left(G_u \frac{dv}{dt} - E_v \frac{du}{dt} \right) - \frac{d\varphi}{dt}$$

where $\varphi = (X_u, W)$.

Proof. We may assume that $X : D \to M$ is an orthogonal patch with X_u spacelike. Then X_v is timelike. Hence $G =< X_v, X_v >< 0$. Since $\mathbf{e}_1 = \dfrac{X_u}{\sqrt{E}}$ and $\mathbf{e}_2 = \dfrac{X_v}{\sqrt{-G}}$ are orthonormal vector fields in the patch $X(u,v)$, by Lemma 5.1.8, and as in the proof of above Proposition 5.1.14, we have

$$\left\langle \frac{dW}{dt}, U \times W \right\rangle = -\frac{d\varphi}{dt} + \frac{1}{2\sqrt{-EG}}\left(G_u \frac{dv}{dt} - E_v \frac{du}{dt}\right).$$

□

Remark 5.1.16 Let $X : D \to M$ be an orthogonal coordinate patch in $\mathbf{L}^3$.

(i) For a unit speed timelike curve $\gamma(s) = X(u(s), v(s))$ with $W(s) = \gamma'(s) = T(s)$ and for $\varphi(s) = (X_u, T(s))$, then

$$\begin{aligned} k_g(s) &= < T'(s), U \times T(s) > \\ &= \frac{d\varphi}{ds} + \frac{1}{2\sqrt{-EG}}\left(G_u \frac{dv}{ds} - E_v \frac{du}{ds}\right). \end{aligned}$$

(ii) For a unit speed spacelike curve $\gamma(s) = X(u(s), v(s))$ with $W(s) = \gamma'(s) = T(s)$ and for $\varphi(s) = (X_u, T(s))$, then

$$\begin{aligned} k_g(s) &= < T'(s), U \times T(s) > \\ &= -\frac{d\varphi}{ds} + \frac{1}{2\sqrt{-EG}}\left(G_u \frac{dv}{ds} - E_v \frac{du}{ds}\right). \end{aligned}$$

Theorem 5.1.17 *In an orthogonal proper patch $X(u,v)$ on a surface M in $\mathbf{E}^3$,*

$$K = -\frac{1}{2\sqrt{EG}}\left[\left(\frac{E_v}{\sqrt{EG}}\right)_v + \left(\frac{G_u}{\sqrt{EG}}\right)_u\right].$$

Proof. Since $F = 0$ for an orthogonal coordinate patch on a surface M in $\mathbf{E}^3$, the Christoffel symbols reduce to

$$\begin{aligned} &\Gamma_{11}^1 = \frac{E_u}{2E}, \ \Gamma_{11}^2 = -\frac{E_v}{2G}, \ \Gamma_{12}^1 = \frac{E_v}{2E}, \\ &\Gamma_{12}^2 = \frac{G_u}{2G}, \ \Gamma_{22}^1 = -\frac{G_u}{2E}, \ \Gamma_{22}^2 = \frac{G_v}{2G}. \end{aligned}$$

Hence,

$$\begin{aligned}
K =& \frac{1}{E}\left\{(\Gamma_{11}^2)_v - (\Gamma_{12}^2)_u + \Gamma_{11}^1\Gamma_{12}^2 + \Gamma_{11}^2\Gamma_{22}^2 - \Gamma_{12}^1\Gamma_{11}^2 - \Gamma_{12}^2\Gamma_{12}^2\right\} \\
=& \frac{1}{E}\left\{-\frac{E_{vv}}{2G} - \frac{G_{uu}}{2G} + \frac{E_vG_v}{4G^2} + \frac{G_u^2}{4G^2} + \frac{E_uG_u}{4EG} + \frac{E_v^2}{4EG}\right\} \\
=& -\frac{1}{2\sqrt{EG}}\left[\left(\frac{E_v}{\sqrt{EG}}\right)_v + \left(\frac{G_u}{\sqrt{EG}}\right)_u\right].
\end{aligned}$$

□

Above, we leave the calculations of the second and third equations in exercises.

Corollary 5.1.18 *In an orthogonal proper patch $X(u,v)$ on a timelike surface M in $\mathbf{L}^3$,*

$$K = \frac{1}{2\sqrt{-EG}}\left[\left(\frac{E_v}{\sqrt{-EG}}\right)_v + \left(\frac{G_u}{\sqrt{-EG}}\right)_u\right].$$

Proof. Since we may assume that $F = 0$ for an orthogonal coordinate patch on a timelike surface M in $\mathbf{L}^3$, we have the same Christoffel symbols as in $\mathbf{E}^3$

$$\Gamma_{11}^1 = \frac{E_u}{2E},\ \Gamma_{11}^2 = -\frac{E_v}{2G},\ \Gamma_{12}^1 = \frac{E_v}{2E},$$
$$\Gamma_{12}^2 = \frac{G_u}{2G},\ \Gamma_{22}^1 = -\frac{G_u}{2E},\ \Gamma_{22}^2 = \frac{G_v}{2G}.$$

Thus, we also have

$$K = \frac{1}{E}\left\{-\frac{E_{vv}}{2G} - \frac{G_{uu}}{2G} + \frac{E_vG_v}{4G^2} + \frac{G_u^2}{4G^2} + \frac{E_uG_u}{4EG} + \frac{E_v^2}{4EG}\right\}.$$

On the other hand,

$$\begin{aligned}
&\frac{1}{2\sqrt{-EG}}\left[\left(\frac{E_v}{\sqrt{-EG}}\right)_v + \left(\frac{G_u}{\sqrt{-EG}}\right)_u\right] \\
=& \frac{1}{2\sqrt{-EG}}\frac{1}{-EG}\left[E_{vv}\sqrt{-EG} - E_v\frac{-E_vG - EG_v}{2\sqrt{-EG}} + G_{uu}\sqrt{-EG}\right. \\
&\left.-G_u\frac{-E_uG - EG_u}{2\sqrt{-EG}}\right] \\
=& -\frac{1}{2EG}(E_{vv} + G_{uu}) + \frac{1}{4E^2G^2}\left(E_v^2G + E_vEG_v + G_uE_uG + EG_u^2\right) \\
=& \frac{1}{E}\left\{-\frac{E_{vv}}{2G} - \frac{G_{uu}}{2G} + \frac{E_vG_v}{4G^2} + \frac{G_u^2}{4G^2} + \frac{E_uG_u}{4EG} + \frac{E_v^2}{4EG}\right\}
\end{aligned}$$

□

The Local Gauss-Bonnet Theorem

Theorem 5.1.19 (The local Gauss-Bonnet theorem in $\mathbf{E}^3$) *Let R be an interior region of a unit speed n-polygonal curve in an orthogonal proper patch on a surface M in $\mathbf{E}^3$ with the unit normal vector field. Then*

$$\sum_{i=1}^{n}\int_{s_{i-1}}^{s_i} k_g(s)ds + \iint_R KdA + \sum_{i=1}^{n}\theta_i = 2\pi$$

where k_g is the geodesic curvature along each regular arc and K is the Gauss curvature of the interior region R.

Proof. Suppose that $\gamma(s) = X(u(s), v(s))$ is a unit speed n-polygonal curve in an orthogonal proper patch. By Remark 4.1.7 and Proposition 5.1.13,

$$k_g = \frac{1}{2\sqrt{EG}}\left(G_u\frac{dv}{ds} - E_v\frac{du}{ds}\right) + \frac{d\varphi_i}{ds}, \quad s \in [s_{i-1}, s_i]$$

and hence

$$\begin{aligned}\sum_{i=1}^{n}\int_{s_{i-1}}^{s_i} k_g(s)ds = &\sum_{i=1}^{n}\int_{s_{i-1}}^{s_i}\frac{1}{2\sqrt{EG}}\left(G_u\frac{dv}{ds} - E_v\frac{du}{ds}\right)ds\\ &+\sum_{i=1}^{n}\int_{s_{i-1}}^{s_i}\frac{d\varphi_i}{ds}ds.\end{aligned}$$

By the Green's Theorem and Theorem 5.1.17,

$$\begin{aligned}\sum_{i=1}^{n}\int_{s_{i-1}}^{s_i} k_g(s)ds &= \iint_{X^{-1}(R)}\left\{\left(\frac{E_v}{2\sqrt{EG}}\right)_v + \left(\frac{G_u}{2\sqrt{EG}}\right)_u\right\}dudv\\ &\quad+\sum_{i=1}^{n}\int_{s_{i-1}}^{s_i}\frac{d\varphi_i}{ds}ds\\ &= -\iint_{X^{-1}(R)} K\sqrt{EG}dudv + \sum_{i=1}^{n}(\varphi_i(s_i) - \varphi_i(s_{i-1}))\\ &= -\iint_R KdA + 2\pi - \sum_{i=1}^{n}\theta_i,\end{aligned}$$

recalling that since $F = 0$, we have $dA = \sqrt{EG}dudv$. □

Definition 5.1.20 A triangle is called the **geodesic triangle** if its edges are all geodesic.

Example 5.1.21 Let T be a geodesic triangle T on a plane P(Figure5.1-13) or on a sphere $S^2(r)$ of radius r(Figure5.1-14). Then

$$\sum_{i=1}^{3} \beta_i = \pi \text{ on } T \subset P$$

and

$$\sum_{i=1}^{3} \beta_i \geq \pi \text{ on } T \subset S^2(r).$$

If T has a positive area in the second case, then $\sum_{i=1}^{3} \beta_i = \pi$.

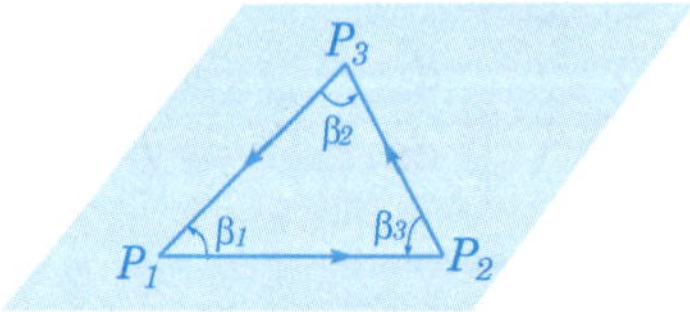

Figure5.1-13

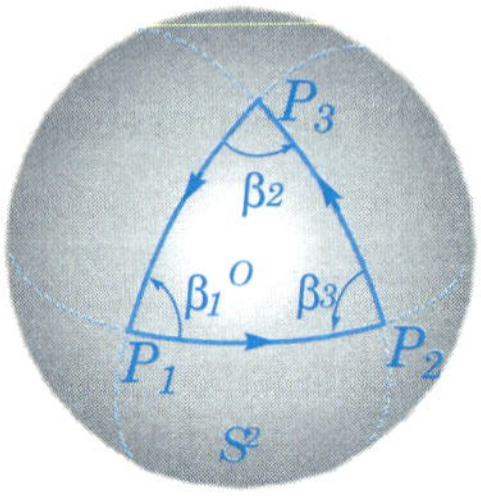

Figure5.1-14

Solution. Since $k_g = 0$, by the local Gauss-Bonnet theorem in E^3,

$$\iint_T K dA + \sum_{i=1}^{3} \theta_i = 2\pi.$$

Since the Gauss curvature $K = \frac{1}{r^2}$ on a sphere $S^2(r)$, the Gauss curvature $K = 0$ on a plane P and since $\beta_i = \pi - \theta_i$, we have $(3\pi - \sum_{i=1}^{3} \beta_i) = 2\pi$. Hence, $\sum_{i=1}^{3} \beta_i = \pi$.

On $S^2(r)$ $\iint_T \frac{1}{r^2} dA + (3\pi - \sum_{i=1}^{3} \beta_i) = 2\pi$. Hence, $\frac{1}{r^2} A = -\pi + \sum_{i=1}^{3} \beta_i$. □

Theorem 5.1.22 (The local Gauss-Bonnet theorem for timelike n-polygons in $\mathbf{L}^3$) *Let R be an interior region of a unit speed timelike n-polygonal curve in a proper patch on a timelike surface M in $\mathbf{L}^3$ with the unit normal vector field. Then*

$$\sum_{i=1}^{n} \int_{s_{i-1}}^{s_i} k_g(s) ds - \iint_R K dA + \sum_{i=1}^{n} \theta_i = 0$$

where k_g is the geodesic curvature along each timelike arc and K is the Gauss curvature of the interior region R.

Proof. Suppose that $\gamma(s) = X(u(s), v(s))$ is a unit speed timelike n-polygonal curve in an orthogonal proper patch.

$$k_g = \frac{1}{2\sqrt{-EG}}\left(G_u \frac{dv}{ds} - E_v \frac{du}{ds}\right) + \frac{d\varphi_i}{ds}, \quad s \in [s_{i-1}, s_i]$$

and

$$\sum_{i=1}^{n} \int_{s_{i-1}}^{s_i} k_g(s)ds = \sum_{i=1}^{n} \int_{s_{i-1}}^{s_i} \frac{1}{2\sqrt{-EG}}\left(G_u \frac{dv}{ds} - E_v \frac{du}{ds}\right) ds + \sum_{i=1}^{n} \int_{s_{i-1}}^{s_i} \frac{d\varphi_i}{ds} ds.$$

By the Green's theorem,

$$\begin{aligned} \sum_{i=1}^{n} \int_{s_{i-1}}^{s_i} k_g(s)ds &= \iint_{X^{-1}(R)} \left\{ \left(\frac{E_v}{2\sqrt{-EG}}\right)_v + \left(\frac{G_u}{2\sqrt{-EG}}\right)_u \right\} dudv \\ &\quad + \sum_{i=1}^{n} \int_{s_{i-1}}^{s_i} \frac{d\varphi_i}{ds} ds \\ &= \iint_{X^{-1}(R)} K\sqrt{-EG} dudv + \sum_{i=1}^{n} (\varphi_i(s_i) - \varphi_i(s_{i-1})) \\ &= \iint_R K dA - \sum_{i=1}^{n} \theta_i. \end{aligned}$$

□

Theorem 5.1.23 (The local Gauss-Bonnet theorem for spacelike n-polygons in $\mathbf{L}^3$) *Let R be an interior region of a unit speed spacelike n-polygonal curve in a proper patch on a timelike surface M in $\mathbf{L}^3$ with the unit normal vector field. Then*

$$\sum_{i=1}^{n} \int_{s_{i-1}}^{s_i} k_g(s)ds - \iint_R K dA - \sum_{i=1}^{n} \theta_i = 0$$

where k_g is the geodesic curvature along each spacelike arc and K is the Gauss curvature of the interior region R.

Proof. Suppose that $\gamma(s) = X(u(s), v(s))$ is a unit speed spacelike n-polygonal curve in an orthogonal proper patch.

$$k_g = \frac{1}{2\sqrt{-EG}}\left(G_u \frac{dv}{ds} - E_v \frac{du}{ds}\right) - \frac{d\varphi_i}{ds}, \quad s \in [s_{i-1}, s_i]$$

and

$$\sum_{i=1}^{n}\int_{s_{i-1}}^{s_i} k_g(s)ds = \sum_{i=1}^{n}\int_{s_{i-1}}^{s_i} \frac{1}{2\sqrt{-EG}}\left(G_u\frac{dv}{ds} - E_v\frac{du}{ds}\right)ds - \sum_{i=1}^{n}\int_{s_{i-1}}^{s_i}\frac{d\varphi_i}{ds}ds.$$

By the Green's theorem,

$$\begin{aligned}\sum_{i=1}^{n}\int_{s_{i-1}}^{s_i} k_g(s)ds &= \iint_{X^{-1}(R)}\left\{\left(\frac{E_v}{2\sqrt{-EG}}\right)_v + \left(\frac{G_u}{2\sqrt{-EG}}\right)_u\right\}dudv \\ &\quad - \sum_{i=1}^{n}\int_{s_{i-1}}^{s_i}\frac{d\varphi_i}{ds}ds \\ &= \iint_{X^{-1}(R)} K\sqrt{-EG}dudv - \sum_{i=1}^{n}(\varphi_i(s_i) - \varphi_i(s_{i-1})) \\ &= \iint_R KdA + \sum_{i=1}^{n}\theta_i.\end{aligned}$$

□

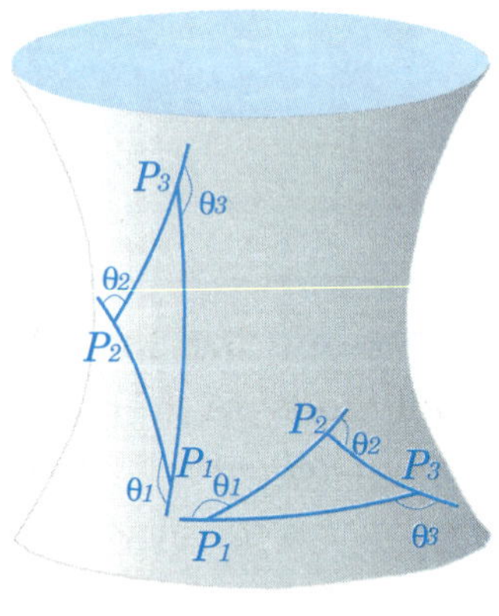

Figure5.1-15

Example 5.1.24 Let $X(u,v) = (r\cosh u\cos v, r\cosh u\sin v, r\sinh u)$ be a coordinate patch of the pseudosphere $S_1^2(r)$ in L^3. Compute $\sum_{i=1}^{3}\theta_i$ on a timelike(or spacelike) geodesic triangle $T \subset S_1^2(r)$.

Solution. Since $k_g = 0$, by the local Gauss-Bonnet theorem for the timelike n-polygon in L^3, $\sum_{i=1}^{3}\theta_i = \iint_T KdA$. Since the Gauss curvature $K = \frac{1}{r^2}$ on a pseudosphere $S_1^2(r)$, we have

$$\sum_{i=1}^{3} \theta_i = \iint_T K dA = \frac{A}{r^2} > 0$$

where A is the area of the timelike triangle $T \subset S_1^2(r)$.

If T is the spacelike triangle on $S_1^2(r)$, by the local Gauss-Bonnet theorem for the spacelike n-polygon,

$$\sum_{i=1}^{3} \theta_i = -\iint_T K dA = -\frac{A}{r^2} < 0.$$

□

For further study, see the following references.

1. Birman, R. and K. Nomizu, The Gauss-Bonnet theorem for two-dimensional space-time, Mich. Math. J. 31(1984) 77-81.

2. do Carmo, Manfredo Perdigan, Differential Geometry of Curves and Surfaces. Englewood Cliffs, NJ: Prentice-Hall, 1976.

3. McCleary, J., Geometry from a Differentiable Viewpoint, Cambridge University Press, 1994.

4. Millman, R. S. and G. D. Parker,Elements of Differential Geometry, Prentice-Hall, Englewood Cliffs, NJ, 1977.

5. O'Neill, B., Elementary Differential Geometry, Academic Press, New York, 1966.

6. Pressley, Andrew, Elementary Differential Geometry, Springer Undergraduate Mathematics Series, 2002.

5.2 The Global Gauss-Bonnet Theorem and its Applications in $\mathbf{E}^3$

The Triangulation or Rectangular Decomposition

Definition 5.2.1 Throughout this section, let R be a compact surface region whose boundary ∂R is is a finite sum of n-polygonal curves which do not intersect unless mentioned. A **triangulation** of R is defined by a family $\mathcal{T} = \{T_i\}_{i=1}^n$ of finite regions called triangles bounded by 3-polygonal curves T_i, $i = 1, 2, 3, ..., n$ satisfying

(a) $R = \bigcup_{i=1}^{n} T_i$

(b) if $T_i \cap T_j \neq$, $T_i \cap T_j$ is a common edge or a common vertex of the triangles T_i and T_j.

A rectangular decomposition of R is similarly defined by a family of finite regions called rectangles bounded by 4-polygonal curves satisfying the above conditions (a) and (b).

Let V, E and F be the number of vertices, edges and triangles(or rectangles) in $\mathcal{T}$ respectively. Then

$$\chi(R) = V - E + F$$

is called the **Euler-Poincare characteristic**.

The Euler-Poincare characteristic $\chi(R)$ of R is a topological invariant and is independent of the triangulation of R. Hence, If $\chi(M) = \chi(M')$ for two topological spaces M and M', M is homeomorphic to M'.

Example 5.2.2 (1) Compute the Euler-Poincare characteristic for the triangulation and the rectangular decomposition of the sphere S^2(Figure5.2-16).
(2) Compute the Euler-Poincare characteristic of the torus T(Figure 5.2-17, 5.2-18).
(3) Compute the Euler-Poincare characteristic of the 2-torus T^2(Figure5.2-19).

Solution. (1) $\chi(S^2) = 4-6+4 = 2$ in a triangulation and $\chi(S^2) = 8-12+6 = 2$ in a rectangular decomposition.
(2) Since each corresponding vertices coincide each other, there are no vertices in a rectangular decomposition of T as in Figure5.2-17. Hence $\chi(T) = 0-8+8 = 0$. Further, since a torus is topologically equivalent to a sphere punctured by two rectangles which are attached by a handle as in Figure5.2-18. since the Euler-Poincare characteristic of the handle is $8 - 12 + 4 = 0$ and since two punctured

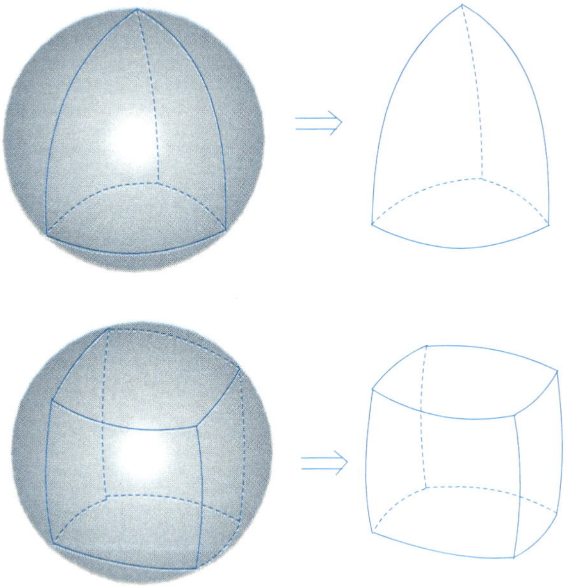

Figure5.2-16 Triangulation and rectangular decomposion

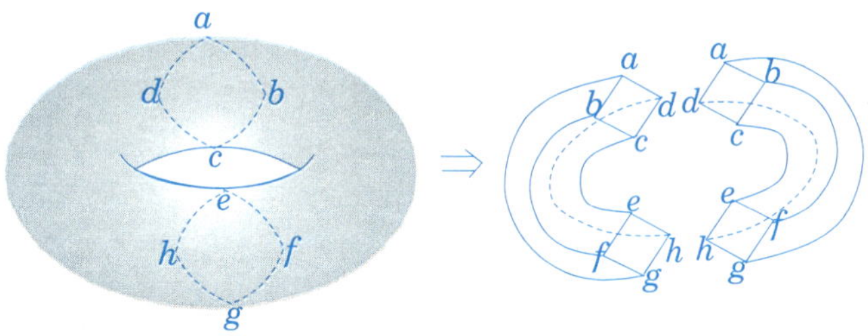

Figure5.2-17 Rectangular decomposition of a torus

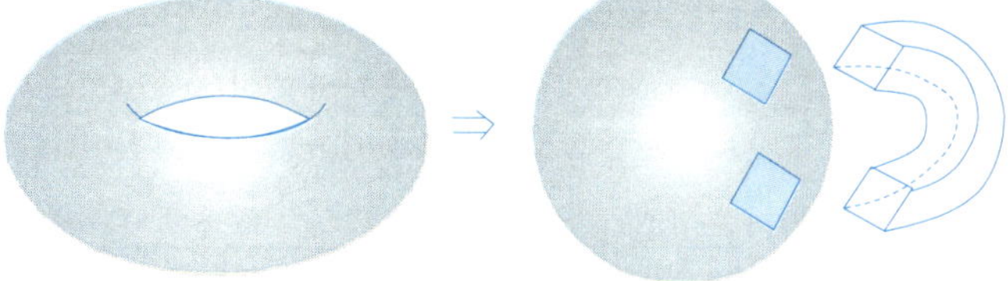

Figure5.2-18 torus represented by a sphere attached with a handle

faces on a sphere minus 2 from $\chi(S^2)$, $\chi(T) = \chi(S^2) - 2 = 0$.
(3) since the 2-torus T^2 is topologically equivalent to a sphere attached with two handles as in Figure5.2-19, $\chi(T^2) = \chi(S^2) - 2 - 2 = -2$. □

If the orientation of the boundary of each triangle $T \in \mathcal{T}$ is positive, the common edge of the adjacent triangles has the opposite orientations as in Figure 5.2-20. The following theorem is presented without proof.

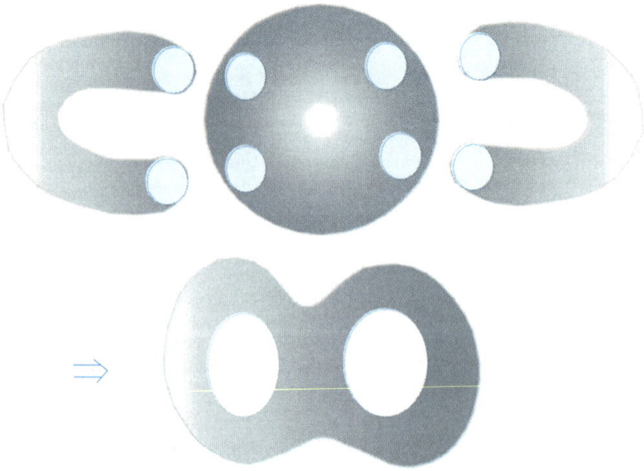

Figure5.2-19 2-torus represented by a sphere attached with two handles

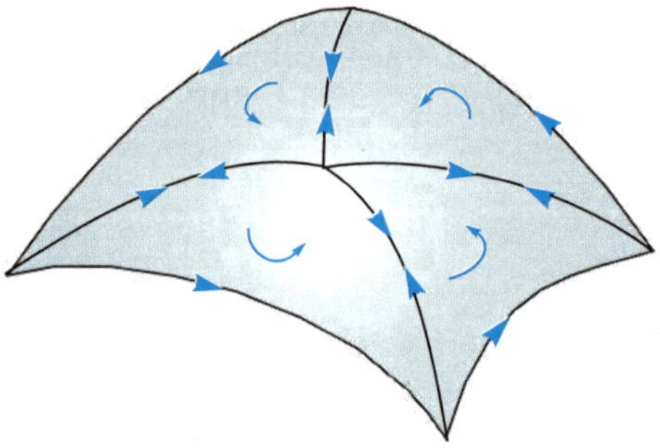

Figure5.2-20

Theorem 5.2.3 *The region R admits a triangulation.*

Now, we prove the global Gauss-Bonnet Theorem for a region of an oriented surface whose boundary is the finite sum of unit speed n-polygonal curves.

Theorem 5.2.4 (The global Gauss-Bonnet theorem) *Suppose that for the interior region R with positive orientation the boundary is the finite sum of unit speed n-polygonal curves C_i for $i = 1, 2, ..., n$, i.e., $\partial R = C_1 + C_2 + \cdots + C_n$ in R and $\theta_1, \theta_2, \cdots, \theta_p$ are the exterior angles of vertices. Then*

$$\sum_{i=1}^{n} \int_{C_i} k_g(s)ds + \iint_R K dA + \sum_{t=1}^{p} \theta_t = 2\pi\chi(R).$$

Proof. Suppose that each triangle T_j in the triangulation $\mathcal{T} = \{T_j\}_{i=1}^F$ of R is contained by subdivision in one of the proper patches of M. Applying the local Gauss-Bonnet theorem to each T_j,

$$\sum_{i=1}^{n} \int_{C_i} k_g(s)ds + \iint_R KdA + \sum_{j,k=1}^{F,3} \theta_{jk} = 2\pi F.$$

Since $\theta_{j1}, \theta_{j2}, \theta_{j3}$ is an exterior angle of the triangle T_j, if $\phi_{j1}, \phi_{j2}, \phi_{j3}$ is an interior angle of the triangle T_j, then $\phi_{jk} = \pi - \theta_{jk}$ and

$$\sum_{j,k}^{F,3} \theta_{jk} = \sum_{j,k}^{F,3} \pi - \sum_{j,k}^{F,3} \theta_{jk} = 3\pi F - \sum_{j,k}^{F,3} \phi_{jk}.$$

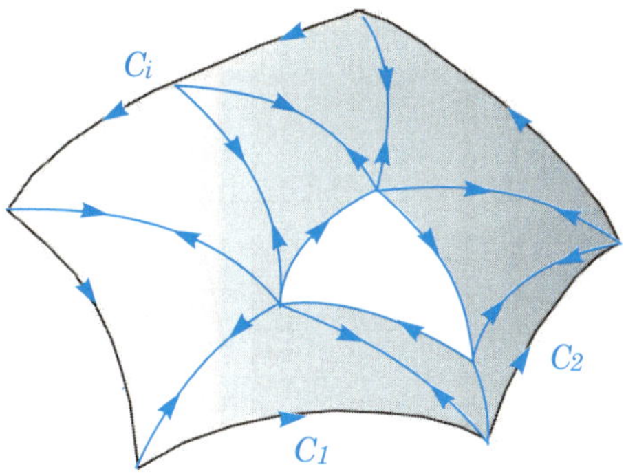

Figure5.2-21

Let E_e=the number of outside edges of $\mathcal{T}$,

E_i=the number of inside edges of $\mathcal{T}$,

V_e=the number of outside vertices of $\mathcal{T}$,

V_i=the number of inside vertices of $\mathcal{T}$,

V_{ec}=the number of the vertices of all C_i,

V_{et}=the number of outside vertices of $\mathcal{T}$ except for vertices of a C_i.

Then

$V_e = V_{ec} + V_{et}$ and $E_e = V_e$.

By the mathematical induction, since

$$
\begin{aligned}
3F &= 2E_i + E_e, \\
\sum_{j,k}^{F,3} \theta_{jk} &= 3\pi F - \sum_{j,k}^{F,3} \phi_{jk} \\
&= 2\pi E_i + \pi E_e - \sum_{j,k}^{F,3} \phi_{jk} \\
&= 2\pi E_i + \pi E_e - \{2\pi V_i + \pi V_{et} + \sum_{t=1}^{p} (\pi - \theta_t)\} \\
&= 2\pi E_i + 2\pi E_e - \pi V_e - 2\pi V_i - \pi V_{et} - \pi V_{ec} + \sum_{t=1}^{p} \theta_t \\
&= 2\pi E - 2\pi V + \sum_{t=1}^{p} \theta_t \\
&= 2\pi(E - V) + \sum_{t=1}^{p} \theta_t.
\end{aligned}
$$

Hence, substituting this result into the first equation,

$$
\begin{aligned}
\sum_{i=1}^{n} \int_{C_i} k_g(s)ds + \iint_R KdA + \sum_{t=1}^{p} \theta_t &= 2\pi(F - E + V) \\
&= 2\pi\chi(R).
\end{aligned}
$$

□

Remark 5.2.5 Similarly, we also may prove this theorem using a rectangular decomposition of the region instead of the triangulation.

If R is an interior region bounded by a n-polygonal curve, then the number of vertices is equal to the number of edges of R. And the number of faces of R is just 1. Hence, $\chi(R) = F - E + V = 1$. Thus, we obtain the local Gauss-Bonnet theorem as follows.

Corollary 5.2.6 *If R is an interior region bounded by a n-polygonal curve, then*

$$
\sum_{i=1}^{n} \int_{s_{i-1}}^{s_i} k_g(s)ds + \iint_R KdA + \sum_{i=1}^{n} \theta_i = 2\pi.
$$

Proof. Since R has only one face, R has the same number of its edges and vertices. Hence, $\chi(R) = 1$. □

If M be a compact surface without boundary, then we have no edges and empty exterior angles. Hence, we obtain the following theorem.

Theorem 5.2.7 *Let M be a compact surface without boundary. Then*

$$\iint_M K dA = 2\pi\chi(M).$$

Using the above Theorem 5.2.7, we may also compute the Euler-Poincare characteristic of the compact surface without boundary from the Gaussian curvature rather than from a triangulation.

Example 5.2.8 Since $\chi(M) = \frac{1}{2\pi}\iint_M KdA$, if $M = S^2(r)$ then

$$\begin{aligned}\chi(S^2(r)) &= \frac{1}{2\pi}\iint_{S^2(r)} \frac{1}{r^2} dA\\ &= \frac{1}{2\pi r^2} Area(S^2(r))\\ &= \frac{1}{2\pi r^2} 4\pi r^2\\ &= 2\end{aligned}$$

Example 5.2.9 (1) $\iint_{S^2} KdA = 2\pi\chi(S^2) = 2\pi(2) = 4\pi$,

(2) $\iint_T KdA = 2\pi\chi(T) = 2\pi(0) = 0$,

(3) $\iint_{T^2} KdA = 2\pi\chi(T^2) = 2\pi(-2) = -4\pi$.

By Corollary 5.2.6, we obtain

Corollary 5.2.10 *Let M be an oriented surface and let $\triangle$ be a triangle on M. Then*

$$\iint_\triangle KdA + \int_{\partial\triangle} k_g ds = \sum_{j=1}^{3} \beta_i - \pi.$$

If K is constant and $\triangle$ is a geodesic triangle on M, then $\iint_\triangle KdA = K$(Area of $\triangle$). By Corollary 5.2.10, K(Area of $\triangle$)$=\sum_{j=1}^{3}\beta_i - \pi$. Hence, the distance between the sum of all interior angles of $\triangle$ and π is proportional to the area of $\triangle$.

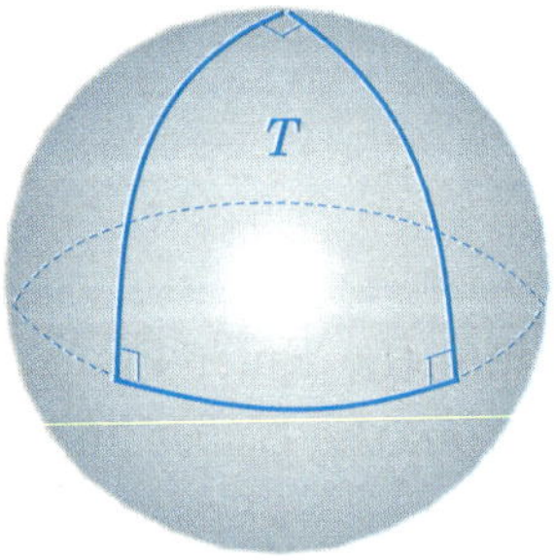

Figure5.2-22

Example 5.2.11 The area of a geodesic triangle T on S^2 is $\sum_{j=1}^{3}\beta_i - \pi$ Hence, we may compute the area of the geodesic triangle with 3 right interior angles as in Figure5.2-22.

Solution. Since $K = 1$, $\sum_{j=1}^{3}\beta_i - \pi = \frac{3\pi}{2} - \pi = \frac{\pi}{2}$. On other hand, since this triangle is $\frac{1}{8}$ portion of S^2, the area of this triangle is $\frac{1}{8}Area(S^2) = \frac{4\pi}{8} = \frac{\pi}{2}$. □

If the Gauss curvature K is positive at every point, then every point is elliptic and hence M is homeomorphic to a sphere. Thus we have

Corollary 5.2.12 *If the Gauss curvature K is positive at every point of the compact surface M without boundary, M is homeomorphic to a sphere.*

Corollary 5.2.13 *Let M be a surface of everywhere negative Gaussian curvature K. Then two geodesics γ_1 and γ_2 starting from a point $p \in M$ can not meet again at a point $q \in M$.*

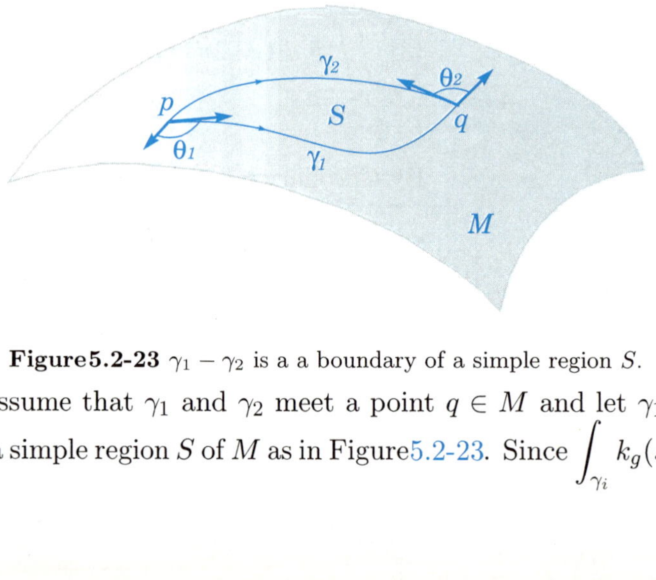

Figure5.2-23 $\gamma_1 - \gamma_2$ is a a boundary of a simple region S.

Solution. Assume that γ_1 and γ_2 meet a point $q \in M$ and let $\gamma_1 - \gamma_2$ be a a boundary of a simple region S of M as in Figure5.2-23. Since $\int_{\gamma_i} k_g(s)ds = 0$, $i =$

1, 2, using the local Gauss-Bonnet theorem,

$$\iint_S K dA + \theta_1 + \theta_2 = 2\pi.$$

Hence, $2\pi - \theta_1 - \theta_2 = \iint_S K dA \leq 0$. Thus, $\theta_1 \geq \pi$ or $\theta_2 \geq \pi$. On the other hand, since γ_1 and γ_2 are not tangent at p and at q, by the uniqueness of geodesics, $\theta_i \leq \pi$, $i = 1, 2$, in contradiction. □

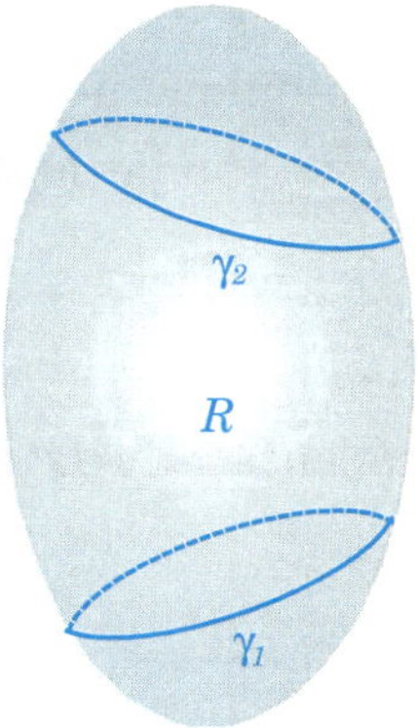

Figure5.2-24

Corollary 5.2.14 *Let M be a compact connected surface without boundary of everywhere positive Gaussian curvature K. Then two simply closed geodesics γ_1 and γ_2 intersect.*

Solution. If γ_1 and γ_2 do not intersect, they are the boundary of a region R homeomorphic to a cylinder with Euler-Poincare characteristic $\chi(R) = 0$ as in Figure5.2-24. By the global Gauss-Bonnet theorem,

$$\iint_R K dA = 0,$$

in contradiction to $K > 0$. □

For further study, see the following reference:

1. do Carmo, Manfredo Perdigan, Differential Geometry of Curves and Surfaces. Englewood Cliffs, NJ: Prentice-Hall, 1976.

2. Millman, R. and G. Parker, Elements of Differential Geometry, Prentice-Hall Inc, 1977.

3. O'Neill, B. Elementary Differential Geometry, Academic Press Inc, 1966.

5.3 The Exponential Map and the Hopf-Rinow Theorem

In this section, we show that the shortest curve $\gamma : [0,1] \to M$ from p to q in a Euclidean surface in $\mathbf{E}^3$ is a geodesic from p to q. In the case of the timelike surface M in $\mathbf{L}^3$, the longest timelike curve $\gamma : [0,1] \to M$ from p to q is also a timelike geodesic from p to q. Moreover, those geodesics are unique up to the reparametrization of the curve γ provided that p and q are not too far apart.

The Exponential Map

Definition 5.3.1 Let M be a surface in $\mathbf{E}^3$ or a timelike surface in $\mathbf{L}^3$. For $p \in M$ and for any $v \in T_pM$,

$$\exp_p(v) = \gamma_v(1)$$

provided that $\gamma_v(1)$ exists. Let $\widetilde{T_pM} = \{v \in T_pM \,|\, \exp_p(v) \text{ exists}\}$. Then $\exp_p : \widetilde{T_pM} \to M$ is called the **exponential map** at p.

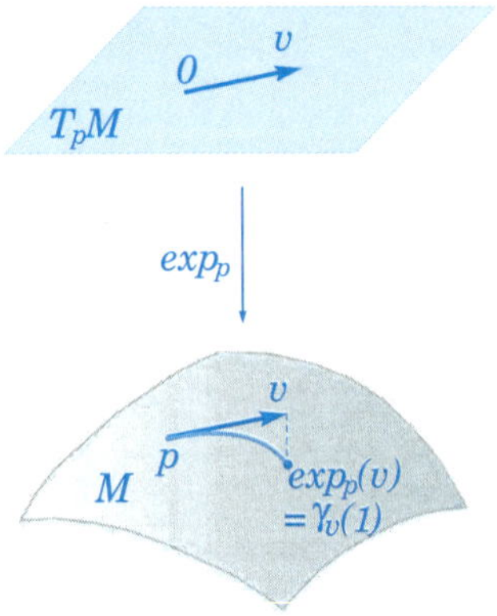

Figure 5.3-25 Exponential map $\exp_p : T_pM \to M$

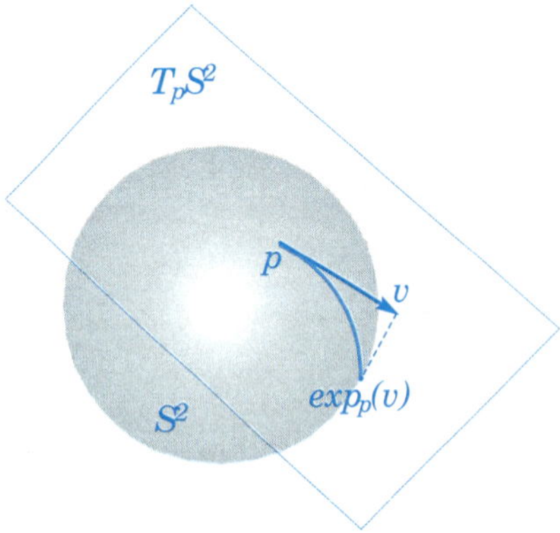

Figure 5.3-26 Exponential map on a sphere

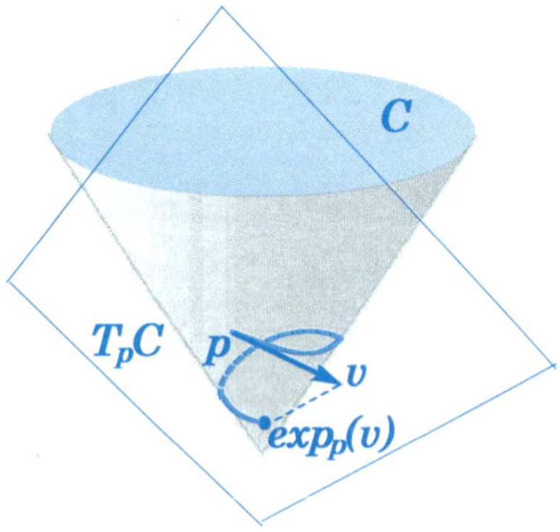

Figure 5.3-27 Exponential map on a cone

Hence, the exponential image of a tangent vector v at a point p of the surface M is a point on a geodesic segment $\gamma_v(t)$ tangent to the vector v after one second as in Figure5.3-25. Examples for spheres and cones in $\mathbf{E}^3$ are shown as in Figure5.3-26, 5.3-27.

Note that $\exp_p(O_p) = p$, so that $O_p \in \widetilde{T_pM}$.

Lemma 5.3.2 *Let M be a surface in $\mathbf{E}^3$ or a timelike surface in $\mathbf{L}^3$. If $\gamma_v(t)$ is defined for $t \in (-\epsilon, \epsilon)$ on M, then for $\lambda \in \mathbb{R}$ nonzero, the geodesic $\gamma_{\lambda v}(t)$ is defined for $t \in (-\epsilon/\lambda, \epsilon/\lambda)$, and $\gamma_{\lambda v}(t) = \gamma_v(\lambda t)$.*

Proof. Let $\alpha : (-\epsilon/\lambda, \epsilon/\lambda) \to M$ be a regular curve defined by $\alpha(t) = \gamma_v(\lambda t)$. Then $\alpha(t)$ is again a geodesic since this is an affine change of parameter. Also

$\alpha(0) = \gamma_v(0) = p$ and $\alpha'(0) = \lambda\gamma_v'(0) = \lambda v$. Thus, $\gamma_{\lambda v}(0) = p$ and $\gamma_{\lambda v}'(0) = \lambda v$. By the uniqueness of geodesics for the initial conditions, $\gamma_{\lambda v}(t) = \alpha(t) = \gamma_v(\lambda t)$. □

Remark 5.3.3 By Lemma 5.3.2, $\exp_p(tv) = \gamma_{tv}(1) = \gamma_v(t)$. Moreover, the exponential map $\exp_p$ is well defined on a ϵ-neighborhood $B_\epsilon = \{v \in T_pM \mid |v| < \epsilon\}$ of the origin $O \in T_pM$ for any $p \in M$.

Theorem 5.3.4 *Let M be a surface in $\mathbf{E}^3$ or a timelike surface in $\mathbf{L}^3$. For some open set $U \subset B_\epsilon$, $\exp_p : U \to \exp_p(U) \subset M$ is a diffeomorphism.*

Proof. Let $O \in B_\epsilon$. Since the tangent space $T_o(T_pM)$ of T_pM at the origin O is linearly isomorphic to T_pM, the differential map $(\exp_p)_* : T_pM \to T_pM$ is well defined. Let v be the initial velocity of the ray $\rho(t) = tv$. By Lemma 5.3.2, $\exp_p(\rho(t)) = \gamma_v(t)$. Since

$$(\exp_p)_*(v) = \frac{d}{dt}\exp_p(\rho(t))|_{t=0} = \frac{d}{dt}\gamma_v(t)|_{t=0} = v,$$

the differential map $(\exp_p)_*$ is a linear isomorphism. By the Inverse Function Theorem, we may take an open neighborhood $U \subset B_\epsilon$ of the origin O so that $\exp_p : U \to \exp_p(U)$ is a diffeomorphism. □

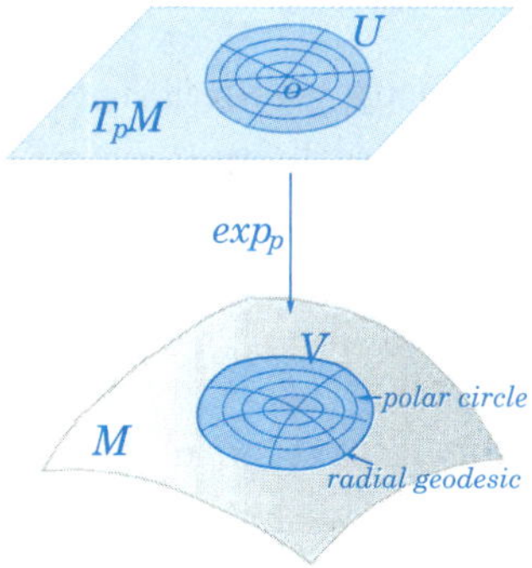

Figure5.3-28 polar coordinates

In Theorem 5.3.4, $V = \exp_p(U)$ is called a **normal neighborhood** of p. Hence, we may take the polar coordinate in the normal neighborhood called the geodesic polar coordinate patch as follows(Figure5.3-28).

Definition 5.3.5 For a surface M in $\mathbf{E}^3$, let $\{e_1, e_2\}$ be a orthonormal basis of T_pM. If $q \in V$, $q = \exp_p(w)$ for $w = u\cos v e_1 + u \sin v e_2 \in U$. Then **geodesic polar coordinates** $X(u,v)$ in the normal neighborhood V of p are defined by

$$X(u,v) = \exp_p(u\cos v e_1 + u\sin v e_2).$$

Then the u-parameter curve $X(u, v_0) = \exp_p(u\cos v_0 e_1 + u \sin v_0 e_2)$ has the initial velocity vector $\cos v_0 e_1 + \sin v_0 e_2$ whose speed is unit and the v-parameter curve $X(u_0, v) = \exp_p(u_0\cos v e_1 + u_0 \sin v e_2)$ is the exponential image of the circle of radius u_0.

Lemma 5.3.6 *Let M be a surface in $\mathbf{E}^3$. For a geodesic polar coordinate $X(u,v)$ with $0 < u < \epsilon$ and $0 < v < 2\pi$, the coefficients of the first fundamental form satisfy*

$$E = 1, \; F = 0, \; G > 0.$$

Proof. Since the u-parameter curve $X(u, v_0)$ is a unit speed geodesic, $E =< X_u, X_u >= 1$, and $X_{uu} = 0$. Since $F_u =< X_u, X_v >_u =< X_{uu}, X_v > + < X_u, X_{vu} >=< X_u, X_{uv} >= \frac{1}{2}E_v = 0$, F is constant on the u-parameter curve. Now, since the v-parameter curve $X(u, v)$ satisfies $X(0, v) = p$ and $X_v(0, v) = 0$, $F(0, v) =< X_u, X_v > |_{u=0} = 0$ for all v. Hence, $F = 0$ for all u and v. Finally, $G = EG - F^2 = |X_u \times X_v|^2 > 0$. □

Remark 5.3.7 $E = 1$ implies that the exponential map preserves the speed of the u-parameter curves called the **radial geodesics** and $F = 0$ implies that the v-parameter curves called the **polar circles** are always orthogonal to the radial geodesics(Figure 5.3-28).

Example 5.3.8 Let $p = (0, 0, 1)$ be the north pole of the unit sphere S^2. Then the geodesic polar coordinates emanating from p are given by

$$X(u, v) = (\sin u \cos v, \sin u \sin v, \cos u).$$

Each u-parameter curve is a unit speed geodesic whose velocity vector is $X_u(0, v) = (\cos v, \sin v, 0) = \cos vE_1 + \sin vE_2$ where $E_1 = \mathbf{i},\ E_2 = \mathbf{j}$. By the uniqueness of geodesics for the initial condition,

$$X(u, v) = \gamma_{\cos vE_1 + \sin vE_2}(u) = \exp_p(u \cos vE_1 + u \sin vE_2),$$

which is the geodesic polar coordinate. Moreover, the largest normal neighborhood of p is $S^2 - \{(0, 0, -1)\}$ where the radial geodesic is the longitudinal semicircle and the polar circles are the circles of latitude(Figure 5.3-29). In the context of Lemma 5.3.6, it may be verified that $E(u, v) = 1, F(u, v) = 0$, and $G(u, v) = \sin^2 u$.

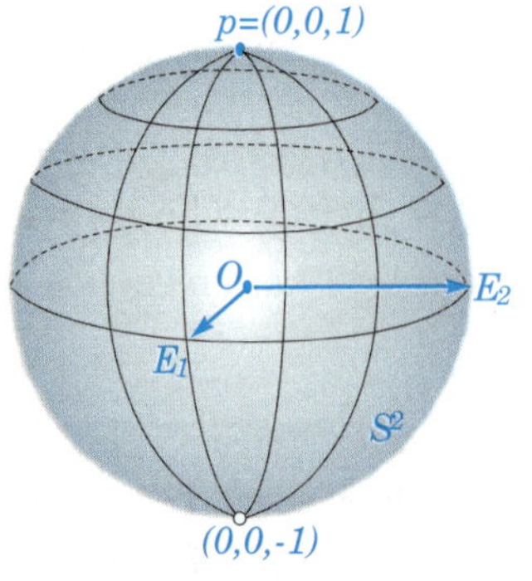

Figure 5.3-29 Polar coordinates on a sphere

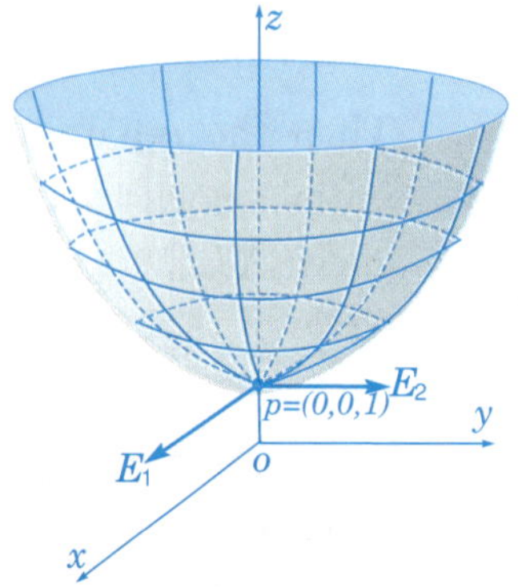

Figure5.3-30 Polar coordinates on a hyperboloid

Example 5.3.9 As in Example 4.1.1 let M be a upper sheet of hyperboloid $x^2+y^2-z^2=-1$ in $\mathbf{L}^3$. Now, for a coordinate patch about $p=(0,0,1)$, let $X(u,v)=(\sinh u\cos v, \sinh u\sin v, \cosh u)$ for $u\geq 0$. Each u-parameter curve is a unit speed geodesic whose velocity vector is $X_u(0,v)=(\cos v, \sin v, 0)=\cos vE_1+\sin vE_2$. Hence, by the uniqueness of geodesics,

$$X(u,v)=\gamma_{\cos vE_1+\sin vE_2}(u)=\exp_p(u\cos vE_1+u\sin vE_2),$$

which is the geodesic polar coordinate(Figure5.3-30). It may be verified for this coordinate patch that $E(u,v)=-1, F(u,v)=0$, and $G(u,v)=\sinh^2 u>0$.

Lemma 5.3.10 *Let M be a timelike surface in $\mathbf{L}^3$. For geodesic polar coordinates $X(u,v)$ with $0<u<\epsilon$ and $0<v<2\pi$, the coefficients of the first fundamental form satisfy*

$$E=-1,\ F=0,\ G>0.$$

Proof. Since the u-parameter curve $X(u,v_0)$ is a unit speed timelike geodesic, $E=<X_u,X_u>=-1$, and $X_{uu}=0$. Since $F_u=<X_u,X_v>_u=<X_{uu},X_v> + <X_u,X_{vu}>=<X_u,X_{vu}>=\frac{1}{2}E_v=0$, F is constant on the u-parameter curve. Now, since the v-parameter curve $X(u,v)$ satisfies $X(0,v)=p$ and $X_v(0,v)=0$, $F(0,v)=<X_u,X_v>|_{u=0}=0$ for all v. Hence, $F=0$ for all u and v. Finally, $G=-(EG-F^2)=|X_u\times X_v|^2>0$ since $X_u\times X_v\neq 0$ for a regular surface patch and this nonzero vector must also be spacelike. □

Theorem 5.3.11 *Let M be a surface in $\mathbf{E}^3$. For each point q of a normal neighborhood $N_\epsilon=\exp_p(B_\epsilon)$ of p in M, the radial geodesic segment of N_ϵ from p to q uniquely minimizes arc length.*

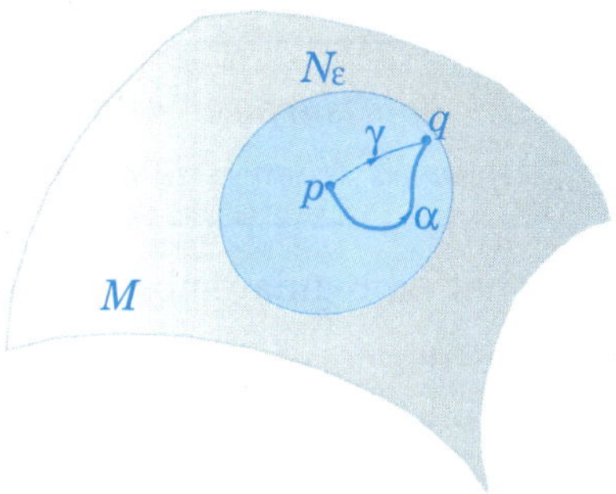

Figure5.3-31 Case1

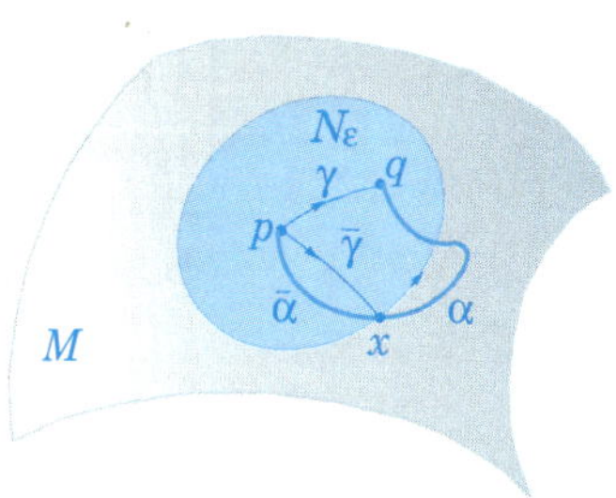

Figure5.3-32 Case2

Proof. Let $X(u, v)$ be geodesic polar coordinates of N_ϵ centered at $p = X(0, 0)$. For $q = X(u_0, v_0)$, the radial geodesic segment from p to q is given by $\gamma(u) = X(u, v_0)$, $(0 \leq u \leq u_0)$. Let $\alpha : [0, u_0] \to M$ be an arbitrary curve segment from p to q.

Case1: we assume that α stays in N_ϵ(Figure5.3-31). Then we can parameterize α by $\alpha(t) = X(u(t), v(t))$, $0 \leq t \leq u_0$ with $\alpha(0) = p$ and $\alpha(u_0) = q$ so that

$$\begin{aligned} u(0) &= 0, \quad u(u_0) = u_0 \\ v(0) &= v_0, \quad v(u_0) = v_0. \end{aligned}$$

Since $E = 1$ and $F = 0$ by Lemma 5.3.6,

$$\begin{aligned} |\alpha'(t)| &= \sqrt{u'(t)^2 + Gv'(t)^2} \\ &\geq |u'(t)| \geq u'(t). \end{aligned}$$

Hence,

$$\begin{aligned} L(\alpha) &= \int_0^{u_0} \sqrt{u'(t)^2 + Gv'(t)^2} dt \\ &\geq \int_0^{u_0} u'(t) dt = u_0. \end{aligned}$$

and $L(\gamma) = \displaystyle\int_0^{u_0} |\gamma'(u)| du = \int_0^{u_0} 1 du = u_0$. Therefore, $L(\alpha) \geq L(\gamma)$.

Case2: If α does not stay in N_ϵ, α must cross at a point x on the polar circle $u = u_0$(Figure5.3-32). Let $\bar{\alpha}$ be the restriction of α to the interval $[0, u_0]$. Then $L(\alpha) \geq L(\bar{\alpha})$. By the previous argument, $L(\bar{\alpha}) \geq L(\bar{\gamma})$ for the radial geodesic segment $\bar{\gamma}$ from p to x and also $L(\bar{\gamma}) = u_0$. Hence $L(\alpha) \geq L(\bar{\gamma}) = u_0$. Therefore, $L(\alpha) \geq L(\gamma)$.

Finally, if $L(\alpha) = L(\gamma)$, then $\sqrt{u'(t)^2 + Gv'(t)^2} = u'(t)$ and $u'(t) \geq 0$. Since $G > 0$, $v'(t) = 0$ and $v(t)$ is constant v_0. Thus, $\alpha(t) = X(u(t), v_0) = \gamma(u(t))$, which implies that α is a monotone reparametrization of γ. □

Corollary 5.3.12 *Let M be a surface in $\mathbf{E}^3$. If α is the shortest piecewise regular curve from p to q in M, then α is an unbroken geodesic.*

Proof. First, we show that all regular subsegments of α are geodesics. Assume that a regular segment of $\alpha|_{[t_{i-1}, t_i]}$ is not a geodesic. Then $\alpha''(s_0) \neq 0$ for some $s_0 \in [t_{i-1}, t_i]$ and by continuity we may suppose $s_0 < t_i$. Thus $\alpha|_{[s_0, s_0+\epsilon]}$ is contained in a normal ϵ-neighborhood N_ϵ of $\alpha(s_0)$(Figure5.3-33). Since $\alpha|_{[s_0, s_0+\epsilon]}$ can not be reparametrized to be geodesic, by Theorem 5.3.11, the radial geodesic segment γ from $\alpha(s_0)$ to $\alpha(s_0 + \epsilon)$ is strictly shorter than $\alpha|_{[s_0, s_0+\epsilon]}$. Replacing

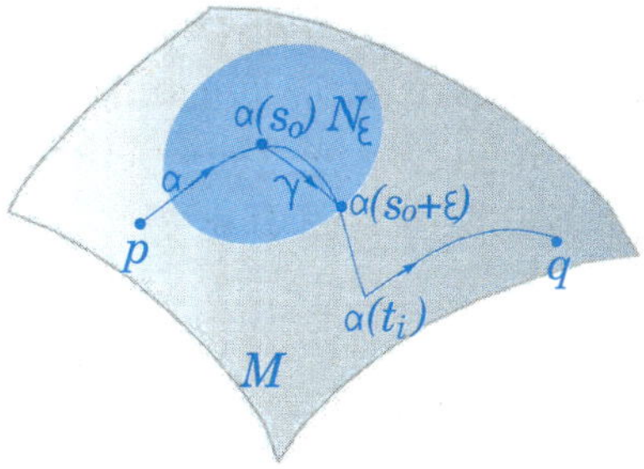

Figure5.3-33

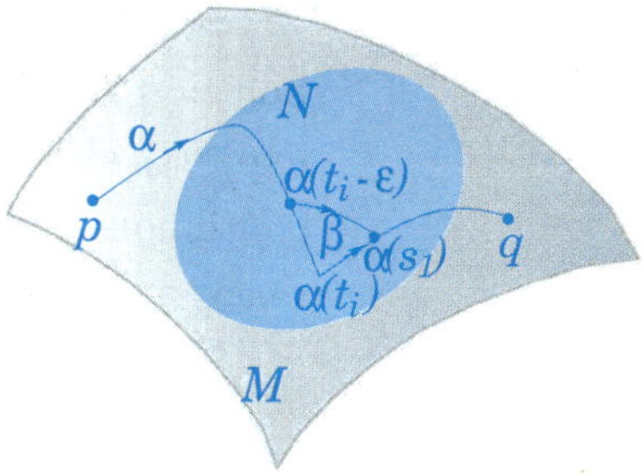

Figure5.3-34

$\alpha|_{[s_0,s_0+\epsilon]}$ by γ, we obtain a curve shorter than α from p to q, which is a contradiction to the assumption that α is the shortest curve from p to q. Thus, α is a broken geodesic.

Now, assume that α has a corner, say $\alpha(t_i)$. Then we may take a normal neighborhood N of $\alpha(t_i - \epsilon)$ for some ϵ containing $\alpha(t_i)$. For some $s_1 \in [t_i, t_{i+1}]$ the subsegment $\alpha|_{[t_i,s_1]}$ is still contained in N(Figure5.3-34). Since the restriction of α from $\alpha(t_i - \epsilon)$ to $\alpha(s_1)$ is contained in N, it is strictly longer than the radial geodesic β from $\alpha(t_i - \epsilon)$ to $\alpha(s_1)$ in N. As before, replacing the restriction of α by β from $\alpha(t_i - \epsilon)$ to $\alpha(s_1)$ shortens α in a contradiction. □

Using Lemma 5.3.10 and similar proofs to Theorem 5.3.11 and Corollary 5.3.12, the following corollary is valid for the timelike surface in $\mathbf{L}^3$.

Corollary 5.3.13 *Let M be a timelike surface in $\mathbf{L}^3$. For each point q of a normal neighborhood $N_\epsilon = \exp_p(B_\epsilon)$ of p in M which is joined to p by a future nonspacelike radial geodesic segment, the radial geodesic segment of N_ϵ from p to q uniquely maximizes arc length. Moreover, if α is the longest piecewise future nonspacelike regular curve from p to q in M, then α is an unbroken geodesic.*

Definition 5.3.14 A geodesic γ joining two points $p, q \in M$ is said to be **minimal**(or **maximal**) if its length $L(\gamma)$ is smaller(or longer) than or equal to the length of any piecewise regular curve(resp., future nonspacelike curve) joining p and q. Thus, $L(\gamma) = d(p, q)$.

A surface M is said to be **complete**(or **geodesically complete**) if all geodesics of M may be defined on the entire real line $\mathbb{R}$(or any geodesic may be extended to be defined for arbitrarily large values of any affine parameter).

Theorem 5.3.15 (Hopf-Rinow) *Let M be a complete surface in $\mathbf{E}^3$. Then, given two points $p, q \in M$, there exists a minimal geodesic in M joining p and q.*

Proof. Let $d(p,q) = r$ and $B_\delta(o)$ be a disk of radius δ contained in a neighborhood U of $o \in T_pM$ where $\exp_p$ is a diffeomorphism on U. Let $B_\delta(p) = \exp_p(B_\delta(o))$. Then $\Sigma = \text{bdry}(B_\delta(p))$ the boundary of $B_\delta(p)$ is compact as the continuous image of a compact set. If $x \in \Sigma$, the continuous function $d(x,q)$ reaches a minimum at $x_0 \in \Sigma$. Hence,

$$x_0 = \exp_p(\delta v), \quad |v| = 1, v \in T_pM.$$

Let γ be a geodesic parameterized by arc length given by $\gamma(t) = \exp_p(tv)$ as in Figure5.3-35. Since M is complete, γ may be defined for all $t \in \mathbb{R}$. In particular,

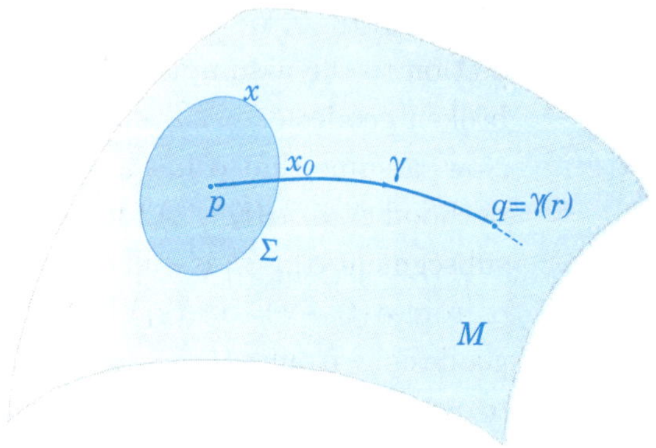

Figure5.3-35

γ is defined on $[0,r]$. We, now, claim that

$$\gamma(r) = q.$$

If so, since γ has unit speed, $L(\gamma) = r = d(p,q)$. Thus, γ is minimal as required. To prove our claim, it suffices to show that

$$d(\gamma(t), q) = r - t \quad \text{for all } t \in [\delta, r].$$

For then setting $t = r$, we obtain $d(\gamma(r), q) = 0$ from which $\gamma(r) = q$ from a basic property of Riemannian distance.

We prove the above equation by two steps.

Step 1: $d(\gamma(\delta), q) = r - \delta$. Every curve α from p to q crosses Σ at a point, say $x \in \Sigma$. Let $\Omega_{p,q}$ denote the set of all piecewise regular curves from p to q. Then

$$\begin{aligned} d(p,q) &= \inf\{L(\alpha)|\ \alpha \in \Omega_{p,q}\} \\ &= \inf_{x\in\Sigma}[\{L(\alpha)|\ \alpha \in \Omega_{p,x}\} + \{L(\alpha)|\ \alpha \in \Omega_{x,q}\}] \\ &= \inf_{x\in\Sigma}[d(p,x) + d(x,q)] = \inf_{x\in\Sigma}[\delta + d(x,q)] \\ &= \delta + d(x_0, q). \end{aligned}$$

Thus, $d(\gamma(\delta), q) = r - \delta$(as $r = d(p,q)$ and $x_0 = \gamma(\delta)$).

Step 2: If our claim holds for $t_0 \in [\delta, r]$ then it holds for $t_0 + \delta'$ where $\delta' > 0$ is sufficiently small.

Let $B_{\delta'}(0) \subset T_{\gamma(t_0)}M$ be a disk contained in a neighborhood U where $\exp_{\gamma(t_0)}$ is a diffeomorphism. Let $B_{\delta'}(\gamma(t_0)) = \exp_{\gamma(t_0)} B_{\delta'}$ and $\Sigma' = \text{bdry}(B_{\delta'}(\gamma(t_0)))$. For $x' \in \Sigma'$ the continuous function $d(x', q)$ has a minimum value at $x_0' \in \Sigma'$ as in Figure5.3-36. Then, with similar reasoning as in Step 1,

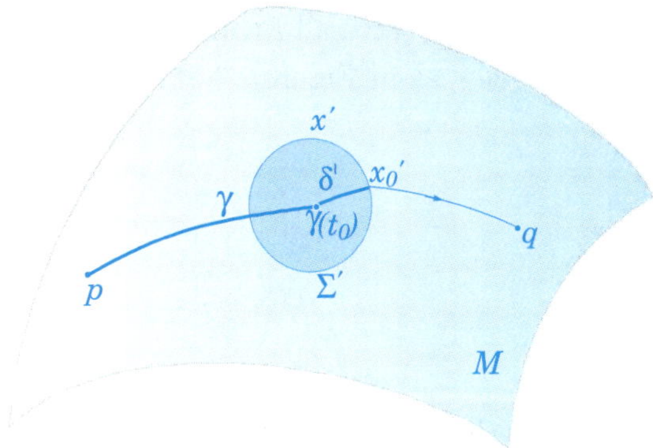

Figure5.3-36

$$d(\gamma(t_0), q) = \inf_{x \in \Sigma'} [d(\gamma(t_0), x') + d(x', q)] = \delta' + d(x_0', q).$$

Since we assumed that $d(\gamma(t), q) = r - t \quad$ for all $t \in [\delta, t_0]$,

$$d(x_0', q) = r - t_0 - \delta'.$$

Moreover, since $d(p, q) \leq d(p, x_0') + d(x_0', q)$,

$$d(p, x_0') \geq d(p, q) - d(x_0', q) = r - (r - t_0 - \delta') = t_0 + \delta'.$$

That is, $d(p, x_0') \geq t_0 + \delta'$.

But the piecewise smooth curve σ , which consists of γ from p to $\gamma(t_0)$ followed by the radial geodesic segment in $B_{\delta'}(\gamma(t_0))$ from $\gamma(t_0)$ to x', has length $t_0 + \delta'$. Thus by Corollary 5.3.12, the curve σ must coincide with $\gamma|[0, t_0 + \delta']$. Hence, $x_0' = \gamma(t_0 + \delta')$. Returning to the equality $d(x_0', q) = r - (t_0 + \delta')$, we now have

$$d(\gamma(t_0 + \delta'), q) = d(x_0', q) = r - (t_0 + \delta')$$

as required.

Now let $A = \{t \in [\delta, r] |\ d(\gamma(t), q) = r - t\}$.

First, A is nonempty as $\delta \in A$ from Step 1. By Step 2, A is open Also, A is closed as the Riemannian distance is continuous. Hence as $[\delta, r]$ is connected, $A = [\delta, r]$. $\square$

Remark 5.3.16 For a timelike surface in $\mathbf{L}^3$, in general, the geodesic completeness does not implies the existence of maximal geodesic segments joining two points with $p << q$. For example, let $M = \{(x,t) \| -\pi/2 < x < \pi/2\}$ with the metric $ds^2 = \sec^2 x(-dt^2 + dx^2)$. Then M is a timelike strip called the **universal cover of anti-de Sitter surface** in $\mathbf{L}^3$. The points p and q are chronologically related in M, but no maximal timelike geodesic in M joins p and q since all future directed timelike geodesics emanating from p are focused at r as in Figure5.3-37.

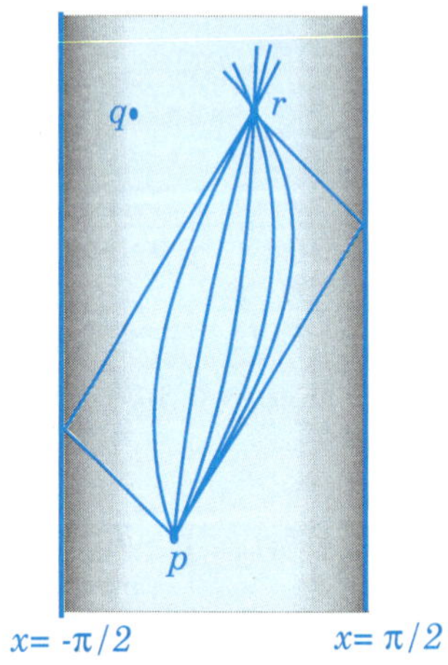

Figure5.3-37

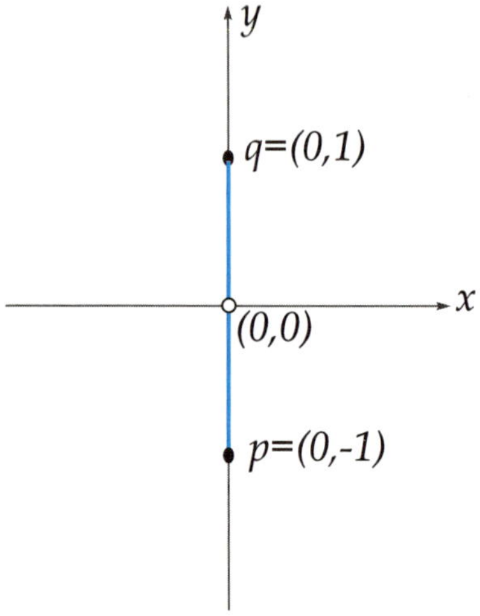

Figure5.3-38

Remark 5.3.17 Let $p = (0,-1)$ and $q = (0,1)$ be two points in a deleted timelike plane $M = \mathbf{R}^2 - (0,0)$ with the metric $ds^2 = dx^2 - dy^2$. Then $d(p,q) = 2$ but p and q can not be joined by a maximal timelike geodesic(cf. Figure5.3-38).

Definition 5.3.18 Let M be a timelike surface in $\mathbf{L}^3$. Given $p, q \in M$, $p \leq q$ means that $p = q$ or there is a piecewise smooth future directed nonspacelike curve from p to q. The **causal future** $J^+(p)$ of p and **causal past** $J^-(p)$ of p are defined by

$$J^+(p) = \{q \in M | p \leq q\}, \quad J^-(p) = \{q \in M | q \leq p\},$$

respectively. A timelike surface M in $\mathbf{L}^3$ is **strongly causal** if M does not contain any point p of M such that there are future-directed nonspacelike curves leaving arbitrarily small neighborhood of p and then returning. Moreover, a strong causal surface M is said to be **globally hyperbolic** if $J^+(p) \cap J^-(q)$ is compact for all $p, q \in M$.

Remark 5.3.19 Avez(1963) and Seifert(1967) obtained a timelike version of Theorem 5.3.15 for globally hyperbolic surface M. We state this result as

Theorem 5.3.20 *Let (M, g) be a globally hyperbolic surface in $\mathbf{L}^3$. Then, given any $p, q \in M$ with $q \in J^+(p)$, there exists a future directed maximal nonspacelike geodesic segment γ from p to q with $L(\gamma) = d(p, q)$.*

Note that if $q \in J^+(p) - I^+(p)$, then γ is lightlike and $L(\gamma) = d(p, q) = 0$.

For further study, see the following references.

1. Beem, J. K., P. E. Ehrlich and K. L. Easley, Global Lorentzian Geometry, Second Edition, Marcel Dekker Inc. New York, 1996.

2. do Carmo, Manfredo Perdigan. Differential Geometry of Curves and Surfaces. Englewood Cliffs, NJ: Prentice-Hall, 1976.

3. Millman, R. and G. Parker, Elements of Differential Geometry, Prentice-Hall Inc, 1977.

4. O'Neill, B. Elementary Differential Geometry, Academic Press Inc, 1966.

5.4 Arc Length Variations and the Bonnet-Myers Theorem

The First Variation Formula

Definition 5.4.1 Let M be a regular surface in $\mathbf{E}^3$ or a timelike surface in $\mathbf{L}^3$ and let $\gamma : [0, b] \to M$ be a unit speed regular curve or a unit speed timelike curve. A **variation** α of γ is a C^∞ map $\alpha : [0, b] \times (-\epsilon, \epsilon) \to M$ such that $\alpha(t, 0) = \gamma(t)$ for $t \in [0, b]$. We also use the notation for neighboring curve of the variation $\alpha(t, s)$ by $\alpha_s(t)$ with $\alpha_0(t) = \alpha(t, 0) = \gamma(t)$. If the variation α of

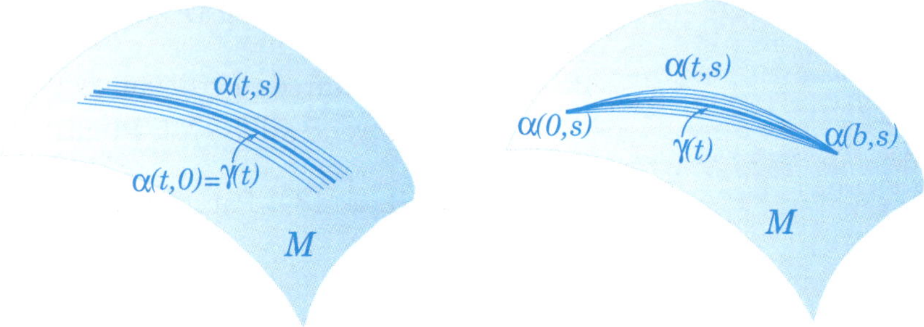

Figure5.4-39 The variation and the proper variation α of γ.

γ satisfies $\alpha(0, s) = \gamma(0)$ and $\alpha(b, s) = \gamma(b)$ for $s \in (-\epsilon, \epsilon)$, then the variation α of γ is called a **proper variation** of γ. Now, consider the partial derivatives of $\alpha(t, s)$ of γ given by $\dfrac{\partial \alpha}{\partial t}$ and $\dfrac{\partial \alpha}{\partial s}$ where especially

$$\left.\frac{\partial \alpha}{\partial t}\right|_{(t,0)} = \gamma'(t) \text{ and we denote } \left.\frac{\partial \alpha}{\partial s}\right|_{(t,0)} = V(t) \text{ for } t \in [0, b].$$

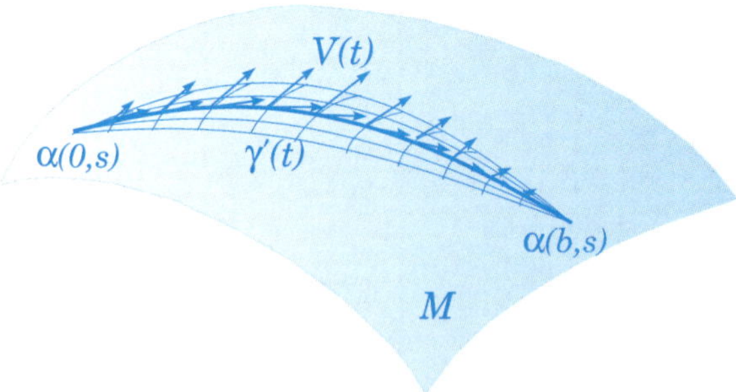

Figure5.4-40 The variation vector field $V(t)$ along $\gamma(t)$.

Here, $V(t)$ is called the **variation vector field** along γ(Figure 5.4-40). If the variation α is proper, then $\left.\dfrac{\partial \alpha}{\partial s}\right|_{(0,0)} = V(0) = 0$ and $\left.\dfrac{\partial \alpha}{\partial s}\right|_{(b,0)} = V(b) = 0$. If the variation vector field $V(t)$ for the variation α satisfies $< V(t), \gamma'(t) >= 0$ along γ, α is called an **orthogonal variation** of α.

Lemma 5.4.2 *Let M be a regular surface in $\mathbf{E}^3$ or a timelike surface in $\mathbf{L}^3$ and let $\gamma : [0,b] \to M$ be a regular curve or a unit speed timelike curve. Then there exists $\delta > 0$ such that $\exp_{\gamma(t)} v$ is defined on the whole interval $[0,b]$ for $v \in T_{\gamma(t)}M$ such that $|v| < \delta$.*

Proof. For each $p \in \gamma([0,b]) \subset M$, let $B_{\delta(p)} = \exp_p \{v \in T_pM; |v| < \delta(p)\}$ where $\delta(p) > 0$ is chosen sufficiently small that $B_{\delta(p)}$ is a normal neighborhood of all of its points (cf. Remark 5.3.3 and Theorem 5.3.4). Then the set $\left\{B_{\delta(p)} | p \in \gamma([0,b])\right\}$ an open covering of $\gamma([0,b])$. Since $[0,b]$ is compact and hence $\gamma([0,b])$ is compact, there are finitely many points $p_1, p_2, \cdots, p_n \in \gamma([0,b])$ such that $\left\{B_{\delta(p_1)}, B_{\delta(p_2)}, \cdots, B_{\delta(p_n)}\right\}$ covers $\gamma([0,b])$. Hence, we take δ for this result by $\delta = \min\{\delta(p_1), \delta(p_2), \cdots, \delta(p_n)\}$. □

Theorem 5.4.3 *Let M be a regular surface in $\mathbf{E}^3$ or a timelike surface in $\mathbf{L}^3$ and let $\gamma : [0,b] \to M$ be a unit speed regular curve or a unit speed timelike curve. Let $V(t)$ be a differentiable vector field along γ, then there is a variation $\alpha : [0,b] \times (-\epsilon, \epsilon) \to M$ such that $V(t)$ is the variation vector field of α. Further, if $V(0) = V(b) = 0$, then α may be chosen to be proper.*

Proof. Let $M = \max\{|V(t)| | t \in [0,b]\}$ and take $\epsilon > 0$ so that $\epsilon < \dfrac{\delta}{M}$. Define

$$\alpha(t,s) = \exp_{\gamma(t)} sV(t),$$

$t \in [0,b]$, $s \in (-\epsilon, \epsilon)$. Then since $\alpha(t,s) = \exp_{\gamma(t)} sV(t) = \gamma_{sV(t)}(1)$ is a geodesic with velocity vector $sV(t)$, $\left.\dfrac{\partial \alpha}{\partial t}\right|_{(t,s)} = sV(t)$, $\alpha(t,0) = \exp_{\gamma(t)} sV(t)\Big|_{(t,0)} = \gamma(t)$, and $\left.\dfrac{\partial \alpha}{\partial s}\right|_{(t,0)} = \dfrac{\partial}{\partial s} \exp_{\gamma(t)} sV(t)\Big|_{(t,0)} = V(t)$. Moreover, if $V(0) = V(b) = 0$, then $\alpha(0,s) = \exp_{\gamma(0)} sV(0) = \gamma(0)$ and $\alpha(b,s) = \exp_{\gamma(b)} sV(b) = \gamma(b)$. Thus, α is proper. □

Lemma 5.4.4 *Let M be a regular surface in $\mathbf{E}^3$ or a timelike surface in $\mathbf{L}^3$. For the C^∞ map $\alpha : [0,b] \times (-\epsilon, \epsilon) \to M$,*

$$\nabla_{\frac{\partial \alpha}{\partial t}} \frac{\partial \alpha}{\partial s} = \nabla_{\frac{\partial \alpha}{\partial s}} \frac{\partial \alpha}{\partial t}.$$

Proof. Let $\alpha(t,s)$ be a C^∞ map in a coordinate patch $X(U) \subset M$ such that $u = u(t,s)$, $v = v(t,s)$. Since

$$\frac{\partial \alpha}{\partial t} = \frac{\partial u}{\partial t} X_u + \frac{\partial v}{\partial t} X_v, \quad \frac{\partial \alpha}{\partial s} = \frac{\partial u}{\partial s} X_u + \frac{\partial v}{\partial s} X_v,$$

and since

$$\begin{aligned}\frac{\partial}{\partial t}\frac{\partial \alpha}{\partial s} &= \frac{\partial}{\partial t}\left(\frac{\partial u}{\partial s}X_u + \frac{\partial v}{\partial s}X_v\right)\\ &= \frac{\partial^2 u}{\partial t\partial s}X_u + \frac{\partial u}{\partial s}\left(\frac{\partial u}{\partial t}X_{uu} + \frac{\partial v}{\partial t}X_{uv}\right)\\ &\quad + \frac{\partial^2 v}{\partial t\partial s}X_v + \frac{\partial v}{\partial s}\left(\frac{\partial u}{\partial t}X_{uv} + \frac{\partial v}{\partial t}X_{vv}\right),\end{aligned}$$

using the Christoffel symbols in section 4.2 we obtain

$$\begin{aligned}\nabla_{\frac{\partial \alpha}{\partial t}}\frac{\partial \alpha}{\partial s} &= \left(\frac{\partial^2 u}{\partial t\partial s} + \Gamma^1_{11}\frac{\partial u}{\partial s}\frac{\partial u}{\partial t} + \Gamma^1_{12}\frac{\partial u}{\partial s}\frac{\partial v}{\partial t} + \Gamma^1_{12}\frac{\partial v}{\partial s}\frac{\partial u}{\partial t}\right.\\ &\quad \left. + \Gamma^1_{22}\frac{\partial v}{\partial s}\frac{\partial v}{\partial t}\right)X_u + \left(\frac{\partial^2 v}{\partial t\partial s} + \Gamma^2_{11}\frac{\partial u}{\partial s}\frac{\partial u}{\partial t} + \Gamma^2_{12}\frac{\partial u}{\partial s}\frac{\partial v}{\partial t}\right.\\ &\quad \left. + \Gamma^2_{12}\frac{\partial v}{\partial s}\frac{\partial u}{\partial t} + \Gamma^2_{22}\frac{\partial v}{\partial s}\frac{\partial v}{\partial t}\right)X_v.\end{aligned}$$

Similarly, we may also compute $\nabla_{\frac{\partial \alpha}{\partial s}}\frac{\partial \alpha}{\partial t}$ and we obtain our result, recalling that $\Gamma^j_{12} = \Gamma^j_{21}$ for $j = 1, 2$. □

Theorem 5.4.5 (The First Variation Formula in $\mathbf{E}^3$) *Let α be a proper variation of a unit speed regular curve γ on a regular surface in $\mathbf{E}^3$ and let $L : (-\epsilon, \epsilon) \to \mathbb{R}$ be the length functional given by*

$$L(s) = \int_0^b \left|\frac{\partial \alpha}{\partial t}\right| dt$$

for $s \in (-\epsilon, \epsilon)$. Then

$$L'(0) = -\int_0^b < V(t), \gamma''(t) > dt,$$

where $V(t) = \frac{\partial \alpha}{\partial s}|_{(t,0)}$.

Proof. Using Lemma 5.4.4 in the 2nd step,

$$L'(s) = \frac{d}{ds}\int_0^b \left|\frac{\partial \alpha}{\partial t}\right| dt = \int_0^b \frac{\partial}{\partial s}\left|\frac{\partial \alpha}{\partial t}\right| dt.$$

$$\begin{aligned}\frac{\partial}{\partial s}\left\langle\frac{\partial \alpha}{\partial t}, \frac{\partial \alpha}{\partial t}\right\rangle^{\frac{1}{2}} &= \left\langle\nabla_{\frac{\partial \alpha}{\partial s}}\frac{\partial \alpha}{\partial t}, \frac{\partial \alpha}{\partial t}\right\rangle \Big/ \left\langle\frac{\partial \alpha}{\partial t}, \frac{\partial \alpha}{\partial t}\right\rangle^{\frac{1}{2}}\\ &= \left\langle\nabla_{\frac{\partial \alpha}{\partial t}}\frac{\partial \alpha}{\partial s}, \frac{\partial \alpha}{\partial t}\right\rangle \Big/ \left\langle\frac{\partial \alpha}{\partial t}, \frac{\partial \alpha}{\partial t}\right\rangle^{\frac{1}{2}}.\end{aligned}$$

Thus

$$L'(s) = \int_0^b \left\langle \nabla_{\frac{\partial\alpha}{\partial t}} \frac{\partial\alpha}{\partial s}, \frac{\partial\alpha}{\partial t} \right\rangle \Big/ \left\langle \frac{\partial\alpha}{\partial t}, \frac{\partial\alpha}{\partial t} \right\rangle^{\frac{1}{2}} dt.$$

Hence

$$\begin{aligned} L'(0) &= \int_0^b \left\langle \nabla_{\frac{\partial\alpha}{\partial t}} \frac{\partial\alpha}{\partial s}, \frac{\partial\alpha}{\partial t} \right\rangle\Big|_{s=0} dt \ \left(\text{since } \left\langle \frac{\partial\alpha}{\partial t}, \frac{\partial\alpha}{\partial t} \right\rangle\Big|_{s=0} = 1\right) \\ &= \int_0^b \frac{\partial}{\partial t}\left\langle \frac{\partial\alpha}{\partial s}, \frac{\partial\alpha}{\partial t} \right\rangle\Big|_{s=0} - \left\langle \frac{\partial\alpha}{\partial s}, \nabla_{\frac{\partial\alpha}{\partial t}} \frac{\partial\alpha}{\partial t} \right\rangle\Big|_{s=0} dt \\ &= \left\langle \frac{\partial\alpha}{\partial s}, \frac{\partial\alpha}{\partial t} \right\rangle\Big|_{t=0}^{b} - \int_0^b \left\langle \frac{\partial\alpha}{\partial s}, \nabla_{\frac{\partial\alpha}{\partial t}} \frac{\partial\alpha}{\partial t} \right\rangle\Big|_{s=0} dt \\ &= -\int_0^b \left\langle \frac{\partial\alpha}{\partial s}, \nabla_{\frac{\partial\alpha}{\partial t}} \frac{\partial\alpha}{\partial t} \right\rangle\Big|_{s=0} dt \ (\text{since } \alpha \text{ is proper}) \\ &= -\int_0^b \langle V(t), \nabla_{\gamma'(t)}\gamma'(t)\rangle\, dt. \end{aligned}$$

□

Since $\left|\frac{\partial\alpha}{\partial t}\right| = \left(-\left\langle \frac{\partial\alpha}{\partial t}, \frac{\partial\alpha}{\partial t} \right\rangle\right)^{\frac{1}{2}}$ on a timelike surface in $\mathbf{L}^3$,

$$\begin{aligned} \frac{\partial}{\partial s}\left(-\left\langle \frac{\partial\alpha}{\partial t}, \frac{\partial\alpha}{\partial t} \right\rangle\right)^{\frac{1}{2}} &= -\left\langle \frac{\partial^2\alpha}{\partial s\partial t}, \frac{\partial\alpha}{\partial t} \right\rangle \Big/ \left(-\left\langle \frac{\partial\alpha}{\partial t}, \frac{\partial\alpha}{\partial t} \right\rangle\right)^{\frac{1}{2}} \\ &= -\left\langle \frac{\partial^2\alpha}{\partial t\partial s}, \frac{\partial\alpha}{\partial t} \right\rangle \Big/ \left(-\left\langle \frac{\partial\alpha}{\partial t}, \frac{\partial\alpha}{\partial t} \right\rangle\right)^{\frac{1}{2}}. \end{aligned}$$

Using the fact that $\left\langle \frac{\partial\alpha}{\partial t}, \frac{\partial\alpha}{\partial t} \right\rangle\Big|_{s=0} = -1$ and in the same manner of the proof of Theorem 5.4.5, we may obtain the following corollary.

Corollary 5.4.6 (The First Variation Formula in $\mathbf{L}^3$) *Let α be a proper variation of a unit speed timelike curve γ on a timelike surface in $\mathbf{L}^3$ and let $L : (-\epsilon, \epsilon) \to \mathbb{R}$ be the length functional given by*

$$L(s) = \int_0^b \left|\frac{\partial\alpha}{\partial t}\right| dt$$

for $s \in (-\epsilon, \epsilon)$. Then

$$L'(0) = \int_0^b < V(t), \gamma''(t) > dt.$$

Corollary 5.4.7 *Let $\gamma : [0, b] \to M$ be a unit speed regular curve on a surface in $\mathbf{E}^3$ or a unit speed timelike curve on a timelike surface in $\mathbf{L}^3$. Then γ is a geodesic if and only if for any proper variation $\alpha : [0, b] \times (-\epsilon, \epsilon) \to M$, $L'(0) = 0$.*

Proof. Suppose γ is a geodesic. Then $\langle V(t), \gamma''(t) \rangle = 0$. Hence, $L'(0) = 0$. Conversely, for any variation α suppose $L'(0) = 0$. Take a variation so that

$$V(t) = f(t) \nabla_{\gamma'(t)} \gamma'(t)$$

where $f : [0, b] \to \mathbb{R}$ is a differentiable function with $f(t) \geq 0$, $f(0) = f(b) = 0$. In $\mathbf{E}^3$,

$$\begin{aligned} 0 = L'(0) &= -\int_0^b \left\langle f(t) \nabla_{\gamma'(t)} \gamma'(t), \nabla_{\gamma'(t)} \gamma'(t) \right\rangle dt \\ &= -\int_0^b f(t) \left| \nabla_{\gamma'(t)} \gamma'(t) \right|^2 dt. \end{aligned}$$

In $\mathbf{L}^3$,

$$\begin{aligned} 0 = L'(0) &= \int_0^b \left\langle f(t) \nabla_{\gamma'(t)} \gamma'(t), \nabla_{\gamma'(t)} \gamma'(t) \right\rangle dt \\ &= \int_0^b f(t) \left| \nabla_{\gamma'(t)} \gamma'(t) \right|^2 dt. \end{aligned}$$

Hence, $f(t) \left| \nabla_{\gamma'(t)} \gamma'(t) \right|^2 = 0$ on $[0, b]$. Thus, $\left| \nabla_{\gamma'(t)} \gamma'(t) \right| = 0$ for any $t \in [0, b]$ since $f(t) > 0$ for some t. Therefore, γ is a geodesic. (In the timelike case, $\nabla_{\gamma'(t)} \gamma'(t)$ can not be null since it is orthogonal to the timelike vector $\gamma'(t)$). □

Lemma 5.4.8 *Let M be a regular surface in $\mathbf{E}^3$ and let $X(u, v)$ be a coordinate patch on an open neighborhood U of $p \in M$. Then*

$$\nabla_{X_v} \nabla_{X_u} X_u - \nabla_{X_u} \nabla_{X_v} X_u = K (X_u \times X_v) \times X_u$$

where K is the Gaussian Curvature of M.

Proof.

$$\begin{aligned} \nabla_{X_v} \nabla_{X_u} X_u &= \nabla_{X_v} \left(\Gamma_{11}^1 X_u + \Gamma_{11}^2 X_v \right) \\ &= \left(\Gamma_{11}^1 \right)_v X_u + \left(\Gamma_{11}^2 \right)_v X_v + \Gamma_{11}^1 \nabla_{X_v} X_u + \Gamma_{11}^2 \nabla_{X_v} X_v \\ &= \left\{ \left(\Gamma_{11}^1 \right)_v + \Gamma_{11}^1 \Gamma_{12}^1 + \Gamma_{11}^2 \Gamma_{22}^1 \right\} X_u \\ &\quad + \left\{ \left(\Gamma_{11}^2 \right)_v + \Gamma_{11}^1 \Gamma_{12}^2 + \Gamma_{11}^2 \Gamma_{22}^2 \right\} X_v. \end{aligned}$$

Similarly, we also obtain

$$\begin{aligned}\nabla_{X_u}\nabla_{X_v}X_u = &\left\{\left(\Gamma^1_{12}\right)_u + \Gamma^1_{11}\Gamma^1_{12} + \Gamma^2_{12}\Gamma^1_{12}\right\} X_u \\ &+ \left\{\left(\Gamma^2_{12}\right)_u + \Gamma^1_{12}\Gamma^2_{11} + \Gamma^2_{12}\Gamma^2_{12}\right\} X_v.\end{aligned}$$

Hence,

$$\begin{aligned}\nabla_{X_v}\nabla_{X_u}X_u - \nabla_{X_u}\nabla_{X_v}X_u = &\left\{\left(\Gamma^1_{11}\right)_v - \left(\Gamma^1_{12}\right)_u + \Gamma^2_{11}\Gamma^1_{22} - \Gamma^2_{12}\Gamma^1_{12}\right\} X_u \\ &+ \left\{\left(\Gamma^2_{11}\right)_v - \left(\Gamma^2_{12}\right)_u + \Gamma^1_{11}\Gamma^2_{12} + \Gamma^2_{11}\Gamma^2_{22}\right. \\ &\left. -\Gamma^1_{12}\Gamma^2_{11} - \Gamma^2_{12}\Gamma^2_{12}\right\} X_v.\end{aligned}$$

By the Gauss equations (4.3-1) and (4.3-4) in Section 4.3,

$$\begin{aligned}\nabla_{X_v}\nabla_{X_u}X_u - \nabla_{X_u}\nabla_{X_v}X_u &= -FKX_u + EKX_v \\ &= K\{< X_u, X_u > X_v - < X_u, X_v > X_u\} \\ &= K(X_u \times X_v) \times X_u.\end{aligned}$$

□

Since $(X_u \times X_v) \times X_u = -(< X_u, X_u > X_v - < X_u, X_v > X_u)$ in the last equation of the above proof in $\mathbf{L}^3$, we may obtain the following corollary

Corollary 5.4.9 *Let M be a timelike surface in $\mathbf{L}^3$ and let $X(u,v)$ be a coordinate patch on an open neighborhood U of $p \in M$. Then*

$$\nabla_{X_v}\nabla_{X_u}X_u - \nabla_{X_u}\nabla_{X_v}X_u = -K(X_u \times X_v) \times X_u$$

where $K(u,v)$ is the Gaussian curvature of M at $X(u,v)$.

Lemma 5.4.10 *Let M be a regular surface in $\mathbf{E}^3$ and let $\alpha(t,s)$ be a C^∞ map on a coordinate patch $X(U) \subset M$ and $V(t,s)$ be a C^∞ vector field on $\alpha(t,s) \subset X(U)$. Then*

$$\nabla_{\frac{\partial\alpha}{\partial s}}\nabla_{\frac{\partial\alpha}{\partial t}}V - \nabla_{\frac{\partial\alpha}{\partial t}}\nabla_{\frac{\partial\alpha}{\partial s}}V = K\left(\frac{\partial\alpha}{\partial t} \times \frac{\partial\alpha}{\partial s}\right) \times V$$

where $K(t,s)$ is the Gaussian curvature of M at $\alpha(t,s)$.

Proof. Express $V(t,s) = a(t,s)X_u + b(t,s)X_v$. Then

$$\begin{aligned}\nabla_{\frac{\partial\alpha}{\partial t}}V &= \nabla_{\frac{\partial\alpha}{\partial t}}(aX_u + bX_v) \\ &= \frac{\partial a}{\partial t}X_u + \frac{\partial b}{\partial t}X_v + a\nabla_{\frac{\partial\alpha}{\partial t}}X_u + b\nabla_{\frac{\partial\alpha}{\partial t}}X_v.\end{aligned}$$

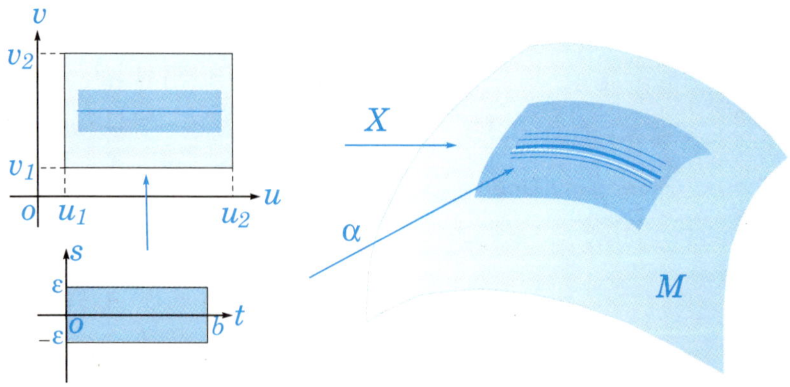

Figure5.4-41

$$\begin{aligned}\nabla_{\frac{\partial\alpha}{\partial s}}\nabla_{\frac{\partial\alpha}{\partial t}}V =&\frac{\partial^2 a}{\partial s\partial t}X_u+\frac{\partial^2 b}{\partial s\partial t}X_v+\frac{\partial a}{\partial s}\nabla_{\frac{\partial\alpha}{\partial t}}X_u+\frac{\partial b}{\partial s}\nabla_{\frac{\partial\alpha}{\partial t}}X_v\\&+\frac{\partial a}{\partial t}\nabla_{\frac{\partial\alpha}{\partial s}}X_u+\frac{\partial b}{\partial t}\nabla_{\frac{\partial\alpha}{\partial s}}X_v+a\nabla_{\frac{\partial\alpha}{\partial s}}\nabla_{\frac{\partial\alpha}{\partial t}}X_u+b\nabla_{\frac{\partial\alpha}{\partial s}}\nabla_{\frac{\partial\alpha}{\partial t}}X_v.\end{aligned}$$

Similarly, exchanging t and s, we also compute $\nabla_{\frac{\partial\alpha}{\partial t}}\nabla_{\frac{\partial\alpha}{\partial s}}V$. Hence,

$$\begin{aligned}\nabla_{\frac{\partial\alpha}{\partial s}}\nabla_{\frac{\partial\alpha}{\partial t}}V-\nabla_{\frac{\partial\alpha}{\partial t}}\nabla_{\frac{\partial\alpha}{\partial s}}V&=a\left(\nabla_{\frac{\partial\alpha}{\partial s}}\nabla_{\frac{\partial\alpha}{\partial t}}X_u-\nabla_{\frac{\partial\alpha}{\partial t}}\nabla_{\frac{\partial\alpha}{\partial s}}X_u\right)\\&+b\left(\nabla_{\frac{\partial\alpha}{\partial s}}\nabla_{\frac{\partial\alpha}{\partial t}}X_v-\nabla_{\frac{\partial\alpha}{\partial t}}\nabla_{\frac{\partial\alpha}{\partial s}}X_v\right)\end{aligned}$$

As in Figure 5.4-41, we parameterize α in $X(U)$ by

$$X(u,v)=X(u(t,s),v(t,s)),$$

whence $\dfrac{\partial\alpha}{\partial t}=\dfrac{\partial u}{\partial t}X_u+\dfrac{\partial v}{\partial t}X_v$ and $\dfrac{\partial\alpha}{\partial s}=\dfrac{\partial u}{\partial s}X_u+\dfrac{\partial v}{\partial s}X_v$. Thus,

$$\nabla_{\frac{\partial\alpha}{\partial t}}X_u=\frac{\partial u}{\partial t}\nabla_{X_u}X_u+\frac{\partial v}{\partial t}\nabla_{X_v}X_u$$

$$\begin{aligned}\nabla_{\frac{\partial\alpha}{\partial s}}\nabla_{\frac{\partial\alpha}{\partial t}}X_u=&\frac{\partial^2 u}{\partial s\partial t}\nabla_{X_u}X_u+\frac{\partial^2 v}{\partial s\partial t}\nabla_{X_v}X_u\\&+\frac{\partial u}{\partial t}\left(\frac{\partial u}{\partial s}\nabla_{X_u}\nabla_{X_u}X_u+\frac{\partial v}{\partial s}\nabla_{X_v}\nabla_{X_u}X_u\right)\\&+\frac{\partial v}{\partial t}\left(\frac{\partial u}{\partial s}\nabla_{X_u}\nabla_{X_v}X_u+\frac{\partial v}{\partial s}\nabla_{X_v}\nabla_{X_v}X_u\right)\end{aligned}$$

By exchanging t and s, we also compute $\nabla_{\frac{\partial\alpha}{\partial t}}\nabla_{\frac{\partial\alpha}{\partial s}}X_u$. Thus, we obtain

$$\begin{aligned}(5.4\text{-}1)\quad&\nabla_{\frac{\partial\alpha}{\partial s}}\nabla_{\frac{\partial\alpha}{\partial t}}X_u-\nabla_{\frac{\partial\alpha}{\partial t}}\nabla_{\frac{\partial\alpha}{\partial s}}X_u\\&=\left(\frac{\partial u}{\partial t}\frac{\partial v}{\partial s}-\frac{\partial v}{\partial t}\frac{\partial u}{\partial s}\right)(\nabla_{X_v}\nabla_{X_u}X_u-\nabla_{X_u}\nabla_{X_v}X_u)\\&=\left(\frac{\partial u}{\partial t}\frac{\partial v}{\partial s}-\frac{\partial v}{\partial t}\frac{\partial u}{\partial s}\right)K(X_u\times X_v)\times X_u.\end{aligned}$$

Similarly,

$$
\begin{aligned}
(5.4\text{-}2)\qquad & \nabla_{\frac{\partial\alpha}{\partial s}}\nabla_{\frac{\partial\alpha}{\partial t}}X_v - \nabla_{\frac{\partial\alpha}{\partial t}}\nabla_{\frac{\partial\alpha}{\partial s}}X_v \\
&= \left(\frac{\partial u}{\partial t}\frac{\partial v}{\partial s} - \frac{\partial v}{\partial t}\frac{\partial u}{\partial s}\right)\left(\nabla_{X_v}\nabla_{X_u}X_v - \nabla_{X_u}\nabla_{X_v}X_v\right) \\
&= \left(\frac{\partial u}{\partial t}\frac{\partial v}{\partial s} - \frac{\partial v}{\partial t}\frac{\partial u}{\partial s}\right)K(X_u \times X_v)\times X_v.
\end{aligned}
$$

Now, recalling our prior calculation

$$
\begin{aligned}
\nabla_{\frac{\partial\alpha}{\partial s}}\nabla_{\frac{\partial\alpha}{\partial t}}V - \nabla_{\frac{\partial\alpha}{\partial t}}\nabla_{\frac{\partial\alpha}{\partial s}}V &= a\left(\nabla_{\frac{\partial\alpha}{\partial s}}\nabla_{\frac{\partial\alpha}{\partial t}}X_u - \nabla_{\frac{\partial\alpha}{\partial t}}\nabla_{\frac{\partial\alpha}{\partial s}}X_u\right) \\
&\quad + b\left(\nabla_{\frac{\partial\alpha}{\partial s}}\nabla_{\frac{\partial\alpha}{\partial t}}X_v - \nabla_{\frac{\partial\alpha}{\partial t}}\nabla_{\frac{\partial\alpha}{\partial s}}X_v\right) \\
&= a\left(\frac{\partial u}{\partial t}\frac{\partial v}{\partial s} - \frac{\partial v}{\partial t}\frac{\partial u}{\partial s}\right)K(X_u \times X_v)\times X_u \\
&\quad + b\left(\frac{\partial u}{\partial t}\frac{\partial v}{\partial s} - \frac{\partial v}{\partial t}\frac{\partial u}{\partial s}\right)K(X_u \times X_v)\times X_v \\
&= K\left(\frac{\partial u}{\partial t}\frac{\partial v}{\partial s} - \frac{\partial v}{\partial t}\frac{\partial u}{\partial s}\right)(X_u \times X_v)\times(aX_u + bX_v).
\end{aligned}
$$

Since

$$
\begin{aligned}
\frac{\partial\alpha}{\partial t}\times\frac{\partial\alpha}{\partial s} &= \left(\frac{\partial u}{\partial t}X_u + \frac{\partial v}{\partial t}X_v\right)\times\left(\frac{\partial u}{\partial s}X_u + \frac{\partial v}{\partial s}X_v\right) \\
&= \left(\frac{\partial u}{\partial t}\frac{\partial v}{\partial s} - \frac{\partial v}{\partial t}\frac{\partial u}{\partial s}\right)(X_u \times X_v),
\end{aligned}
$$

we obtain

$$
\nabla_{\frac{\partial\alpha}{\partial s}}\nabla_{\frac{\partial\alpha}{\partial t}}V - \nabla_{\frac{\partial\alpha}{\partial t}}\nabla_{\frac{\partial\alpha}{\partial s}}V = K\left(\frac{\partial\alpha}{\partial t}\times\frac{\partial\alpha}{\partial s}\right)\times V.
$$

□

In $\mathbf{L^3}$, we add the the minus sign to the Gauss curvature K in the equations (5.4-1) and (5.4-2) as in Corollary 5.4.9. Thus we also obtain the following corollary.

Corollary 5.4.11 *Let M be a timelike surface in $\mathbf{L^3}$ and let $\alpha(t,s)$ be a C^∞ map on a coordinate patch $X(U)\subset M$ and $V(t,s)$ be a C^∞ vector field on $\alpha(t,s)\subset X(U)$. Then*

$$
\nabla_{\frac{\partial\alpha}{\partial s}}\nabla_{\frac{\partial\alpha}{\partial t}}V - \nabla_{\frac{\partial\alpha}{\partial t}}\nabla_{\frac{\partial\alpha}{\partial s}}V = -K\left(\frac{\partial\alpha}{\partial t}\times\frac{\partial\alpha}{\partial s}\right)\times V
$$

where $K(t,s)$ is the Gaussian curvature of M at $\alpha(t,s)$.

Theorem 5.4.12 (The Second Variation Formula in $\mathbf{E}^3$) *Let α be a proper orthogonal variation of a unit speed geodesic γ and $V(t) = \left.\dfrac{\partial \alpha}{\partial s}\right|_{(t,0)}$ be the associated variation vector field along γ on a regular surface M in $\mathbf{E}^3$. Then*

$$L''(0) = \int_0^b < V'(t), V'(t) > -K(t) < V(t), V(t) > dt$$

where $V'(t) = \nabla_{\gamma'(t)} V(t)$ and $K(t) = K(t,0)$ is the Gaussian curvature on $\gamma(t) = \alpha(t,0)$.

Proof. From the proof of Theorem 5.4.5,

$$L'(s) = \int_0^b \left\langle \nabla_{\frac{\partial \alpha}{\partial t}} \frac{\partial \alpha}{\partial s}, \frac{\partial \alpha}{\partial t} \right\rangle \Big/ \left\langle \frac{\partial \alpha}{\partial t}, \frac{\partial \alpha}{\partial t} \right\rangle^{\frac{1}{2}} dt.$$

Hence, using Lemma 5.4.4 for the second term

$$\begin{aligned} L''(s) = & \int_0^b \left(\frac{\partial}{\partial s} \left\langle \nabla_{\frac{\partial \alpha}{\partial t}} \frac{\partial \alpha}{\partial s}, \frac{\partial \alpha}{\partial t} \right\rangle \right) \left\langle \frac{\partial \alpha}{\partial t}, \frac{\partial \alpha}{\partial t} \right\rangle^{\frac{1}{2}} \Big/ \left\langle \frac{\partial \alpha}{\partial t}, \frac{\partial \alpha}{\partial t} \right\rangle dt \\ & - \int_0^b \left(\left\langle \nabla_{\frac{\partial \alpha}{\partial t}} \frac{\partial \alpha}{\partial s}, \frac{\partial \alpha}{\partial t} \right\rangle \right)^2 \Big/ \left\langle \frac{\partial \alpha}{\partial t}, \frac{\partial \alpha}{\partial t} \right\rangle^{\frac{3}{2}} dt. \end{aligned}$$

At $s = 0$, since $\gamma(t)$ is a unit speed orthogonal geodesic, we have

$$\left.\left\langle \frac{\partial \alpha}{\partial t}, \frac{\partial \alpha}{\partial t} \right\rangle\right|_{(t,0)} = < \gamma'(t), \gamma'(t) >= 1,$$

$$\left.\left\langle \frac{\partial \alpha}{\partial t}, \frac{\partial \alpha}{\partial s} \right\rangle\right|_{(t,0)} = < \gamma'(t), V(t) >= 0, \text{ and } \nabla_{\gamma'(t)} \gamma'(t) = 0.$$

Thus,

$$\begin{aligned} \left.\left\langle \nabla_{\frac{\partial \alpha}{\partial t}} \frac{\partial \alpha}{\partial s}, \frac{\partial \alpha}{\partial t} \right\rangle\right|_{(t,0)} & =< \nabla_{\gamma'(t)} V(t), \gamma'(t) > \\ & = \gamma'(t) \left(< V(t), \gamma'(t) > \right) - < V(t), \nabla_{\gamma'(t)} \gamma'(t) >= 0. \end{aligned}$$

Then

$$L''(0) = \int_0^b \frac{\partial}{\partial s} \left.\left\langle \nabla_{\frac{\partial \alpha}{\partial t}} \frac{\partial \alpha}{\partial s}, \frac{\partial \alpha}{\partial t} \right\rangle\right|_{(t,0)} dt.$$

Here,

$$\begin{aligned} \frac{\partial}{\partial s} \left.\left\langle \nabla_{\frac{\partial \alpha}{\partial t}} \frac{\partial \alpha}{\partial s}, \frac{\partial \alpha}{\partial t} \right\rangle\right|_{(t,0)} = & \left.\left\langle \nabla_{\frac{\partial \alpha}{\partial s}} \nabla_{\frac{\partial \alpha}{\partial t}} \frac{\partial \alpha}{\partial s}, \frac{\partial \alpha}{\partial t} \right\rangle\right|_{(t,0)} \\ & + \left.\left\langle \nabla_{\frac{\partial \alpha}{\partial t}} \frac{\partial \alpha}{\partial s}, \nabla_{\frac{\partial \alpha}{\partial s}} \frac{\partial \alpha}{\partial t} \right\rangle\right|_{(t,0)}. \end{aligned}$$

By Lemma 5.4.4,

$$\left\langle \nabla_{\frac{\partial\alpha}{\partial t}} \frac{\partial\alpha}{\partial s}, \nabla_{\frac{\partial\alpha}{\partial s}} \frac{\partial\alpha}{\partial t} \right\rangle \Big|_{(t,0)} = \left\langle \nabla_{\frac{\partial\alpha}{\partial t}} \frac{\partial\alpha}{\partial s}, \nabla_{\frac{\partial\alpha}{\partial t}} \frac{\partial\alpha}{\partial s} \right\rangle \Big|_{(t,0)}$$
$$= \langle \nabla_{\gamma'(t)} V(t), \nabla_{\gamma'(t)} V(t) \rangle =< V'(t), V'(t) > .$$

Further, by Lemma 5.4.10, $\left(\text{with } V = \frac{\partial\alpha}{\partial s}\right)$

$$\left\langle \nabla_{\frac{\partial\alpha}{\partial s}} \nabla_{\frac{\partial\alpha}{\partial t}} \frac{\partial\alpha}{\partial s}, \frac{\partial\alpha}{\partial t} \right\rangle$$
$$= \left\langle K\left(\frac{\partial\alpha}{\partial t} \times \frac{\partial\alpha}{\partial s}\right) \times \frac{\partial\alpha}{\partial s}, \frac{\partial\alpha}{\partial t} \right\rangle + \left\langle \nabla_{\frac{\partial\alpha}{\partial t}} \nabla_{\frac{\partial\alpha}{\partial s}} \frac{\partial\alpha}{\partial s}, \frac{\partial\alpha}{\partial t} \right\rangle$$

Now, using Theorem 1.3.1

$$\left\langle K\left(\frac{\partial\alpha}{\partial t} \times \frac{\partial\alpha}{\partial s}\right) \times \frac{\partial\alpha}{\partial s}, \frac{\partial\alpha}{\partial t} \right\rangle \Big|_{(t,0)}$$
$$=< K(t)(\gamma'(t) \times V(t)) \times V(t), \gamma'(t) >$$
$$= K(t)(< \gamma'(t), V(t) >< V(t), \gamma'(t) > - < V(t), V(t) >< \gamma'(t), \gamma'(t) >$$
$$= -K(t) < V(t), V(t) >$$

and

$$\left\langle \nabla_{\frac{\partial\alpha}{\partial t}} \nabla_{\frac{\partial\alpha}{\partial s}} \frac{\partial\alpha}{\partial s}, \frac{\partial\alpha}{\partial t} \right\rangle \Big|_{(t,0)} = \frac{\partial}{\partial t} \left\langle \nabla_{\frac{\partial\alpha}{\partial s}} \frac{\partial\alpha}{\partial s}, \frac{\partial\alpha}{\partial t} \right\rangle \Big|_{(t,0)}$$
$$- \left\langle \nabla_{\frac{\partial\alpha}{\partial s}} \frac{\partial\alpha}{\partial s}, \nabla_{\frac{\partial\alpha}{\partial t}} \frac{\partial\alpha}{\partial t} \right\rangle \Big|_{(t,0)}$$
$$= \frac{\partial}{\partial t} \left\langle \nabla_{\frac{\partial\alpha}{\partial s}} \frac{\partial\alpha}{\partial s}, \frac{\partial\alpha}{\partial t} \right\rangle \Big|_{(t,0)}$$

since $\nabla_{\frac{\partial\alpha}{\partial t}} \frac{\partial\alpha}{\partial t}\Big|_{(t,0)} = \nabla_{\gamma'(t)}\gamma'(t) = 0$. Hence,

$$L''(0) = \int_0^b \left(< V'(t), V'(t) > -K(t) < V(t), V(t) >\right) dt$$
$$+ \left\langle \nabla_{\frac{\partial\alpha}{\partial s}} \frac{\partial\alpha}{\partial s}, \frac{\partial\alpha}{\partial t} \right\rangle (b,0) - \left\langle \nabla_{\frac{\partial\alpha}{\partial s}} \frac{\partial\alpha}{\partial s}, \frac{\partial\alpha}{\partial t} \right\rangle (0,0).$$

Since α is a proper variation, we have $\nabla_{\frac{\partial\alpha}{\partial s}} \frac{\partial\alpha}{\partial s}(b,0) = \nabla_{\frac{\partial\alpha}{\partial s}} \frac{\partial\alpha}{\partial s}(0,0) = 0$. Therefore,

$$L''(0) = \int_0^b < V'(t), V'(t) > -K(t) < V(t), V(t) > dt.$$

□

Corollary 5.4.13 *For the variation vector field $V(t)$ of the proper orthogonal variation along a unit speed geodesic γ on a regular surface M in $\mathbf{E}^3$,*

$$L''(0) = -\int_0^b \langle V''(t) + K(t)V(t), V(t)\rangle \, dt.$$

Proof. Since $\dfrac{d}{dt} < V(t), V'(t) >=< V'(t), V'(t) > + < V(t), V''(t) >$,

$$\begin{aligned} L''(0) &= \int_0^b < V'(t), V'(t) > -K(t) < V(t), V(t) > dt \\ &= \langle V(t), V'(t)\rangle\big|_0^b - \int_0^b \langle V''(t) + K(t)V(t), V(t)\rangle \, dt \\ &= -\int_0^b \langle V''(t) + K(t)V(t), V(t)\rangle \, dt, \end{aligned}$$

using $V(0) = V(b) = 0$. □

Theorem 5.4.14 (The Second Variation Formula in $\mathbf{L}^3$) *Let α be a proper orthogonal variation of a unit speed timelike geodesic γ and $V(t) = \left.\dfrac{\partial\alpha}{\partial s}\right|_{(t,0)}$ be the associated variation vector field along γ on a timelike surface M in $\mathbf{L}^3$. Then*

$$L''(0) = -\int_0^b (< V'(t), V'(t) > +K(t) < V(t), V(t) >) \, dt$$

where $V'(t) = \nabla_{\gamma'(t)} V(t)$ and $K(t) = K(t,0)$ is the Gaussian curvature at $\gamma(t) = \alpha(t,0)$.

Proof. From the proof of Theorem 5.4.5 and Corollary 5.4.6, recalling Lemma 5.4.4,

$$L'(s) = \int_0^b -\left\langle \nabla_{\frac{\partial\alpha}{\partial t}} \frac{\partial\alpha}{\partial s}, \frac{\partial\alpha}{\partial t}\right\rangle \Big/ \left(-\left\langle \frac{\partial\alpha}{\partial t}, \frac{\partial\alpha}{\partial t}\right\rangle\right)^{\frac{1}{2}} dt.$$

Hence,

$$\begin{aligned} &L''(s) \\ &= \int_0^b -\left(\frac{\partial}{\partial s}\left\langle \nabla_{\frac{\partial\alpha}{\partial t}} \frac{\partial\alpha}{\partial s}, \frac{\partial\alpha}{\partial t}\right\rangle\right)\left(-\left\langle \frac{\partial\alpha}{\partial t}, \frac{\partial\alpha}{\partial t}\right\rangle\right)^{\frac{1}{2}} \Big/ -\left\langle \frac{\partial\alpha}{\partial t}, \frac{\partial\alpha}{\partial t}\right\rangle dt \\ &\quad -\int_0^b \left(\left\langle \nabla_{\frac{\partial\alpha}{\partial t}} \frac{\partial\alpha}{\partial s}, \frac{\partial\alpha}{\partial t}\right\rangle\right)^2 \Big/ \left(-\left\langle \frac{\partial\alpha}{\partial t}, \frac{\partial\alpha}{\partial t}\right\rangle\right)^{\frac{3}{2}} dt. \end{aligned}$$

At $s = 0$, since $\gamma(t)$ is a unit speed orthogonal geodesic, we have

$$\left\langle \frac{\partial \alpha}{\partial t}, \frac{\partial \alpha}{\partial t} \right\rangle\Big|_{(t,0)} = < \gamma'(t), \gamma'(t) >= -1,$$

$$\left\langle \frac{\partial \alpha}{\partial t}, \frac{\partial \alpha}{\partial s} \right\rangle\Big|_{(t,0)} = < \gamma'(t), V(t) >= 0, \text{ and } \nabla_{\gamma'(t)}\gamma'(t) = 0.$$

Thus,

$$\begin{aligned}\left\langle \nabla_{\frac{\partial \alpha}{\partial t}} \frac{\partial \alpha}{\partial s}, \frac{\partial \alpha}{\partial t} \right\rangle\Big|_{(t,0)} &=< \nabla_{\gamma'(t)}V(t), \gamma'(t) > \\ &= \gamma'(t)(< V(t), \gamma'(t) >) - < V(t), \nabla_{\gamma'(t)}\gamma'(t) >= 0.\end{aligned}$$

Then

$$L''(0) = -\int_0^b \frac{\partial}{\partial s} \left\langle \nabla_{\frac{\partial \alpha}{\partial t}} \frac{\partial \alpha}{\partial s}, \frac{\partial \alpha}{\partial t} \right\rangle\Big|_{(t,0)} dt.$$

Here,

$$\begin{aligned}\frac{\partial}{\partial s}\left\langle \nabla_{\frac{\partial \alpha}{\partial t}} \frac{\partial \alpha}{\partial s}, \frac{\partial \alpha}{\partial t} \right\rangle\Big|_{(t,0)} &= \left\langle \nabla_{\frac{\partial \alpha}{\partial s}} \nabla_{\frac{\partial \alpha}{\partial t}} \frac{\partial \alpha}{\partial s}, \frac{\partial \alpha}{\partial t} \right\rangle\Big|_{(t,0)} \\ &\quad + \left\langle \nabla_{\frac{\partial \alpha}{\partial t}} \frac{\partial \alpha}{\partial s}, \nabla_{\frac{\partial \alpha}{\partial s}} \frac{\partial \alpha}{\partial t} \right\rangle\Big|_{(t,0)}.\end{aligned}$$

By Lemma 5.4.4,

$$\begin{aligned}\left\langle \nabla_{\frac{\partial \alpha}{\partial t}} \frac{\partial \alpha}{\partial s}, \nabla_{\frac{\partial \alpha}{\partial s}} \frac{\partial \alpha}{\partial t} \right\rangle\Big|_{(t,0)} &= \left\langle \nabla_{\frac{\partial \alpha}{\partial t}} \frac{\partial \alpha}{\partial s}, \nabla_{\frac{\partial \alpha}{\partial t}} \frac{\partial \alpha}{\partial s} \right\rangle\Big|_{(t,0)} \\ &= \langle \nabla_{\gamma'(t)}V(t), \nabla_{\gamma'(t)}V(t) \rangle =< V'(t), V'(t) > .\end{aligned}$$

Further, by Corollary 5.4.11, and using the Lorentz cross product

$$\begin{aligned}&\left\langle \nabla_{\frac{\partial \alpha}{\partial s}} \nabla_{\frac{\partial \alpha}{\partial t}} \frac{\partial \alpha}{\partial s}, \frac{\partial \alpha}{\partial t} \right\rangle \\ &= \left\langle -K\left(\frac{\partial \alpha}{\partial t} \times \frac{\partial \alpha}{\partial s}\right) \times \frac{\partial \alpha}{\partial s}, \frac{\partial \alpha}{\partial t} \right\rangle + \left\langle \nabla_{\frac{\partial \alpha}{\partial t}} \nabla_{\frac{\partial \alpha}{\partial s}} \frac{\partial \alpha}{\partial s}, \frac{\partial \alpha}{\partial t} \right\rangle\end{aligned}$$

Now, using Theorem 1.3.10

$$\begin{aligned}&\left\langle K\left(\frac{\partial \alpha}{\partial t} \times \frac{\partial \alpha}{\partial s}\right) \times \frac{\partial \alpha}{\partial s}, \frac{\partial \alpha}{\partial t} \right\rangle\Big|_{(t,0)} \\ &=< K(t)(\gamma'(t) \times V(t)) \times V(t), \gamma'(t) > \\ &= -K(t)\left(< V(t), \gamma'(t) >^2 - < V(t), V(t) >< \gamma'(t), \gamma'(t) >\right) \\ &= -K(t) < V(t), V(t) >\end{aligned}$$

and

$$\begin{aligned}\left\langle \nabla_{\frac{\partial \alpha}{\partial t}} \nabla_{\frac{\partial \alpha}{\partial s}} \frac{\partial \alpha}{\partial s}, \frac{\partial \alpha}{\partial t} \right\rangle\Big|_{(t,0)} &= \frac{\partial}{\partial t}\left\langle \nabla_{\frac{\partial \alpha}{\partial s}} \frac{\partial \alpha}{\partial s}, \frac{\partial \alpha}{\partial t} \right\rangle\Big|_{(t,0)} \\ &\quad - \left\langle \nabla_{\frac{\partial \alpha}{\partial s}} \frac{\partial \alpha}{\partial s}, \nabla_{\frac{\partial \alpha}{\partial t}} \frac{\partial \alpha}{\partial t} \right\rangle\Big|_{(t,0)} \\ &= \frac{\partial}{\partial t}\left\langle \nabla_{\frac{\partial \alpha}{\partial s}} \frac{\partial \alpha}{\partial s}, \frac{\partial \alpha}{\partial t} \right\rangle\Big|_{(t,0)}\end{aligned}$$

since $\nabla_{\frac{\partial\alpha}{\partial t}} \frac{\partial\alpha}{\partial t}\Big|_{(t,0)} = \nabla_{\gamma'(t)}\gamma'(t) = 0$. Hence,

$$\begin{aligned} L''(0) = &- \int_0^b \left(< V'(t), V'(t) > + K(t) < V(t), V(t) >\right) dt \\ &- \left\langle \nabla_{\frac{\partial\alpha}{\partial s}} \frac{\partial\alpha}{\partial s}, \frac{\partial\alpha}{\partial t} \right\rangle (b,0) + \left\langle \nabla_{\frac{\partial\alpha}{\partial s}} \frac{\partial\alpha}{\partial s}, \frac{\partial\alpha}{\partial t} \right\rangle (0,0). \end{aligned}$$

Since α is a proper variation, we have $\nabla_{\frac{\partial\alpha}{\partial s}} \frac{\partial\alpha}{\partial s}(b,0) = \nabla_{\frac{\partial\alpha}{\partial s}} \frac{\partial\alpha}{\partial s}(0,0) = 0$. Therefore,

$$L''(0) = - \int_0^b \left(< V'(t), V'(t) > + K(t) < V(t), V(t) >\right) dt.$$

□

Corollary 5.4.15 *For the variation vector field $V(t)$ of a proper orthogonal variation along a unit speed timelike geodesic γ on a timelike surface M in $\mathbf{L}^3$,*

$$L''(0) = \int_0^b \langle V''(t) - K(t)V(t), V(t) \rangle \, dt.$$

Proof. Since $\frac{d}{dt} < V(t), V'(t) > = < V'(t), V'(t) > + < V(t), V''(t) >$,

$$\begin{aligned} L''(0) &= - \int_0^b < V'(t), V'(t) > + K(t) < V(t), V(t) > dt \\ &= - \langle V(t), V'(t) \rangle \big|_0^b + \int_0^b \langle V''(t) - K(t)V(t), V(t) \rangle \, dt \\ &= \int_0^b \langle V''(t) - K(t)V(t), V(t) \rangle \, dt. \end{aligned}$$

□

Remark 5.4.16 The second variation formula only depends on the particular variation through its variation vector field. Moreover, the first and second variation formulas in $\mathbf{E}^3$ imply that if $\gamma : [0,b] \to M$ is a minimizing geodesic segment, then for any variation of γ, $L'(0) = 0$ and $L''(0) \geq 0$. On the other hand, in $\mathbf{L}^3$ they imply that if $\gamma : [0,b] \to M$ is a maximizing geodesic segment, then for any variation of γ, $L'(0) = 0$ and $L''(0) \leq 0$.

Definition 5.4.17 The **diameter** of a regular surface M in $\mathbf{E}^3$ or **timelike diameter** of a timelike surface M is defined by

$$\mathrm{diam(M)} = \sup\{\rho(p,q) | \; p, q \in M\}$$

where $\rho(p,q)=\rho_E(p,q)$ given in Definition 4.3.1 for the regular surface M in $\mathbf{E}^3$ and $\rho(p,q)=\rho_L(p,q)$ given in Definition 4.3.3 for the timelike surface M in $\mathbf{L}^3$.

If $S^2(r)$ is a sphere of radius r, its Gaussian curvature K is $\frac{1}{r^2}$ and its diameter diam(S^2) is $r\pi$. The following theorem is the 2-dimensional version of **Bonnet-Myers Diameter Theorem** which is a comparison of the diameters between the complete surface with Gaussian curvature K and the sphere of radius $\frac{1}{\sqrt{\delta}}$.

Theorem 5.4.18 (**Bonnet-Myers**) *Let M be a complete surface with the Gaussian curvature $K \geq \delta > 0$. Then M has the diameter*

$$\mathrm{diam}(M) \leq \frac{\pi}{\sqrt{\delta}}$$

and hence M is compact.

Proof. Since M is complete, by the Hopf-Rinow theorem, for any two points $p, q \in M$ there is a minimizing geodesic segment γ of length $l = \rho_E(p,q)$ joining p and q. Assume that $\frac{\pi}{\sqrt{\delta}} < l$ and consider variations of the geodesic $\gamma : [0,l] \to M$. For a unit vector $w \in T_{\gamma(0)}M$ such that $< w, \gamma'(0) >= 0$, let $W(t)$ be a parallel translation of w along γ. Clearly, $|W(t)| = 1$ and $< W(t), \gamma'(t) >= 0$ for $t \in [0,l]$. Now define a vector field $V(t)$ by

$$V(t) = \sin\left(\frac{\pi t}{l}\right) W(t), \quad t \in [0,l].$$

Since $V(0) = V(l) = 0$ and $< V(t), \gamma'(t) >= 0$, by Theorem 5.4.3, we may take a proper orthogonal variation of γ with the variation vector field $V(t)$. By the second variation formula,

$$L''(0) = \int_0^l \left(< V'(t), V'(t) > -K < V(t), V(t) >\right) dt.$$

Since $W(t)$ is a parallel vector field along γ, $V'(t) = \frac{\pi}{l}\cos\left(\frac{\pi t}{l}\right) W(t)$. Since $l > \frac{\pi}{\sqrt{\delta}}$ is assumed, $K \geq \delta > \frac{\pi^2}{l^2}$. Hence,

$$\begin{aligned} L''(0) &= \int_0^l \left(\frac{\pi^2}{l^2}\cos^2\left(\frac{\pi t}{l}\right) - K\sin^2\left(\frac{\pi t}{l}\right)\right) dt \\ &< \int_0^l \left(\frac{\pi^2}{l^2}\cos^2\left(\frac{\pi t}{l}\right) - \frac{\pi^2}{l^2}\sin^2\left(\frac{\pi t}{l}\right)\right) dt \\ &= \frac{\pi^2}{l^2}\int_0^l \cos\left(\frac{2\pi t}{l}\right) dt = 0. \end{aligned}$$

This is a contradiction to the fact that $L'(0) = 0$ and $L''(0) \geq 0$ for any variation of γ since γ is a minimizing geodesic segment. Thus, $l \leq \frac{\pi}{\sqrt{\delta}}$. Since p, q are any two points in M, $diamM \leq \frac{\pi}{\sqrt{\delta}}$ and hence M is bounded. Since M is complete and bounded, M is compact. □

Corollary 5.4.19 (**Timelike Bonnet-Myers Theorem**) *Let M be a globally hyperbolic timelike surface with the Gaussian curvature $K \leq -\delta < 0$. Then M has the timelike diameter*

$$\mathrm{diam}(M) \leq \frac{\pi}{\sqrt{\delta}}.$$

Proof. By Theorem 5.3.20, for any two points $p, q \in M$ with $p \ll q$ there is a maximal geodesic segment γ of length $l = \rho_L(p, q)$ joining p and q. Assume that $\frac{\pi}{\sqrt{\delta}} < l$ and consider a variation of the unit speed timelike geodesic $\gamma : [0, l] \rightarrow M$. For a unit spacelike vector $w \in T_{\gamma(0)}M$ such that $< w, \gamma'(0) >= 0$, let $W(t)$ be a parallel translation of w along γ. Clearly, $|W(t)| = 1$ and $< W(t), \gamma'(t) >= 0$ for $t \in [0, l]$. Now define a vector field $V(t)$ by

$$V(t) = \sin\left(\frac{\pi t}{l}\right) W(t), \ t \in [0, l].$$

Since $V(0) = V(l) = 0$ and $< V(t), \gamma'(t) >= 0$, by Theorem 5.4.3, we may take a proper orthogonal variation of γ with the variation vector field $V(t)$. By the second variation formula,

$$L''(0) = -\int_0^l (< V'(t), V'(t) > +K < V(t), V(t) >)\, dt.$$

Since $W(t)$ is a parallel vector field along γ, $V'(t) = \frac{\pi}{l}\cos\left(\frac{\pi t}{l}\right) W(t)$. Since $\frac{\pi}{\sqrt{\delta}} < l$ is assumed, $K \leq -\delta < -\frac{\pi^2}{l^2}$. Hence,

$$\begin{aligned} L''(0) &= -\int_0^l \left(\frac{\pi^2}{l^2}\cos^2\left(\frac{\pi t}{l}\right) + K\sin^2\left(\frac{\pi t}{l}\right)\right) dt \\ &> -\int_0^l \left(\frac{\pi^2}{l^2}\cos^2\left(\frac{\pi t}{l}\right) - \frac{\pi^2}{l^2}\sin^2\left(\frac{\pi t}{l}\right)\right) dt \\ &= -\frac{\pi^2}{l^2}\int_0^l \left(\cos\left(\frac{2\pi t}{l}\right)\right) dt = 0, \end{aligned}$$

so that $L''(0) > 0$ for this variation. This is a contradiction to the fact that $L'(0) = 0$ and $L''(0) \leq 0$ for any variation of γ since γ is a maximal geodesic segment. Thus, $l \leq \frac{\pi}{\sqrt{\delta}}$. Since p, q are any two chronologically related points in M, $diam M \leq \frac{\pi}{\sqrt{\delta}}$. □

For further study, see the following references.

1. Beem, J. K., P. E. Ehrlich and K. L. Easley, Global Lorentzian Geometry, Second Edition, Marcel Dekker Inc. New York, 1996.

2. do Carmo, Manfredo Perdigan, Differential Geometry of Curves and Surfaces. Englewood Cliffs, NJ: Prentice-Hall, 1976.

3. Millman, R. and G. Parker, Elements of Differential Geometry, Prentice-Hall Inc, 1977.

4. O'Neill, B. Elementary Differential Geometry, Academic Press Inc, 1966.

5. O'Neill, B. Semi-Riemannian Geometry, Academic Press, 1983.

Index

Paul E. Ehrlich

Professor of Mathematics
The University of Florida
Gainesville FL 32611-8105

Seon-Bu Kim

Professor of Mathematics
Chonnam National University
Gwangju 500-757 Korea

Differential Geometry in Euclidean
and Minkowskian Spaces

ISBN 978-89-6105-640-3

KM KYUNG MOON SA

174, Wausan-ro Mapo-gu Seoul, Korea
Tel: (82) 2 332 2004
Fax: (82) 2 336 5193

Printed in Korea

Price: ₩ 18,000